Although animals are widely employed as research subjects, only recently have we acknowledged the bond that frequently, perhaps inevitably, develops between subject and researcher. Whatever the qualities of this relationship, an increasing body of evidence suggests that it may result in profound behavioral and physiological changes in the animal subject. Such effects are apparent in behavioral studies conducted in both laboratory and field settings. They also appear in physiological studies ranging from biomedical (e.g., heart rate, blood pressure, and immunological changes) to animal science studies (e.g., growth and production). Such effects are not confined to obvious cases involving primates or dogs, but are found with unexpected animals like chickens, reptiles, and even octopuses.

Despite the fact that most researchers are trained to minimize or avoid interactions with animal subjects, they continue to occur. This book is the first of its kind to address the issue systematically, describing many examples of this "inevitable bond" between scientist and animal. The discussion will allow researchers to anticipate these potentially confounding effects and take advantage of them in designing more effective and humane test environments for animal subjects.

The inevitable bond

Dr. Sarah T. Boysen displaying *The Inevitable Bond* with her subject, Sheba. (Photo by Paul Wilkins.)

The inevitable bond
Examining scientist–animal interactions

Edited by

Hank Davis
University of Guelph

Dianne Balfour
Ministry of Health
British Columbia

CAMBRIDGE
UNIVERSITY PRESS

Published by the Press Syndicate of the University of Cambridge
The Pitt Building, Trumpington Street, Cambridge CB2 1RP
40 West 20th Street, New York, NY 10011–4211, USA
10 Stamford Road, Oakleigh, Victoria 3166, Australia

First published 1992

Printed in the United States of America

Library of Congress Cataloging-in-Publication Data
The inevitable bond : examining scientist–animal interactions / edited
by Hank Davis, A. Dianne Balfour.
p.   cm.
Includes index.
ISBN 0–521–40510–6 (hardcover)
1. Animal experimentation – Effect of experimenters on.   2. Human–
animal relationships.   I. Davis, Hank, 1941– .   II. Balfour, A.
Dianne.
QL55.I44   1992
591′.0724 – dc20   91-27509
                                                                           CIP

A catalog record for this book is available from the British Library.

ISBN 0-521-40510-6 hardback

To all the rats, mice, squirrels, and ducks
who have enriched my life
and to Susan Simmons, who enhanced
those experiences by sharing them with me   (H.D.)

To Garp (D.B.)

Contents

List of contributors   *page* ix
Acknowledgments   xi

1   The inevitable bond   *Hank Davis and A. Dianne Balfour*   1

2   Interactions, relationships, and bonds: the conceptual basis for
    scientist–animal relations   *Daniel Q. Estep and Suzanne Hetts*   6

3   Studies of rodent–human interactions in animal psychology
    Donald A. Dewsbury   27

4   The covalent animal: on bonds and their boundaries in
    behavioral research   *John C. Fentress*   44

5   The phenomenon of attachment in human–nonhuman
    relationships   *John Paul Scott*   72

6   Humanity's "best friend": the origins of our inevitable bond
    with dogs   *Benson E. Ginsburg and Laurie Hiestand*   93

7   The use of dog–human interaction as a reward in instrumental
    conditioning and its impact on dogs' cardiac regulation
    Ewa Kostarczyk   109

8   Behavioral arousal and its effect on the experimental animal
    and the experimenter   *Alastair J. S. Summerlee*   132

9   Practice makes predictable: the differential effect
    of repeated sampling on behavioral and physiological
    responses in monkeys   *Maria L. Boccia, Christy Broussard,
    James Scanlan, and Mark L. Laudenslager*   153

10 Improved handling of experimental rhesus monkeys
    Viktor Reinhardt   171

Contents

11 Social interaction as a condition for learning in avian species: a synthesis of the disciplines of ethology and psychology
Irene M. Pepperberg 178

12 Pongid pedagogy: the contribution of human–chimpanzee interactions to the study of ape cognition *Sarah T. Boysen* 205

13 The role of social bonds in motivating chimpanzee cognition
David L. Oden and Roger K. R. Thompson 218

14 Minimizing an inevitable bond: the study of automated avoidance in rats *Morrie Baum and Laurie Hiestand* 232

15 Underestimating the octopus *Jennifer Mather* 240

16 The scientist and the snake: relationships with reptiles
Bonnie B. Bowers and Gordon M. Burghardt 250

17 Fear of humans and its consequences for the domestic pig
P. H. Hemsworth, J. L. Barnett, and G. J. Coleman 264

18 The effect of the researcher on the behavior of poultry
Ian J. H. Duncan 285

19 Early human–animal relationships and temperament differences among domestic dairy goats *David M. Lyons* 295

20 The effect of the researcher on the behavior of horses
Sharon L. Crowell-Davis 316

21 Imprinting and other aspects of pinniped–human interactions
Ronald J. Schusterman, Robert Gisiner, and Evelyn B. Hanggi 334

22 Humans as predators: observational studies and the risk of pseudohabituation *Nancy G. Caine* 357

23 Human–bear bonding in research on black bear behavior
Gordon M. Burghardt 365

24 Scientist–animal bonding: some philosophical reflections
Hugh Lehman 383

Index 397

Contributors

A. Dianne Balfour, *Ministry of Health, Victoria, British Columbia*

J. L. Barnett, *Animal Research Institute, Department of Agriculture and Rural Affairs, Werribbee, Victoria, Australia*

Morrie Baum (deceased), *Department of Psychology, Rensslaer Polytechnic Institute, Troy, New York*

Maria L. Boccia, *Department of Psychiatry, Developmental Psychobiology Research Group, University of Colorado Health Sciences Center, Denver*

Bonnie B. Bowers, *Graduate Program in Ethology, University of Tennessee, Knoxville*

Sarah T. Boysen, *Primate Cognition Project, Ohio State University, and Yerkes Regional Primate Research Center, Emory University, Atlanta, Georgia*

Christy Broussard, *Department of Psychiatry, Developmental Psychobiology Research Group, University of Colorado Health Sciences Center, Denver*

Gordon M. Burghardt, *Department of Psychology, University of Tennessee, Knoxville*

Nancy G. Caine, *Department of Psychology, Bucknell University, Lewisburg, Pennsylvania*

G. J. Coleman, *Department of Psychology, La Trobe University, Bundoora, Victoria, Australia*

Hank Davis, *Department of Psychology, University of Guelph, Guelph, Ontario*

Sharon L. Crowell-Davis, *Department of Anatomy, College of Veterinary Medicine, University of Georgia, Athens*

Donald A. Dewsbury, *Department of Psychology, University of Florida, Gainesville*

Ian J. H. Duncan, *Department of Animal Science, University of Guelph, Guelph, Ontario*

Contributors

Daniel Q. Estep, *Animal Behavior Associates, Littletown, Colorado*

John C. Fentress, *Department of Psychology, Dalhousie University, Halifax, Nova Scotia*

Benson E. Ginsburg, *Department of Biobehavioral Sciences, University of Connecticut, Storrs*

Robert Gisiner, *Long Marine Laboratory, University of California at Santa Cruz*

Evelyn B. Hanggi, *Long Marine Laboratory, University of California at Santa Cruz*

P. H. Hemsworth, *Animal Research Institute, Department of Agriculture and Rural Affairs, Werribee, Victoria, Australia*

Suzanne Hetts, *Department of Clinical Sciences, Veterinary Teaching Hospital, Colorado State University, Fort Collins*

Laurie Hiestand, *Department of Biology, Wesleyan Collage, Middletown, Connecticut*

Ewa Kostarczyk, *Department of Neurophysiology, Nencki Institute of Experimental Biology, Warsaw, Poland*

Mark L. Laudenslager, *Department of Psychiatry, Developmental Psychobiology Research Group, University of Colorado Health Sciences Center, Denver*

Hugh Lehman, *Department of Philosophy, University of Guelph, Guelph, Ontario*

David M. Lyons, *Behavioral Biology, California Primate Research Center, University of California at Davis*

Jennifer Mather, *Department of Psychology, University of Lethbridge, Lethbridge, Alberta*

David L. Oden, *Psychology Department, LaSalle University, Philadelphia*

Irene M. Pepperberg, *Departments of Psychology, Ecology, and Evolutionary Biology, University of Arizona, Tucson*

Viktor Reinhardt, *Wisconsin Regional Primate Research Center, University of Wisconsin, Madison*

James Scanlan, *Department of Psychiatry, Developmental Psychobiology Research Group, University of Colorado Health Sciences Center, Denver*

Ronald J. Schusterman, *Departments of Psychology and Biology, California State University, Hayward, and Long Marine Laboratory, University of California at Santa Cruz*

John Paul Scott, *Department of Psychology, Bowling Green State University, Bowling Green, Ohio*

Alastair J. S. Summerlee, *Department of Biomedical Sciences, Ontario Veterinary College University of Guelph, Guelph, Ontario*

Roger K. R. Thompson, *Departments of Biology and Psychology, Franklin and Marshall College, Lancaster, Pennsylvania*

Acknowledgments

A project of this magnitude requires the help of many people. We are grateful to Alan Crowden and Robin Smith of Cambridge University Press for their encouragement; to Mary Racine for her superb copyediting; to Craig Hawryshyn for moral and material support; to the Natural Sciences and Engineering Research Council of Canada for its support (Grant A0673 to H.D.) of all aspects of the preparation of this book; and especially to Laurie Hiestand, whose involvement, in capacities far beyond what is reflected in authorship, was essential to the success of the book. Finally, we are grateful to our contributors for all the obvious reasons, but also for some more subtle. Many of these persons took considerable risks and wrote about material they had avoided or not addressed directly before.

—1—

The inevitable bond

Hank Davis and A. Dianne Balfour

Scientists in many disciplines work with animals every day. To various degrees, that statement has been true for more than a century, and despite the best efforts of animal rights activists, it is likely to remain true for some time. Perhaps the number of animals used and the manner in which they are used will change, but it is likely that scientists and animals will continue to interact.

Many of these interactions are far from superficial. They constitute what, in many other settings, would be called a *relationship*, perhaps even a *bond*. Although it has not been fashionable to do so, we would like to label this relationship for what it is and underscore its importance. The advantages of doing so will become obvious.

Twenty years ago this book would not have been well received. Indeed, many scientists in the forefront of their fields would have avoided associating themselves with the project. A decade later the zeitgeist was changing. An Animal Behavior Society symposium on the scientist–animal bond (Davis, 1985) drew an enthusiastic response, although a number of researchers continued to express reservations about the topic. A typical comment was: "This is exactly the kind of stuff we were trained to avoid. It taints our research. It compromises the conclusions we draw. This is why we *number* our animals instead of naming them. The last thing we need is to raise the specter of this kind of contamination."

This reaction deserves comment. We agree that most of us were trained explicitly to avoid excessive interactions with our research animals. The reasons fell into two general categories which reflect the fact that the scientist–animal bond is reciprocal: (1) effects on the researcher and (2) effects on the animal. The first concern can be summarized as: "If you bond with your animal, you won't be able to dissect/shock/euthanize the

1

animal. Save yourself the conflict." This concern is beyond the scope of our book.

The second concern lies at the center of our agenda. It is: "If you bond with your animal, the behavior or physiology of the animal may be changed. In short, the subject matter of your experiment may be altered by this relationship." We believe this is a real possibility and a legitimate concern. However, unlike our hypothetical critic, we do not believe such an outcome *must* be a source of contamination. It is precisely for this reason that we have written this book.

If we realize beforehand that the scientist–animal relationship is (1) likely to occur and (2) likely to modulate our data in measurable ways, we can accommodate this outcome. As a number of our contributors point out, we might even put the effects of such bonding to good use. But we can do nothing if we continue to minimize our awareness of such effects or deny their existence altogether in the name of "pure science."

Many colleagues have confused the agenda of this book with the "Clever Hans" effect. There is a fundamental difference between the sorts of issues we are talking about and the famous episode involving a putative counting horse. Hans, as we now know, was not performing long division or converting fractions to decimals. He was responding to minimal cues from his interrogator. From this memorable incident in the history of experimental psychology, we learned to avoid cuing our research animals. This is unquestionably an important message, but it is not the point of our book. In the Clever Hans situation, there was a direct question ("What is the square root of sixteen?") and a correct answer ("Four"). The experimenter knew the answer and inadvertently conveyed it to his animal subject. By treating the animal's performance uncritically, we reached an erroneous conclusion about his abilities.

In the chapters that follow, our contributors refer to more than just test situations involving correct responses. They discuss the imparting of more general information from human to animal, and they examine the motivational and emotional effects of a *relationship* on the animal and its performance. Relationships allow the parties to make predictions about each other's behavior. Thus, the animal may make reasonable inferences about the scientist, such as "This person feeds me" or "This person hurts me." If such cognitive attributions offend some readers, the case can be simplified to "I feel good/fearful in the presence of this person." Such emotional expectancies require only the simplest kind of Pavlovian conditioning involving the well-established phenomenon of "person as CS" (conditioned stimulus).

Obviously, what an animal learns from its relationship with a scientist may be considerably more complex. It may involve highly refined discriminative cues based on particular sensory stimuli or temporal factors. In any case, the effects will yield measurable changes in behavior or physiology. Thus, one can construe the processes underlying the inevitable bond in complex cognitive terms or in simple Pavlovian terms. Neither choice of vocabulary allows one to sidestep the fundamental point of this book. Even the simplest Pavlovian conditioning may have profound motivational effects. When these effects intrude on the researcher's domain of interest, we are facing the subject matter at hand.

Since we have excluded from our coverage the "Clever Hans" class of scientist–animal problems, we should identify yet a second type of event that may be confused with our book's topic. It is known that experimenter expectancies or biases can have profound, albeit unintended, effects on the outcomes of research. For example, a scientist's theory or point of view often affects his or her conclusions about the subject's behavior. These effects are typically caused by biased reporting; that is, behavior that is inconsistent with the scientist's hypothesis may be undervalued or ignored altogether. For example, if a scientist believes a particular strain of rats is brighter than another, the collection and analysis of maze performance data are likely to support this conclusion, regardless of how the animals actually perform. This is one reason that so-called double-blind procedures are widely used. Despite the ubiquity of the problem, like the Clever Hans effect it lies beyond the scope of our book. The critical feature missing from the concept of experimenter bias is the possibility that the animal's behavior itself is affected. Presumably the only thing that is altered is the *scientist's* perception.

However, psychologist Robert Rosenthal has argued that interactions between scientist and subject may actually alter the subject's behavior. Such effects do lie within the framework of our book and are not confined to the use of human subjects. Rosenthal and Fode (1963) reported an interesting experiment in which purportedly "bright" rats showed superior performance that did not stem from biased recordings but from "superior" treatment by the experimenters. They took pains to demonstrate that the markedly different ways the scientists treated rats they believed to be superior led to observably different behavior and measurably improved performance. It is unimportant *why* the scientists treated the animals in a better way. What matters is that the relationship between the scientists and the animals had measurable effects on the animals' performance. Had Rosenthal and Fode not identified the importance of the scientist–animal relationship, conclusions about the maze performance of rats would have failed to identify an underlying determinant of performance.

This result is an instance of what has been called a "Pygmalian effect," a fairly commonplace principle in psychological research on humans. The

3

Rosenthal and Fode example is unusual primarily because it involves an interspecies situation. Indeed, many of the effects reported in this book would not be newsworthy in cases of social psychological research with humans. Scientists working with humans generally assume that the subject is not a passive source of data, but rather an active organism with needs and motives that may have an impact on the data being collected. Thus, the design of experiments and experimenter–subject interactions typically take this into account. We propose that no less consideration be given to the case of animal subjects.

The critical reader will note that we have not offered a single or binding definition of key terms in our thesis such as *bond, interaction*, and *relationship*. Nor have we attempted to restrict our contributors' use of these terms. Indeed, some contributors have taken pains to define exactly what they mean by each of these words. It is revealing in a Wittgensteinian sense that even such attempts at precision have resulted in differences. This is perhaps as it should be. There is variation even within the small bandwidth of meaning these terms convey. To understand the central point of this book, it is not essential to establish the exact point at which a relationship ends and a bond begins.

Some readers may react with surprise to the range of species covered in our book. Bonding with dogs or other primates is one thing. But a scientist–animal relationship involving reptiles? poultry? octopuses? We are pleased to include these less orthodox subjects in our coverage. Although the effects described in these chapters are arguably smaller or more subtle, they are nonetheless real. More important, they leave unresolved the question of whether their reduced magnitude reflects less bonding per se or less sensitivity on our part in detecting it. This question is no less relevant to the case of animal groups (e.g., insects, fish, crustaceans) for which we were unable to solicit a contribution.

It will become obvious that much of the information reported in this book is anecdotal. This is not surprising or unreasonable given that the issue is in its infancy. With the publication of the book, however, we pose two challenges to the field: (1) Continue to generate more examples in as wide a range of contexts as possible, and (2) test them directly. There is no reason that the scientist–animal relationship cannot be given the status of a conventional independent variable and treated with the same rigor and explanatory power as more conventional variables.

In summary, our case is the following: Although animals are widely employed as research subjects, only recently have we acknowledged the bond that frequently, perhaps inevitably, develops between subject and researcher. Whatever the qualities of this relationship, an increasing body of evidence suggests that it may result in profound behavioral and physiological changes in the animal. Such effects are apparent in behavioral studies conducted in both laboratory and field settings. They also appear in physiological studies ranging from biomedical (e.g., heart rate, blood pressure, and immunological changes) to animal science studies (e.g., growth and production).

Our primary goal in this volume is to document the extent and diversity of such effects. While most of the contributors discuss the behavioral and physiological effects of the scientist–animal bond, others emphasize methodological accommodations that were necessitated by foreknowledge of their occurrence. To ignore the existence and power of these effects or to minimize the interactions that created them would reflect considerable myopia. At the least, it would reveal insensitivity to the complexity of the animal subject. At the most, it might distort our understanding of the very processes we intend to study.

Acknowledgments

The authors thank Loraleigh Keashly, David Piggins, and Ian Duncan for their thoughtful comments. The preparation of this chapter was supported in part by Grant A0673 to Hank Davis from the Natural Sciences and Engineering Research Council of Canada.

References

Davis, H. (1985). The importance of experimenter–animal interactions in the development and measurement of behavior: A cross-species analysis. Invited paper session, Animal Behavior Society, Raleigh, NC.

Rosenthal, R. A., & Fode, K. L. (1963). Three experiments in experimenter bias. *Psychological Reports, 12,* 491–511.

—2—

Interactions, relationships, and bonds: the conceptual basis for scientist–animal relations

Daniel Q. Estep and Suzanne Hetts

Every animal psychology experiment is necessarily an encounter between animal and man.

Hediger (1968, p. 131)

Editors' introduction

Estep and Hetts explore the variety of relationships scientists can develop with their research animals. They cite such possibilities as dominant–subordinate, parent–offspring, predator–prey, potential sexual partner, and combinations or variations of these. Estep and Hetts argue that each of these relationships, whether intentionally or unintentionally developed, can have important consequences for research. In some types of research, certain relationships may facilitate performance or data collection. In other cases, particular relationships may be unnecessary and even deterimental.

Most of the book's basic concerns and issues are covered in this chapter. Virtually every point raised by Estep and Hetts is elaborated later in a variety of contexts with different species.

Introduction

By their very nature, almost all scientific studies involving nonhuman animals involve some human contact with the subjects. In some cases the interactions are intense, frequent, and a carefully planned part of the study, such as in an investigation of the cognitive abilities of captive chim-

6

panzees. In other cases human contacts may be brief and infrequent, such as in certain biomedical investigations of the anatomy or physiology of the animal. The only exceptions to this generalization may be some field studies in which the investigator conceals him- or herself from the animal subject and the two never come into direct contact.

This interaction between the scientist and the nonhuman animal (simply "animal" after this) is, by most definitions, communication. Communication can be defined as an action performed by one organism that alters the behavior of another (Dewsbury, 1978). Any communication that passes between the two is likely to influence not only their immediate behavior but subsequent behavior as well. Even the olfactory, auditory, or visual cues left by a concealed field-worker may influence the behavior of the animal, which in turn can alter the behavior of the investigator. In some laboratory studies the interactions between the investigator and the animal may be minimal, but interactions between the animal and other members of the research team, especially animal caretakers, may be much more frequent. The results of these interactions with other team members could have a considerable influence on the interactions the animal has with the investigator and cannot be ignored. The consequences of this communication between animal subject and all the members of the research team are the subject of this book. Whether they are planned or unplanned, extensive or brief, they can have an important impact on the conduct of animal research.

The other contributors to this volume will provide ample documentation of the diversity of interactions that can arise between scientists and a wide variety of animal species. The goal of this chapter is to outline the basic kinds of interactions, relationships, and bonds that can form between scientists and their animals and to explore the theoretical basis for such phenomena. From this knowledge, practical suggestions will be made for improving research.

A conceptual framework for discussing scientist–animal relations

The conceptual framework adopted here is that of Hinde (1976), which was developed to facilitate the study of intraspecific social structure. While the interactions between scientist and animal are necessarily interspecific, the framework is still of considerable value. The essence of Hinde's view is that social structures are built up of relationships and that relationships are built up of interactions. Although the system is hierarchical, it is also dynamic and interactive across levels. For example, previous relationships can affect present interactions and some aspects of social structure may constrain or limit certain relationships. Relationships and social structure can change over time as a result of interactions and other experiences.

In Hinde's system, interactions between two or more organisms are the basic elements. They usually take the form of "*A* does *X* to *B*." It is implied that social interactions involve mutual influence and thus involve communication. Interindividual relationships are built up from a series of interactions between two individuals that are known to each other. If two male baboons have a series of agonistic interactions over a period of time that result in a reliable pattern of responding, then it can be said that they have an agonistic relationship. Within a given relationship, several kinds of interactions may be seen: affiliative, agonistic, care giving, and so on. Social structure emerges from the nature, quality, and patterning of all the relationships among all the animals that are part of the system.

Hinde's system does not directly address the concept of the social bond that can arise between animals, but this most important concept should be defined. The term "bond" seems to be used interchangeably with the term "attachment" by most investigators. Even a cursory reading of the literature reveals that these two terms are rarely defined, and when they are, there are considerable differences among definitions and in the phenomena that are included. For example, some investigators view bonding as a mutual state involving mutual influence, interactions, or relationships (e.g., Wolfle, 1985). For others, bonding may be one-sided, as when an organism becomes attached to an object or a place (e.g., Cairns, 1966b). Cairns (1972) points out that the term "attachment" is often used in two ways. First, it can refer to a class of behaviors that maintain proximity between the organism and another, or to the behavioral disruption that results from involuntary separation. Second, it can refer to an internal process that energizes, regulates, and directs the behaviors just described. For the purposes of this chapter, Cairns's (1972) second definition will be employed (i.e., referring to an internal process), and the terms "bond" and "attachment" will be used interchangeably. Furthermore, it will be assumed that a bond or attachment is not necessarily a type of interaction or relationship, since it need not be mutual. That is, an organism can be bonded to an inanimate object, or it can be bonded to another organism that is not bonded to it. Other aspects of social attachment will be considered in a subsequent section ("Scientist as conspecific").

Communication and interactions of one sort or another between the research team and the animal subject are inevitable in almost all scientific studies involving animals. However, there has been a long-standing tradition in animal research that scientist–subject interactions should be held to a minimum and should be standardized. The purpose is to achieve better experimental control by reducing the variance introduced by such interactions. Once such attempts are made, it is assumed that whatever scientist–subject interactions remain are relatively unimportant for the research outcome and can be ignored. In a word, scientist–animal interactions are viewed as a nuisance variable to be controlled, not studied. This attitude has tended to blind investigators to the importance and variety of interactions that can occur in the research environment – feeding interactions, handling, training to a particular task, and so on. Interindividual relationships, bonds, and even complex social structures may arise between scientists and animals. All of these can have profound effects on the way the research is conducted, on the way the results are interpreted, and on the welfare of both the scientists and the animals.

With the exception of this book, there have been few systematic attempts to describe the interactions, interindividual relationships, or social structures that arise between scientists and animals. At best, brief mention is usually made in the methods section of research reports describing training procedures, animal care, or attempts to habituate the animal to the investigator. These descriptions rarely provide details about how the animal is approached and handled, who contacts the animal and how often, and in what ways the animal responds to these contacts. Also frequently neglected is the history of the animal with humans in general and the investigator and others of the research team in particular. Such interactions and history can have an extremely important influence on the way in which the animal responds to the investigator and ultimately can affect the outcome of the study.

Systematic descriptions of scientist–animal relations are important because they make explicit the full range of interactions that the animal has had with humans in the past and in the present investigation. From these interactions the important interindividual relationships can be derived and the social structure, if any, described. By providing such detail, important sources of experimental variance are revealed that otherwise might be overlooked. Differences in human–animal interactions may help to explain why some studies "work" and other do not.

Animals' perceptions of scientists

This section focuses on the kinds of interactions and relationships that can arise between scientists and animals, chiefly from the perspective of the animal. All relationships involve communication among participants by use of species-typical communication signals. However, each participant may or may not have the same perception of the signal, and therefore of the interaction or relationship. When perceptions are congruent, communication is effective and likely results in predictable behavior patterns. When perceptions are not congruent, miscommunication and misinterpretation of behavior among the participants can occur. Then the actions of each participant may not be easily explained or predicted by the other. Thus, to understand the interactions and relationships that arise in the research setting, it is important to understand how all the participants (both human and nonhuman) perceive the others. While it is not possible to gain direct knowledge about the perceptions of nonhumans, their perceptions can be reasonably inferred from their behavior. For example, an animal that perceives a human as a predator would be unlikely to respond to this human with affiliative behavior patterns normally directed toward conspecifics.

The Swiss zoologist Heini Hediger has made important contributions to the conceptualization and understanding of the ways that animals can perceive humans and vice versa. Hediger (1965) has stated that, to humans, nonhumans can take on innumerable roles, as he puts it "from dead merchandise up to a deity" (p. 298). Humans may also be perceived by the animal in a variety of ways. Hediger describes five different kinds of perceptions or roles that are most frequently observed. They are the human as predator, the human as prey, the human as a part of the environment without social significance, the human as a symbiont, and the human as a member of the animal's own species. It is probably rare for any animal to perceive a human as belonging exclusively to only one of these five categories. The categories instead form a continuum, and an animal probably perceives any given human as a member of a combination of them. These categories are simply convenient referents that are useful in describing an animal's behavior toward humans and thereby inferring its perceptions. It will become clear that research animals' perceptions of humans in the research environment ultimately affect the kinds of interactions and relationships that arise between the scientist and the animal.

Scientist as predator

Hediger points out that wild (or untamed domestic) animals generally treat humans as predators (or, to use his term, "enemies"). For many animals this perception of humans is unlearned and does not depend on specific experiences, although experiences with humans can greatly modify it. While it is most common to think of this perception arising between field researchers and wild animals, it can also exist between captive wild animals and laboratory workers and even between domestic laboratory animals and their human investigators. For example, it is common knowledge among laboratory workers that even domestic laboratory rodents such as rats, mice, and hamsters should be handled before experimentation (i.e., "gentled") to reduce escape behavior, biting, and other fearful responses

(Short, 1967). Such responses are consistent with the interpretation that the animal perceives the scientist as a predator.

Scientist as prey

It is rare for most animal species to respond to humans as prey, and particularly so in research environments. It is not unheard of, however, as attested by occasional reports of attacks by wild carnivores on the field researchers studying them (e.g., Scott, Bentley, & Warren, 1985).

Scientist as socially insignificant part of the environment

In most research, the scientist aspires to have the animal behave toward the investigator as if he or she were a socially insignificant part of the environment. This reduces communication between the two to a minimum. Many field workers and some laboratory investigators go to great lengths to either conceal themselves from their subjects with blinds or use remote sensing devices (binoculars, radiotelemetry devices, etc.) to accomplish this goal. Others spend enormous amounts of time and energy habituating the animal to the presence of the investigator. How well these attempts succeed in reducing the reactivity of the animal to the researcher is difficult to assess and is rarely addressed directly. Investigators do not often describe how their subjects react to them.

An interesting and important exception to this generalization is seen in a recent study by Caine (1990; see also Chapter 22, this volume). Her research questions a general assumption made by many investigators that if the animal does not respond to the human observer in dramatic and easily observable ways, then it must be unaware of or habituated to the observer's presence. While studying a captive colony of red-bellied tamarins, Caine discovered that the animals significantly delayed entering their sleeping nests at dusk in the presence of human observers (an antipredator behavior), despite the fact that the animals had been under observation for a considerable period of time and seemed habituated to humans in other ways. The implications of this are that different aspects of an animal's behavior may habituate at different rates and that some very subtle forms of antipredatory behavior may never habituate. As Caine points out, such unrecognized antipredator responses can bias the results of studies in important and unexpected ways.

Scientist as symbiont

Symbiosis involves the living together of members of different species to their mutual benefit (McFarland, 1987b). Communication occurs between the two species, but the organisms, in general, do not respond to each other as either predator and prey or as conspecifics. In some situations, conspecific signals may be modified and used for interspecific communication; in other situations, new signals may arise through evolutionary modification and/or learning that allow for effective communication. As an example of the latter, macaques living in large social groups at the Yerkes Regional Primate Research Center Field Station have been trained to respond to human hand and voice signals to carry out certain research tasks. Individual animals have learned to separate themselves from the group, to enter transport boxes and individual cages, to present an arm, a leg, and/or their anogenital regions, and to passively allow the collection of blood, the collection of vaginal secretions, and the execution of minor veterinary medical procedures. The human signals and the animals' responses are not typical of either species' predator–prey or conspecific communication but have developed through mutual adjustment and learning.

In many laboratory settings the animal and members of the research team engage in a symbiotic relationship. The research team provides the animal with food, water, shelter, health care, and other necessities. In turn the animal provides data and, for some researchers, intellectual stimulation and emotional gratification from working with animals. The degree to which the scientist and the animal each benefit from the relationship varies and is frequently difficult to quantify. In research settings it may be difficult to distinguish a symbiotic relationship from a conspecific one or even from a predator–prey relationship. Potential benefits to the animal, when weighed against the costs, may be a matter of debate. Benefits to both scientist and animal presumably also accrue in conspecific relationships. The use of modified conspecific signals by the interactants in symbiotic relationships also may blur the distinction between conspecific and symbiotic relationships. Such blurring can occur, too, when relationships change over time.

Scientist as conspecific

Perhaps the most interesting, and arguably the most influential, interaction that can occur between human and nonhuman is that in which both of the interactants perceive and respond to each other as members of their own species, or one of the interactants does so. Hediger (1965) has termed such a relation a social partnership and, as discussed in subsequent sections, has made important contributions to the understanding of how such relations develop. Unfortunately, the use of the term "social partnership" is misleading in two ways. First, the term "partnership" implies some sort of mutual agreement or perception between the participants concerning how they will interact. However, Hediger himself points out that the participants' perceptions of each other are not always congruent. Thus, a seemingly paradoxical situation can arise in which a human perceives an animal as a social partner, but the human does not become a social partner to the animal.

Second, the term "social partnership" seems to imply positive social relations based on affiliative interactions. However, conspecifics do not always behave in affiliative ways. Agonistic behavior is common among conspecifics in a variety of situations, especially those involving competition for dominance status or for limited resources such as food, nest sites, or mates. Hediger provides several examples of zoo animals responding aggressively to humans whom they perceived to be conspecific rivals. It also appears that some problems of aggressive behavior among companion and domestic animals (e.g., dominance aggression in dogs, aberrant behavior syndrome in llamas) occur because the animals perceive the humans as conspecific rivals (Ebel, 1989; Marder & Marder, 1985; Voith & Borchelt, 1982).

How conspecific perceptions, conspecific-like behavior, and social attachment all influence one another is complex and not always clear. It is obvious from the writings of Cairns (1966a), Scott (1962, 1971), and others that the formation of social bonds can lead to conspecific perceptions and conspecific-like behavior. However, social attachments may not be necessary for the occurrence of some kinds of conspecific-like behavior. Such behavior can be observed in at least three different kinds of social context. The simplest context is one in which an organism *occasionally* directs conspecific signals toward inanimate objects or heterospecifics, but the objects or heterospecifics are not *consistently* responded to as conspecifics. This usually happens when some element of the structure of the object or the anatomy or behavior of the heterospecific resembles a conspecific social signal or releaser that would normally trigger a social response. For example, Tinbergen (1951, p. 87) has described how the food-begging responses of nestling thrushes can be elicited by the visual stimulation of a human hand waved above the nest.

The second context is one in which an object (or class of objects) or a heterospecific (or a class of heterospecifics) is consistently responded to as a conspecific, but no interindividual relationship or social structure can be observed. Hediger's (1965) concept of social partnership, Roy's (1980) concept of species identity, and Scott's (1971) concept of socialization all seem to be relevant in that they describe processes by which organisms come to respond preferentially to classes or groups of objects or organisms. This preferential response is described as conspecific behavior (or conspecific-like if the preferred object is a heterospecific).

The third social context is similar to, but more complex than, the one just described. In this context, the preference takes the form of an attachment or bond whereby the organism acts to maintain proximity with the attachment object and shows signs of separation distress at involuntary separation. The attachment behavior typically directed toward a conspecific is directed toward a heterospecific. In this context, interindividual relationships will likely form, which may be affiliative, agonistic, sexual, care giving, care seeking, or a combination thereof.

The distinction between the last two contexts can be seen in domestic dogs that have been well socialized to humans. Such dogs respond to unfamiliar humans as they would to unfamiliar dogs, showing species-typical greetings, play-invitations, and even agonistic displays. However, the quality, intensity, and complexity of such interactions are very different from those seen between dogs and humans that have social attachments to each other. Here the human is responded to not as an unfamiliar dog but as a member of the dog's social group – its pack. These three phenomena clearly overlap and most properly should be considered different levels of conspecific-like behavior that can be directed toward heterospecifics. However, distinguishing among the levels is important, especially when one is considering the mechanisms that give rise to the different kinds of perceptions of heterospecifics (see next section).

Some researchers have intentionally set out to develop conspecific perceptions in animals in order to study the processes by which conspecific identity or attachments form or to learn something about an animal that otherwise would be difficult to learn. Lorenz's (1937) work on imprinting in birds and the work by Gardner and Gardner (1971) and Hayes and Nissen (1971) on the cognitive abilities of chimpanzees are examples. However, it should be noted that if conditions are right (see next section) such perceptions can form unintentionally and can have important effects on the research.

Factors that affect the kind of relationships formed between scientists and animals

Assimilation tendency

Hediger, as one of the first scientists to theorize about the establishment of human–animal bonds and interspecific relationships, believed that the most important factor was the assimilation tendency. He defined it as "the deep-rooted tendency in all higher living beings – man included – to see in creatures of a different species, with whom there exists a certain familiarity, creatures of their own kind and to treat them accordingly" (1981, p. 2). The behavioral manifestations of the assimilation tendency were anthropomorphic behavior in humans and zoomorphic behavior in animals. Anthropomorphism occurred when humans perceived and behaved toward other species as if they were human. Hediger (1965) saw zoomorphism as a homologous characteristic whereby animals perceived and behaved toward humans or other heterospecifics as if they were conspecifics.

In an important paper, Hediger (1965) argued that this assimilation tendency, which he thought was restricted to "higher vertebrates," could result in the formation of social partnerships between humans and animals, provided that certain conditions were met. The major condition for the formation of a partnership was a "certain intimacy" between the two organisms. Regrettably, Hediger never defined what he meant by "intimacy." It is unclear whether it refers simply to a prolonged sensory contact (or familiarity) between the two or something more. Subsequent work on social attachment has indeed demonstrated that prolonged sensory contact with the object is necessary for the formation of an attachment – a factor discussed in a subsequent section ("Sensory contact or familiarity").

While the notion of assimilation tendency is appealing and has been pursued by other investigators (e.g., Walther, 1984), there are serious questions about its utility as an explanatory concept. First, assimilation tendency appears to be little more than an intervening variable with no real explanatory power. Assimilation tendency cannot be measured directly; it can only be inferred by observations of anthropomorphic or zoomorphic behavior. The tendency is not equally present in all species or even in all members of a species. Its expression depends on a variety of variables such as similarities between the organisms in sensory systems, body size, and taxonomic position, as well as the amount of intimate contact between them. Because all of these conditions must be specified in order to understand why an organism is behaving anthropomorphically or zoomorphically, nothing seems to be gained by invoking the concept. In these respects, the concept of assimilation tendency is similar to such drive concepts as sex drive, hunger, and thirst. They all have little explanatory power in and of themselves, tend to oversimplify more complex relationships, and actually hinder research (see Hinde, 1970, for a more complete critique of drive concepts).

Second, anthropomorphism and zoomorphism are little more than descriptions of the behavior of organisms toward one another. Even in this Hediger is inconsistent in his use of the terms. In some writings they refer to attachments or bonds involving complex interindividual relationships in which the animal or human has incorporated the other into its own social system (Hediger, 1964). In other writings the terms refer to interactions in which the other organism could not possibly be incorporated into a

15

heterospecific social system, but only where one organism interprets and responds to the behavior of the heterospecific as if it came from a conspecific (Hediger, 1965). An example of the latter is Hediger's discussion of the human-oriented behavior and language that zoo visitors direct toward animals. Most zoo visitors do not have true attachments or bonds with the animals, yet they do behave anthropomorphically. Zoo visitor behavior is closer to the occasional conspecific-like behavior discussed previously. Such looseness of definitions undermines the descriptive power of the terms and makes investigation of underlying mechanisms more difficult.

Although the concept of assimilation tendency explains nothing in and of itself, Hediger's ideas about factors that affect the assimilation tendency have suggested other variables that may be important in determining an animal's perception of humans and human–animal relationships. Those variables as well as others postulated by other investigators will now be examined. It should be noted that most of the thinking about the causes and correlates of human–animal relationships have focused on conspecific perceptions and human–animal bonds. However, as will be seen, most of these factors are important in establishing other kinds of perceptions of humans by animals, and thus other human–animal relationships as well.

Similarities in communication systems

Hediger (1965) thought that bonds between species were more likely to form when the organisms were closely related taxonomically and similar in body size, sensory systems, and modes of communication. In general, these factors are all correlated with one another. Similarities in one factor make it more likely that there will be similarities in the others.

Perhaps the most important of these is similarity in communication systems. The greater the overlap in communication signals between two organisms, the more likely they are to influence each other, to behave as symbionts or conspecifics, and to form social attachments. The less the overlap, the more likely the organisms are to respond to each other as predator and prey or not to respond socially at all. Hediger's (1965) observation that human–animal bonding tends to occur most often with "higher vertebrates" is probably due in part to the greater overlap in communication systems between humans and these other species.

Perhaps one of the most interesting findings in the literature on social attachment is that attachments can be made to organisms not sharing common communication systems and even to inanimate objects (Scott, 1971). The imprinting literature is full of examples of birds and mammals becoming attached to flashing lights, moving objects, and cloth-covered dummies (see Rajecki, Lamb, & Obmasher, 1978). Such attachments can be found in humans as well as animals. Anecdotes of humans becoming attached to objects, clothing, and places are widely known. However, it is also true that the greater the attractivity of an object (i.e., its salience) and the more likely the object is to produce behavioral responses by the organism, such as following or clinging, then the greater the likelihood of attachment to the object (Cairns, 1966a). Clearly, the more similar two organisms are in their communication systems, the greater the likelihood they will find signals from the other salient and the greater the probability that they will respond to them. This in turn should increase the likelihood that symbiotic or conspecific perceptions will develop for one or both of them.

Sensory contact or familiarity

Recent work on social attachment in vertebrates has established that one of the most important factors, if not *the* most important factor, in establishing a social attachment or bond to some other organism or object is prolonged sensory contact with it (Cairns, 1966a; Scott, 1962). Reinforcing factors such as direct tactile contact or feeding, punishment (or the lack of punishment), as well as an organism's age may all influence the speed and likelihood of attachment formation, but they are not necessary for it to occur. Thus, animals that spend significant amounts of time near other animals or humans, or in a given cage, or with a particular ball should be expected to form attachments to these things, regardless of the kinds of interactions experienced. Organisms form attachments to the familiar. Familiarity influences the formation of other sorts of human–animal perceptions and relationships as well. The more familiar organisms are, the more likely they are to respond to each other as symbionts or as conspecifics and the less likely they are to respond as predator and prey. It has been shown, for example, that domestic cats reared with rats from an early age do not respond to the rats as prey but rather as symbionts or conspecifics (Kuo, 1930).

Reinforcement and punishment

Physical contact, feeding, other reinforcers, and the absence of punishment seem to facilitate symbiotic and conspecific perceptions in most species. Punishment, in general, tends to facilitate predator–prey perceptions; however, the effects are complex and may depend on the animal's species, age, and prior experience. Punishment is thought to diminish positive social perceptions by increasing the distance between the animal and the punishing object. Paradoxically, punishment need not deter social bonding if the organism and attachment object are kept in proximity (Cairns, 1966a). Scott (1962) has argued that punishment can even facilitate the attachment process under some circumstances. He cited an unpublished doctoral dissertation by Fisher (1955) which showed that puppies receiving only punishing contacts with humans responded with quicker approaches to humans once the punishment stopped than other puppies that received only affiliative contacts. These results are in marked contrast to those of Hemsworth, Barnett, and Hansen (1987), who used a similar design but found that young pigs that had received punishing stimuli from humans approached them less readily than those receiving only affiliative contacts.

Age

For most species of birds and mammals (e.g., ducks, dogs, cats, and rhesus monkeys), conspecific perceptions and attachments are established more quickly during sensitive periods early in the organism's life (Scott, 1963). However, as Hediger (1965) so keenly observed, they can be formed at almost any time in the animal's life. Anecdotes abound about wild-trapped captive animals forming strong attachments to their keepers (Hediger, 1968). Woolpy and Ginsburg (1967) have described the processes involved in socializing adult wolves to humans. Hediger saw that the social learning process that occurred early in life was not fundamentally different from adult social learning, but rather that the two were extremes of a continuum. Age appears to be important in the formation of other kinds of human–animal perceptions, but has received little direct attention. As mentioned previously, some species are less likely to perceive others as prey if they have had social contact from an early age. Studies of social attachment have also shown that organisms that have had little or no social contact with others during a sensitive period early in life are more likely to perceive others as predators or as social insignificants (Rajecki et al., 1978).

Fear

Fear is another major factor in determining animals' perceptions of humans. In the presence of humans, most wild and many domestic animals show avoidance, escape, or fearful behavior if they have not had experience with humans (Hediger, 1965; Hemsworth & Barnett, 1987). Thus, many species that have not had experience with humans treat them as potential predators. Limited evidence suggests that genotype can influence fearfulness and that domestic species are less likely to show fearful behavior than are wild species (reviewed in Hemsworth & Barnett, 1987). As described previously, nonthreatening experiences with humans during early sensitive periods or later in life can change an animal's perception of the human from one of predator to one of neutral part of the environment, symbiont, or even conspecific. Similarly, punishing or threatening experiences can alter the perception from a more positive to a more negative one.

Most scientists try to reduce fear of humans as much as possible in their animals to improve experimental control. However, even procedures that seem innocuous to the investigator may produce fear in the animal (Hemsworth, Gonyou, & Dziuk, 1986; Hemsworth et al., 1987; see also Hemsworth, Barnett, & Coleman, Chapter 17, this volume). Thus, without a thorough knowledge of the animal's behavior the scientist may be inadvertently promoting a perception that is not wanted.

Genetic predispositions

Genetic predispositions may affect the formation of specific perceptions in other ways. Voith (1985) has pointed out that certain aspects of an animal's social system, which presumably are influenced by genetic predispositions, may make social bonding with humans more likely. Among these is a prolonged period of parental care and living in a large and complex social group. Animals not having such social systems may be less likely to treat humans as conspecifics. Messent and Serpell (1981) have even argued that humans have selectively bred for anthropomorphic and neotenic physical and behavioral traits in dogs to facilitate human–animal bonding. Infantile traits such as flattened faces, large round eyes, and playfulness in adult animals may increase the overlap in the communication systems of animals and humans. This may be because these characteristics resemble those in human infants that stimulate parental care-giving responses (McFarland, 1987a).

Finally, implicit in several of the factors already discussed, such as proximity, sensitive periods and fear, is the notion that prior experience between humans and animals can influence later perceptions and relationships. Extensive positive interactions with humans can facilitate the formation of conspecific-like behavior and bonds between scientists and their animals. Little experience with humans, negative experiences, or inconsistent experiences can lead to the animal treating the human as a predator.

To summarize, several factors seem to be important in determining animals' perceptions of humans. First are genetic predispositions to live in complex groups, show extended parental care, and show fear of humans. Second is the degree of overlap in the communication systems of the scientist and the animal. The greater the degree of overlap, the more likely that one or both will perceive the other as a symbiont or conspecific. Third is the animal's experience with humans. Prolonged sensory contact, especially during sensitive periods, promotes positive perceptions and reduces negative ones. Extensive positive reinforcing experiences reduce fear and facilitate positive perceptions. Finally, fear, whether learned or unlearned, can affect the animal's perceptions of the scientist.

Implications for scientific studies involving animals

Perhaps the most important implication of this review of human–animal relations is that when scientist and animal come into contact, some relationship between them is inevitable. The animal's perception of the scientist will affect its behavior, and this in turn will influence the kind of relationship that forms between the two. This relationship can be ignored, or studied and used to advantage, but it will not cease to exist. The scientist who acknowledges the existence of these relationships and understands how they are formed can use this information to produce better, more efficient, and more humane research.

The scientist's first step in utilizing information about animal perceptions and about relationships is to decide ahead of time what kind of relationship is needed for the specific study that is planned. It is doubtful that a pure predator relationship would ever be called for in the laboratory; however, in the field it might be desirable for the animal to respond to some humans (or all humans some of the time) as predators. For example, some field researchers have promoted their wild animal's perceptions of humans as predators to study antipredator behavior (e.g., Bildstein, 1983; Knight, 1984; Kruuk, 1964).

Similarly, it may not always be desirable for the animal to perceive humans as conspecifics or to form bonds with them. Hediger (1964) has described how animals with attachments to humans sometimes treat humans as sexual or dominance rivals, creating dangerous situations for both. Kilgour (1984) and Grandin (1987) have pointed out that bonding to humans can disrupt the performance of livestock on certain handling tasks. It seems reasonable that such bonds may disrupt some research tasks as well. In contrast, Wolfle (1985) has argued that laboratory animal caretakers *should* strive to form strong mutual bonds with research animals. This, he says, will improve the animal's ability to cope with stress and improve its health, well-being, and quality. Thus, it may be advantageous to have different members of the research team perceived differently by the animal and to promote different kinds of relationships with the animal. For example, it may be best to have the animal form a bond with the caretaker but to respond to the experimenter as a social insignificant or as a symbiont.

20

While it is important for researchers to decide what kinds of animal perceptions and relationships are needed for a given study, such decisions do not always ensure that such perceptions and relationships will come to pass. Knowledge about the formation of heterospecific perceptions and relationships is still incomplete, and control over the relevent variables, such as prior experience, is not always possible.

The second implication of the research reviewed here is that the more one knows about the animal being worked with, the better. Knowledge of the animal's behavior with conspecifics is important in determining the animal's capacities for developing certain kinds of perceptions and relationships. If a close bond with the animal is desirable, then species that live in large, complex social groups and have prolonged periods of parental care such as chimpanzees, rhesus monkeys, dogs, and horses should be considered. If bonding is undesirable, then species that are solitary and have very short periods of parental care, such as golden hamsters, should be considered.

Equally important is knowledge about the communication systems and social signals of the animal. As emphasized previously, the greater the overlap in communication signals between animal and scientist, the more likely it is that the animal will perceive the scientist as a symbiont or a conspecific.

The degree of communication overlap can affect the training of animals and the manipulation of their behavior. A wonderful example of how humans can use this overlap is the herding of cattle by the Fulani tribesmen of Africa (Lott & Hart, 1979). The Fulani use knowledge about cattle dominance signals and affiliative behavior to insert themselves into their herds as social dominants and leaders. In this way they can control the activities and movements of the cattle without fences, gates, or danger to themselves or the cattle. However, as Bouissac (1981) points out, this overlap between human and animal is imperfect, and a lack of knowledge about the communication system of an animal can led to miscommunication and misinterpretation by the scientist of the animal's behavior.

Bouissac (1981), Pryor (1981), and Voith and Borchelt (1982) provide examples of how circus trainers, porpoise trainers, and pet-dog owners, respectively, have easily misinterpreted the behavior of these animals when they did not have accurate knowledge of their species-typical social signals. Pryor describes how the opened-mouth, gaping-jaw display of the bottle-nosed dolphin is often interpreted as smiling or an affiliative display, but in reality it is often a dominance challenge. Such mistakes on the part of human handlers have led to dangerous encounters with animals.

Prescott (1981) described how inadequate knowledge about the sensory abilities of dolphins resulted in inadequate research designs and the misinterpretation of research results. Inadvertent auditory cues were given to animals, which led to false conclusions about their sensory abilities. Greater knowledge about their auditory capacities would have enabled the researchers to use better controls and draw more valid conclusions.

Knowledge about the normal signals used in agonistic, affiliative, sexual, parental, and predator–prey interactions is extremely useful in manipulating the animal's behavior and in helping to create a desirable relationship with the animal. Particularly important in this respect is knowledge about fearful behavior. Since many species are fearful of humans, recognition of fear-related behavior is useful in determining what aspects of the research protocol or scientist's behavior might produce fear. Steps can then be taken to reduce or eliminate fear-producing stimuli.

The third implication of the literature reviewed here is that the scientist should recognize that, all other things being equal, the longer the scientist or other humans spend with the animal, the more likely the animal is to form positive perceptions of humans. If the goal is to form an attachment, the animal must spend time with humans. If a bond is not desirable, the less time spent with the animal, the better. It should be remembered, however, that during sensitive periods, bond formation can occur very rapidly, so that even brief contacts with the animal could result in inadvertent bond formation. Thus, sensitive periods, reinforcement, punishment, and intense physical contact can all modulate the likelihood and rate of bonding and can be used to facilitate or inhibit the formation of bonds and to manipulate the animal's perceptions of the scientist.

The fourth implication is the recognition that the scientist, other members of the research team, and all other humans that come into contact with the animal will form perceptions of and relationships with the animal as well. Humans who come into contact with the animal will respond to it as socially insignificant, predator or prey, symbiont, conspecific, or in some other way. Humans, like the animal, must form some perception. Humans who participate in human–animal interactions and relationships are subject to many of the same processes described for their animal counterparts.

The social system of humans seems to predispose them to form attachments with members of their own species and other species as well. Voith (1985) has pointed out that humans become attached to pets in part because pets' behavior taps into the human intraspecific attachment system, frequently mimicking the behavior of children or other adult humans and eliciting care-giving and affiliative behavior.

Clearly, the previous experience of the scientist and other members of the research team with animals will greatly influence the kinds of perceptions and ultimately the kinds of relationships that they will form with the animal. Negative experiences may promote fear and predator perceptions; positive experiences may promote symbioses or bonding. The more prolonged the contact between scientist and animal, the more likely that the scientist will form an attachment to the animal.

Finally, the dangers of anthropomorphism in science are widely known (Hediger, 1981; but see Lockwood, 1985, for an opposing view). It is commonly believed that the more knowledge that scientists have about the natural behavior and behavioral capacities of the animal, the less likely they will be to anthropomorphize. Conversely, it has also been observed by some scientists studying the human–animal bond that the less a human owner knows about a pet's species-typical behavior, the more likely she or he is to be anthropomorphic. If a pet owner frequently interprets the animal's species-typical behavior as human behavior, the owner will be more likely to view the animal as "human-like" and thus form a stronger attachment to it (S. Hetts, unpublished observations). In other words, because the owner *perceives* that mutual communication is occurring, regardless of the animal's perception, the owner will become more attached to the pet, even if the pet is not as strongly attached to the owner. The strength of attachment need not be mutual. This hypothesis does not, therefore, contradict the concept that increased knowledge of species-typical behaviors will increase the degree of two-way communication and thus increase mutual attachment between human and animal. In the former case, a bond forms because of the human *misperception* of shared communication and shared emotions. In the latter case, a bond forms because of truly shared communication.

The implications of this for the scientist and the rest of the research team are that a lack of knowledge about the animal may predispose them not only to anthropomorphize but also to form stronger bonds with the animal, if other conditions for bond formation are present. This may occur regardless of whether the animal in turn bonds to them. Scientists must keep a constant vigil against anthropomorphic thinking and interpretation when performing animal research.

Throughout this chapter the formation of relationships between scientist and animal has been discussed from the point of view either of the animal or of the scientist. However, it should be recognized that a relationship involves the actions and reactions of both parties. They will mutually influence each other, and this can alter all future interactions. The animal that comes into the research situation perceiving the human as a predator (or a conspecific) will alter the behavior of the scientist regardless of the kind of perception the scientist was predisposed to form. What ultimately transpires in the research will be, in part, a function of the kind of relationships that develop between the scientist, other members of the research team, and the animal.

Acknowledgments

We thank Dr. Sharon Crowell-Davis and Dr. Laurie Hiestand and the editors for their very useful comments on earlier drafts of this chapter.

References

Bildstein, K. L. (1983). Why white-tailed deer flag their tails. *American Naturalist*, *121*, 709–15.

Bouissac, P. (1981). Behavior in context: In what sense is a circus animal performing? In *The Clever Hans Phenomenon: Communication with Horses, Whales, Apes, and People*, ed. T. A. Sebeok & R. Rosenthal. *Annals of the New York Academy of Science*, *364*, 18–25.

Caine, N. G. (1990). Unrecognized anti-predator behavior can bias observational data. *Animal Behaviour*, *39*, 195–7.

Cairns, R. B. (1966a). Attachment behavior of mammals. *Psychological Review*, *73*, 409–26.

(1966b). Development, maintenance, and extinction of social attachment behavior in sheep. *Journal of Comparative and Physiological Psychology*, *62*, 298–306.

(1972). Attachment and dependency: A psychobiological and social-learning synthesis. In *Attachment and Dependency*, ed. J. L. Gewirtz, pp. 29–80. Washington, DC: Winston.

Dewsbury, D. A. (1978). *Comparative Animal Behavior*. New York: McGraw-Hill.

Ebel, S. (1989). The llamma industry in the United States. In *Llama Medicine*, ed. L. W. Johnson. *Veterinary Clinics of North America: Food Animal Practice*, *5*, 1–20.

Fisher, A. E. (1955). The effects of differential early treatment on the social and exploratory behavior of puppies. Unpublished doctoral dissertation, Pennsylvania State University.

Gardner, B. T., & Gardner, R. A. (1971). Two way communication with an infant chimpanzee. In *Behavior of Nonhuman Primates*, ed. A. M. Schrier & F. Stollnitz, Vol. 4., pp. 117–84. New York: Academic Press.

Grandin, T. (1987). Animal handling. In *Farm Animal Behavior*, ed. E. O. Price. *Veterinary Clinics of North America: Food Animal Practice*, *3*, 323–38.

Hayes, K. J., & Nissen, C. H. (1971). Higher mental functions of a home-raised chimpanzee. In *Behavior of Nonhuman Primates*, ed. A. M. Schrier & F. Stollnitz, Vol. 4, pp. 60–115. New York: Academic Press.

Hediger, H. (1964). *Wild Animals in Captivity*. New York: Dover.

(1965). Man as a social partner of animals and vice-versa. In *Social Organization of Animal Communities*, ed. P. E. Ellis. *Symposia of the Zoological Society of London*, *14*, 291–300.

(1968). *The Psychology and Behavior of Animals in Zoos and Circuses*. New York: Dover.

(1981). The Clever Hans phenomenon from an animal psychologist's point of view. In *The Clever Hans Phenomenon: Communication with Horses, Whales, Apes and People*, ed. T. A. Sebeok & R. Rosenthal. *Annals of the New York Academy of Science*, *364*, 1–17.

Hemsworth, P. H., & Barnett, J. L. (1987). Human–animal interactions. In *Farm

Animal Behavior, ed. E. O. Price. *Veterinary Clinics of North America: Food Animal Practice*, *3*, 339–56.

Hemsworth, P. H., Barnett, J. L., & Hansen, C. (1987). The influence of inconsistent handling by humans on the behaviour, growth and corticosteroids of young pigs. *Applied Animal Behaviour Science*, *17*, 245–52.

Hemsworth, P. H., Gonyou, H. W., & Dziuk, P. J. (1986). Human communication with pigs: The behavioural response of pigs to specific human signals. *Applied Animal Behaviour Science*, *15*, 45–54.

Hinde, R. A. (1970). *Animal Behaviour* (2d ed.). New York: McGraw-Hill.

  (1976). Interactions, relationships and social structure. *Man*, *11*, 1–17.

Kilgour, R. (1984). The role of human–animal bonds in farm animals and welfare issues. In *The Pet Connection*, ed. R. K. Anderson, B. L. Hart, & L. A. Hart, pp. 58–74. Minneapolis: University of Minnesota Press.

Knight, R. L. (1984). Responses of nesting ravens to people in areas of different human densities. *Condor*, *86*, 345–6.

Kruuk, H. (1964). Predators and anti-predator behaviour of the black-headed gull (*Larus ridibundus*). *Behaviour*, Suppl. 11, 1–129.

Kuo, Z. Y. (1930). The genesis of the cat's response to the rat. *Journal of Comparative Psychology*, *11*, 1–35.

Lockwood, R. (1985). Anthropomorphism is not a four letter word. In *Advances in Animal Welfare Science, 1985/1986*, ed. M. W. Fox & L. D. Mickley, pp. 185–99. Washington, DC: Humane Society of the United States.

Lorenz, K. Z. (1937). The companion in the bird's world. *Auk*, *54*, 245–73.

Lott, D. F., & Hart, B. L. (1979). Applied ethology in a nomadic cattle culture. *Applied Animal Ethology*, *5*, 309–19.

Marder, A. R., & Marder, L. R. (1985). Human–companion animal relationships and animal behavior problems. In *The Human–Companion Animal Bond*, ed. J. Quackenbush & V. L. Voith. *Veterinary Clinics of North America: Small Animal Practice*, *15*, 411–21.

McFarland, D. (1987a). Anthropomorphism. In *The Oxford Companion to Animal Behavior*, ed. D. McFarland, pp. 16–17. New York: Oxford University Press.

  (1987b). Symbiosis. In *The Oxford Companion to Animal Behavior*, ed. D. McFarland, pp. 539–41. New York: Oxford University Press.

Messent, P. R., & Serpell, J. A. (1981). An historical and biological view of the pet–owner bond. In *Interrelations Between People and Pets*, ed. B. Fogle, pp. 5–22. Springfield, IL: Thomas.

Prescott, J. H. (1981). Clever Hans: Training the trainers, or the potential for misinterpreting the results of dolphin research. In *The Clever Hans Phenomenon: Communication with Horses, Whales, Apes and People*, ed. T. A. Sebeok & R. Rosenthal. *Annals of the New York Academy of Science*, *364*, 130–6.

Pryor, K. (1981). Why porpoise trainers are not dolphin lovers: Real and false communication in the operant setting. In *The Clever Hans Phenomenon: Communication with Horses, Whales, Apes and People*, ed. T. A. Sebeok & R. Rosenthal. *Annals of the New York Academy of Science*, *364*, 137–43.

Rajecki, D. W., Lamb, M. E., & Obmascher, P. (1978). Toward a general theory of infantile attachment: A comparative review of aspects of the social bond. *Behavioral and Brain Sciences*, *3*, 417–64.

Roy, M. A. (1980). An introduction to the concept of species identity. In *Species Identity and Attachment*, ed. M. A. Roy, pp. 3–21. New York: Garland STPM Press.

Scott, J. P. (1962). Critical periods in behavioral development. *Science, 138,* 948–58.
    (1963). The process of primary socialization in canine and human infants. *Society for Research in Child Development Monographs, 28,* 1–47.
    (1971). Attachment and separation in dogs and man: Theoretical propositions. In *The Origins of Human Social Relations,* ed. R. Schaffer, pp. 227–46. New York: Academic Press.

Scott, P. A., Bentley, C. V., & Warren, J. J. (1985). Aggressive behavior by wolves towards humans. *Journal of Mammalogy, 66,* 807–9.

Short, D. J. (1967). Handling laboratory animals. In *UFAW Handbook on the Care and Management of Laboratory Animals,* (3d ed.), pp. 114–33. London: Livingstone.

Tinbergen, N. (1951). *The Study of Instinct.* New York: Oxford University Press.

Voith, V. L. (1985). Attachment of people to companion animals. In *Symposium on the Human–Companion Animal Bond,* ed. J. Quackenbush & V. L. Voith. *Veterinary Clinics of North America: Small Animal Practice, 15,* 289–95.

Voith, V. L., & Borchelt, P. L. (1982). Diagnosis and treatment of dominance aggression in dogs. In *Symposium on Animal Behavior,* ed. V. L. Voith & P. L. Borchelt. *Veterinary Clinics of North America: Small Animal Practice, 12,* 655–63.

Walther, F. R. (1984). *Communication and Expression in Hoofed Mammals.* Bloomington: Indiana University Press.

Wolfle, T. (1985). Laboratory animal technicians. In *Symposuim on the Human–Companion Animal Bond,* ed. J. Quackenbush & V. L. Voith. *Veterinary Clinics of North America: Small Animal Practice, 15,* 449–54.

Woolpy, J. H., & Ginsburg, B. E. (1967). Wolf socialization: A study of temperament in a wild social species. *American Zoologist, 7,* 357–63.

—3—
Studies of rodent–human interactions in animal psychology

Donald A. Dewsbury

Editors' introduction

As Dewsbury notes, rodents have been widely studied as a matter of convenience, rather than because of any intense interest in them per se. Given the extensive amount of contact between scientists and rodent subjects, it is not surprising that these animals have been affected in measurable ways. Such effects have been both obvious and subtle, immediate as well as long term.

Dewsbury introduces a number of themes that recur in other chapters. For example, human social contact is most effective when it mimics rodent intraspecific behavior. Dewsbury also notes that rodents are capable of differentiating among human handlers. Such discriminations may allow particular individuals to be a source of either positive or negative motivation and reinforcement. Finally, Dewsbury notes that intrinsic interest in scientist–rodent interactions may now be replacing earlier, more exploitive attitudes.

Rodents, especially laboratory rats (domesticated forms of *Rattus norvegicus*), have been the subjects in an inordinate proportion of studies in animal psychology. Often, rodents have been studied because of convenience; sometimes they have been treated as surrogate human subjects. We now know, however, that rat–human relationships are more complex than often implicitly assumed by researchers. Conducted in the 1950s and 1960s, much of the relevant research can be viewed somewhat differently today. I shall thus discuss rodent–human relationships in broad social and historical context.

Three traditions in animal psychology

Although they were not always clearly differentiated, one can discern three primary traditions in animal psychology (Dewsbury, 1984). These correspond to the three journals in the field now published by the American Psychological Association: *Behavioral Neuroscience*, the *Journal of Experimental Psychology: Animal Behavior Processes*, and the *Journal of Comparative Psychology*. The tradition in physiological psychology has been concerned primarily with the analysis of physiological processes underlying behavior and will not be dealt with here. One can then contrast a process-oriented tradition with a comparative psychological tradition. In the former, the emphasis is on discovering processes of psychological importance. There is relatively little interest in the animal subject per se; the focus is on experiments revealing processes of relevance to human behavior. Studies of learning have been especially prominent. In the comparative tradition, by contrast, there is more focus on the naturally occurring behavior of the species in question and more emphasis is placed on the ways in which particular species adapt to situations of biological relevance in the natural lives of animals. A long-standing interest in the study of behavioral development exists within this tradition. The comparative tradition developed around the turn of the twentieth century and has been continuous throughout the century. The process-oriented tradition grew rapidly during the middle part of the century, often obscuring activity in the comparative tradition. A balance has been restored in recent years, and there has been something of a blending, as with research on foraging, learning in natural contexts, and animal cognition.

Social behavior

Social behavior can be defined as "behavior having to do with interactions between conspecifics or between animals of different species where some form of symbiosis exists in which the behavioral communication is similar to that within a species" (Immelmann & Beer, 1989, p. 274). The study of social behavior has been especially important in the comparative psychological tradition (see Dewsbury, 1984). Hebb and Thompson (1954) reviewed a considerable number of studies on "animal social psychology"; Zajonc (1969) edited a book with that title.

Psychologists have long recognized the social tendencies of nonhuman animals, often postulating some form of social drive or instinct. Both Galton (1907) and McDougall (1908) wrote of gregarious instincts; James (1890) discussed sociability in his chapter on instinct. Comparative psychologists have long recognized that the opportunity to engage in social interactions can be rewarding to both human and nonhuman animals. Thus, Ligon (1929) found some indication that rats could serve as positive incentives for approach by conspecifics. Zajonc (1969) devoted a whole section of his book to "affiliation," reprinting papers by Ashida (1964) and Shelley and Hoyenga (1966) on gregariousness in rats. Latané conducted an extensive research program on social attraction in rats (e.g., Latané & Hothersall, 1972). Affiliative behavior is present even in very young rats, as in their huddling behavior (Alberts, 1978a).

Interspecific social behavior

The focus in most research in the comparative tradition has been on intraspecific social behavior. Indeed, social behavior has often been defined so as to be limited to a single species. For Drever (1952) "social" refers to the "relation of an individual to others of the same species" (p. 268). Similarly, in treating social behavior, Warren (1934) emphasized responses "directed toward other organisms, usually of the same species" (p. 252), and Heymer (1977) stressed "behaviour patterns the frequency of which is changed by the presence of a conspecific" (p. 166).

Despite definitional restrictions and a research emphasis on social behavior within species, some research has been directed at interspecific social interactions. Perhaps the best known is that of Scott and Fuller (1965) on dog–human interactions, but the work of Mason (1983) on dogs as companions for monkeys and that of Cairns and Johnson (1965) on attachments between lambs and dogs is also relevant. As reported in the rodent literature, rats have been reared with cats (e.g., Kuo, 1938) and with mice (Denenberg, Hudgens, & Zarrow, 1964).

There has been special interest in human–nonhuman social interactions. Again, the work of Scott and Fuller (1965) stands out. In much literature on canines, the assumption has been that dogs treat humans as a member of the pack and display social responses accordingly. Among the reasons for special interest in animal–human relationships are the extensive domestication of nonhuman animals by humans (e.g., Boice, 1973; Hale, 1969) and the extent of interactions between nonhuman animals and our own species. More recently, various forces have combined to yield a broader consideration of human–nonhuman bonding (e.g., Katcher & Beck, 1983). Questions of the ethics of human–nonhuman interactions have been important in engendering this shift in emphasis and increased sensitivity to human–nonhuman relationships and the associated responsibilities of humans (e.g., Sperling, 1988).

Developments from the 1950s to the present

The 1950s and 1960s

Several developments combined to generate a spurt of interest in behavioral development and social behavior in the 1950s and 1960s. Much of the impetus came from process-oriented animal psychology. Thus, laboratory rats became the species of choice, often with little consideration as to appropriateness.

Learning theory was dominant in animal psychology, and motivation was generally viewed as being built via a process of secondary conditioning on a few primary drives (hunger, thirst, sex, etc.). In the 1950s psychologists began dismantling primary drive theory, as they discovered evidence that learning could occur in the absence of a reduction in what were regarded as primary drives. They found that the opportunity to explore and to interact with objects yielding "contact comfort" was reinforcing to young rhesus macaques (e.g., Harlow, 1976). Copulation without drive reduction was found to be reinforcing for rats (Sheffield, Wulff, & Backer, 1951). The fact that in neither case was there primary drive reduction led to research interest in a variety of motivational problems and in the reinforcing effects of various stimuli, including social interactions.

Even more substantial was the developmental focus; a large literature developed on the effects of early experience on adult behavior. Critical in the development of this trend were the theories of Freud, Hebb (1949), and Bowlby (1958). The general view was that events early in life were critical for later development. The study that initiated this vast literature, that of Hall and Whiteman (1951), was done in response to a challenge to the view that early trauma had profound effects on later personality. The "critical period hypothesis" (e.g., Scott, 1978) became focal.

The intent in this research was to study processes rather than animals. Because rats were the common species in psychological vivaria, they became the de facto choice of subjects. The easiest way for experimenters to stimulate and traumatize rats was to do things to them. Indeed, in some cases stimulation by humans produced effects similar to the delivery of electric shocks (e.g., Levine, 1962). Thus, a large literature on human–rat interactions developed, not out of intrinsic interest in such interactions, but as a matter of convenience because treatment by humans was the easiest way to provide stimulation and rats were the easiest subjects to use. As a natural adjunct to these developmental studies, psychologists studied the reactions of rats (e.g., reinforcing value) to the stimulation patterns being used when humans handled rats in studies of effects of stimulation on development (e.g., Candland, Faulds, Thomas, & Candland, 1960).

Parallel and interacting developments occurred in the comparative tradition. This was the time of interaction between comparative psychologists and European ethologists. European ethology had a strong tradition in the study of social behavior (e.g., Tinbergen, 1953), and the best-known ethological research was the imprinting work of Lorenz (e.g., 1935/1970). Beach and Jaynes (1954) and King (1968) tried to steer the study of early experience toward the comparative tradition, emphasizing a wide range of species and naturally occurring behavioral patterns. Although there was some effort in this direction, most of the work remained process-oriented and designed more to reveal processes than to provide information about the natural lives of animals.

More recent developments

In the 1950s and 1960s there developed a large literature on human–rat interactions related to handling and early experience, with hundreds of similar studies appearing in the psychological literature. As the comparative and process-oriented traditions blended, substantive criticisms of this literature were one factor in its decline. In the typical early-experience study, a treatment was given to one group of rats but not the other, the rats were then returned to their cages, and testing occurred at a later age. The emphasis was on the experimental study of discrete treatments rather than on developmental processes, which are continuous. Schneirla and Rosenblatt (1963) argued that this research missed the important aspects of the developmental process. They also criticized the study of social interactions between, rather than within, species.

Perhaps the most telling criticism of the developmental literature was that of Daly (1973), who critically reviewed the literature on the presumed benefits of early stimulation. Daly, like others, pointed to the considerable number of contradictory results reported in this literature, in many cases even with apparently similar treatments. He then questioned the whole conceptual framework of this literature, pointing out that because the very young of these animals develop in burrows, the laboratory may already be overstimulative and that the presumed benefits of increased stimulation might not be benefits at all if viewed in relation to the natural habitat. Daly was critical of the anthropocentric focus of this literature.

The open-field test, in which an animal is placed in a featureless environment and its activity and eliminative behavior measured, became the method of choice in many studies of early experience. Performance in the open field was thought to be related to "emotionality" or an "exploratory drive," depending on the context of the study. This method came under strong criticism (e.g., Walsh & Cummins, 1976). Gallup and Suarez (1980)

31

and Suarez and Gallup (1982) analyzed the open-field test situation with chickens and rodents in a more ethological framework, now viewing the experimenter as a "predator" of the subject in the experiment.

These telling criticisms led to the gradual decline of research in this tradition, although occasional studies continued to appear. Animal psychologists tended to emphasize problems of naturally occurring behavior on the one hand and animal cognition on the other. More recently, however, researchers have become more interested in human–animal relationships for their own sake (e.g., Katcher & Beck, 1983). The data collected during the earlier spurt of research may now be of interest in the context of rodent–human relationships and their implications for research. I shall review some of this research in this perspective.

A brief overview of research on the reinforcing value of rodent–human interactions

The reinforcing value of intraspecific social interactions

The early study of Ligon (1929) was followed in studies such as those of Ashida (1964) and Shelley and Hoyenga (1966) on rats and Lindzey, Winston, and Roberts (1965) on house mice. All tended to indicate that social stimuli were reinforcing to rodents. The most extensive analyses were those of Latané and his colleagues (e.g., Latané & Hothersall, 1972), who not only showed that rats tend to aggregate, but analyzed stimulus determinants by studying the reactions of rats to a variety of objects. Huddling was also studied in spiny mice, *Acomys cahirinus*, in which it is dependent on familiarity (Porter, Wyrick, & Pankey, 1978).

The reinforcing value of social interaction was strong enough to be viewed as a potential artifact in studies of the reward value of different aspects of copulation with and without "drive reduction" in males and for females (e.g., Bolles, Rapp, & White, 1968; Drewett, 1973; Ware, 1968). It became necessary to demonstrate reinforcing value over and above that of some control for "mere" social interaction. Interestingly, in the literature on social attraction it became necessary to demonstrate that attraction was not due merely to sexuality (Sloan & Latané, 1974).

The reinforcing value of nonhuman interspecific social interactions

Relatively little work was done on interspecific contact other than that with humans. Shelley and Hoyenga (1966) showed that rats would approach a caged chick. Latané and Hothersall (1972) showed that rats and gerbils would aggregate, although the extent of aggregation was less than half as much as that within species.

The reinforcing value of interactions with humans for rats

The reactions of rats to contact with humans depend on the nature of the contact. The initial reason for undertaking studies of animals seeking contact with humans was to examine the reinforcing value of the pattern of stimulation used in the early-experience literature. Weininger (1953) developed a method of "gentling," in which the rat was held in one hand and stroked from head to tail at a rate of approximately 50 strokes per minute. In the initial study Weininger noted that during gentling rats made a clicking noise with their teeth, which was interpreted as a "sign of happiness or content" (Weininger, 1953, p. 112). Candland et al. (1960) assessed the preference of rats for 30 seconds of gentling versus 30 seconds of confinement in the goal box of a U-shaped discrimination apparatus. Regardless of their previous experience with gentling or with other rats, the rats learned to avoid gentling when given a choice. In a subsequent study, Candland, Horowitz, and Culbertson (1962) showed that the negative reinforcing value of gentling was sufficient to support the learning of a new discrimination response. Wong (1972) obtained similar results with male rats and unhandled females, but found no negative reinforcement in handled females. Although it was thus shown that the pattern of gentling used in early-experience studies could be aversive to rats, Candland et al. (1960) were careful to point out that "a different rate of stroking, or a different manner of gentling, might yield different results" (p. 57).

In a study by Sperling and Valle (1964), neither rats with handling experience and without food deprivation nor a group with no pre-experiment experience showed any change in preference for entering a compartment associated with 5 minutes of gentling. However, in a third group, gentling was paired with feeding. Before the experiment the rats had free access to water but were fed just once per day, with an experimenter stroking them during the feeding period. When later tested in the two-compartment apparatus, these rats displayed acquisition positively reinforced by gentling. The results were interpreted as being due to secondary reinforcement, because gentling had been paired with feeding.

Positive reinforcing effects, without overt pairing with feeding, have also been reported. As part of their systematic program to determine the stimulus parameters underlying intraspecific social attraction in rats, Latané and his associates used stimuli delivered by humans. Werner and Latané (1974) argued that what are critical about the reinforcing value of intraspecific social interactions are the complexity and unpredictability of the the interactions, rather than any particular tactile sensation or odor per se. They argued that a human hand "is just about rat size, soft, warm, and flexible, easily manipulated, and clearly not a rat" (p. 329). The stimulation, termed "fondling," was designed to mimic intraspecific interactions, as the experimenter poked, lifted, rubbed, scratched, tapped, and tumbled the rat, but never picked it up. The results of three experiments showed that rats that had been familiar with the human hand were attracted to it and interacted with it, sniffing it, nibbling it, following it, and crawling on it, in a manner similar to that of intraspecific play. Werner and Anderson (1976) compared the reactions of rats to fondling and 50-stroke-per-minute gentling. Whereas fondling served as a positive reinforcer, gentling had negative reinforcing effects.

Another indication of positive responses to human stimulation was found in the work of Boice, Harmon, and Boice (1967). They gave rats free trials in an E-shaped maze to determine their preference for removal by the experimenter or via a mechanical handling box. Untamed rats preferred the mechanical removal, whereas prehandled rats preferred removal by humans.

More recently, Davis and Pérusse (1988) studied the acquisition of a lever-press response in rats, with interaction with a human contingent on the operant response. "The experimenter picked up, petted, spoke to, and allowed the rat to continuously climb on her person" (p. 90). With this pattern of interaction, acquisition (i.e., evidence of positive reinforcement) was shown in half of the animals tested. The course of acquisition and extinction was not unlike that found with conventional rewards, such as food and water.

Do rodents discriminate among individual humans?

Within many animal societies there is individual recognition, the discrimination of one conspecific from another. One may question whether rodents discriminate among individual humans. Kintz, Delprato, Mettee, Persons, and Schappe (1965a) studied the ability of rats to use an experimenter as a discriminative stimulus for choice in tests in a T-maze with food reward. Rats were tested by two experimenters varying in random order. When one experimenter handled the animals, the food was on the left; when the other did the handling, the food was on the right. The rats learned to discriminate between the two experimenters as a cue to the location of food. The amount of handling between trials was varied and had little effect.

An even more dramatic individual discrimination can be found in the "caretaker effect" of McCall, Lester, and Dolan (1969). They studied the behavior of rats in different parts of an open field as a function of the presence of objects in the open field and of experimenters outside the open field. Because several factors were varied and some treatment interactions obtained, the results are somewhat complex. However, the rats generally spent more time on the side of the open field near their caretaker, a male dressed in a familiar white laboratory coat, as opposed to a stranger, a female unfamiliar to the rats and dressed in street clothes. McCall, Lester, and Corter (1969) manipulated the caretaker effect systematically, studying two groups of rats, each reared by a different caretaker. In each group the rats spent more time near their caretaker than near the caretaker of the other rats. In addition, when special tops were fitted to the open field in order to determine the nature of cues used in making the discrimination, the results suggested that olfactory, not visual, cues were important.

Morlock, McCormick, and Meyer (1971) designed a study to determine whether the caretaker effect might be due to approach toward the care-taker or avoidance of the stranger. Rats were tested in a five-path elevated maze with just one of the two humans present at a time. The rats explored the maze more in the presence of the caretaker than with the stranger. Analysis of the pattern of path entries suggested that the rats were avoid-ing the stranger but not differentially approaching the caretaker.

Anecdotally, I recall a caretaker at the University of Michigan who handled rats with the effect that they came to follow him as he moved about the periphery of the table on which they were placed. The phenom-enon was not studied systematically.

Contemporaneous influences of humans on the behavior of rodents

In addition to being reinforcers for approach and choice behavior in rats, humans can influence the behavior of rats in a number of ways, some of them subtle. Those discussed in the present section are contemporaneous, rather than developmental, effects.

That the presence of conspecifics can influence the rate at which behavioral patterns are displayed has been emphasized by Zajonc (1965, 1969) in reviews of social facilitation. Among the paradigms studied were audience effects and coaction effects. An example of the coaction effect can be found in research of Harlow (1932) showing that rats eat more when grouped than when alone.

Some effects of human presence and interaction are fairly straight-forward. For example, Hughes (1978) compared activity in an automated open field with and without a human present, finding that the presence of a human inhibits ambulation. The extent of familiarity between experimenter and rats was not manipulated. Walsh and Cummins (1976) advocated studying how the ways in which animals are transported to the apparatus affect open-field behavior. Extra handling can have a beneficial effect on performance in learning tasks. Bernstein (1952) found that performance in a T-maze discrimination task with food deprivation and reward improved in rats receiving extra handling. Barry (1957) found fewer errors during retraining among rats that were handled daily during a posttraining interval in a study of escape learning in a water maze than among rats not handled. West and Michael (1987) found that rats prehandled for 7 days acquired a lever-pressing response motivated by brain self-stimulation more quickly than nonhandled rats.

Other effects can be more subtle. The term "experimenter effect" was coined to cover cases of more subtle and unintentional effects of humans on the behavior of rodent subjects in experiments (Kintz, Delprato, Metee, Persons, & Schappe, 1965b; McGuigan, 1963). Rosenthal and Fode (1963) studied undergraduate students testing Sprague–Dawley rats that they had been told were of maze-bright and maze-dull lines. The rats believed by the experimenters to be of the maze-bright line learned more rapidly. The authors attributed this difference to the different patterns the experimenters used in handling the two groups of rats, thus facilitating performance. Maier (1956) reported a case of two research assistants working in adjacent rooms, one of whom obtained the expected number of "fixated" responses, whereas the other did not. Maier reported that the one "felt sorry for the rats, and this may have caused him to pet the rats between trials somewhat more than other researchers" (p. 36).

In the early-experience literature, to be discussed in the next section, it was found that some "treatment effects" might be due not to the treatment itself, but to variations in the behavior of the young and of the mother toward the young after the treatment had been imposed (Lee & Williams, 1974; Richards, 1966; Young, 1965). Thus, human handling alters intraspecific interactions, which may in turn produce the effect.

Effects of early experience

The literature on the effects of early experience on rodent behavior that developed in the 1950s and 1960s is massive, and its review is well beyond the scope of this chapter (see Daly, 1973; King, 1958; Newton & Levine, 1968). In general, these were studies in which humans stimulated rodents, but the fact that it was humans who were stimulating the animals was generally incidental to the objectives of the research. Although many treatments were used and the terminology differs somewhat from study to study, we can simplify by adopting two terms, "handling" and "gentling." In the handling literature (e.g., Denenberg, 1968; Levine, 1962) preweaning rats typically were removed daily from the home cages and placed in containers for a few minutes before they were returned to their mothers.

In some cases electric shocks were delivered to the young; these often produced effects similar to handling. The "gentling" manipulation generally was used with postweaning animals. As noted earlier, animals typically were picked up and stroked from head to tail at a rate of approximately 50 strokes per minute for a period of about 10 minutes (e.g., Gertz, 1957; Weininger McClelland, & Arima, 1954).

In various studies, extra stimulation produced (1) increased growth and accelerated development, (2) reduced emotionality, as measured by activity and defecation in the open field, (3) improved performance in learning tasks, and (4) more appropriate endocrine responses to adult stress. The results were interpreted as being due to (1) direct stimulative effects, (2) hypothermia produced when pups were removed from the warm nest, (3) altered maternal behavior, or (4) early stress. However, as this literature grew, an increasing number of contradictory effects and failures to obtain effects were reported. Overall, it is clear that early experience can influence development, with effects on endocrine responsivity perhaps a bit more robust than others. Other effects, such as those on open-field behavior, were reported in enough studies that they must be regarded as real, replicable phenomena, even though not obtained in all studies. The inference that these observed effects are beneficial appears strained, however (see Daly, 1973).

An additional example of the complexity of the subtle interactions among treatments can be found in results showing that handled rats weighed more than controls if reared in a heavily trafficked room before weaning, whereas controls weighed more if reared in quiet rooms (Denenberg, Schell, Karas, & Haltmeyer, 1966; McMichael, 1966). Since much of the background stimulation was from humans, this represents another subtle rat–human interaction.

Conclusions and the need for further research

The interactions between rodents and humans in psychological research are more extensive and more subtle than often recognized. Various apparently trivial actions by the humans can have consequences for the rodents and their behavior. Furthermore, contact with humans can be either rewarding or punishing for the rodent depending on its past experience and the kind of stimulation used. There is some indication that rodents can and do discriminate among humans, with odors appearing to provide the primary cue. Surely there are many anecdotes of close interactions between individuals and their pet rodents. However, rarely have such interactions been subjected to systematic investigation.

Implications for research

These results have several implications for future research with rodents. First, it seems worthwhile to study these phenomena in their own right. However, in addition, it is clear that humans can influence the behavior of rodents in ways rarely recognized by researchers in this field. The nature of the human–rodent interaction as one important factor affecting the results of experiments needs recognition. It should be recognized that differing results from different laboratories or individuals within laboratories may be produced not only by differences in strain, maintenance conditions, and so on, but also as a result of the ways in which humans interact with the rodents. Parametric designs in which the nature of these interactions is varied systematically as one variable would pay dividends in revealing some of the more subtle interactions. This was effective in studies of the caretaker effect and may work equally well in other contexts.

Are the effects adaptive for rats?

In contemporary comparative psychology there is often a search for the adaptive significance of behavior – its role in promoting survival and successful reproduction for the individual. It is difficult to see why the formation of positive bonds with humans should have evolved among rodents. Humans have generally preyed upon rodents and acted to eliminate them. Because, in captivity, individuals that are favored by humans may survive longer and be chosen for breeding, related behavioral patterns would appear adaptive. However, it seems unlikely that such patterns could have evolved de novo during the time rodents have been in captivity. Rather, it seems more likely that humans are able to mimic stimuli the approach to which is adaptive to rodents in conspecific interactions. This is most obvious in the work of Latané and his associates, in which a human hand was manipulated so as to resemble the behavior of a rat as closely as possible. However, stroking and handling in other situations may have similar elements. Thus, in studies where rodents approach humans, it may reflect the action of response systems evolved to function in intraspecific interactions.

Is it a true bond?

The question of whether the cases of apparent rodent–human bonding are true bonds or merely reactions to stimulation mimicking that in intraspecific interactions may be indeed a pseudoquestion. Since early ethological studies, such as Lorenz's *Kumpan* work (Lorenz, 1935/1970), it has been recognized that many prolonged social interactions can be based on mutual exchange of appropriate stimuli and reciprocal feedback as interactions progress. It is thus possible that the bases of all social interactions may be reduced to stimulus control of behavior not intrinsically different in social and nonsocial situations.

The cues for affiliation

Further study of the stimulus aspects of human–rodent interactions may be rewarding. Latané and associates (e.g., Latané & Hothersall, 1972) have undertaken such analyses for intraspecific interactions, and Alberts (1978b) has analyzed the multisensory control of huddling by infant rat pups. Similar analyses might reveal more clearly the nature of the cues relevant to rodents both in responding directly to humans and as human behavior produces more subtle effects on rodent behavior.

Why persist?

It is not clear whether a major research effort directed at human–rodent interactions will yield major insights into rodent behavior or its control, development, evolutionary history, or adaptive significance – the primary foci in contemporary comparative psychology. It is always possible, however, that analyses of unusual stimuli will reveal more about rodents than will studies of natural situations – an aspect of ethological analyses more common than is often recognized. One reason for continued study is that rodents continue to be used as research subjects, and we must understand the nature of the whole experimental situation as clearly as possible. A primary reason for continued research is to gain a better understanding of human–nonhuman interactions and their implications for animal use in experimentation. Furthermore, as more researchers have come to treat their subjects as something more than a mere vehicle for research, there has developed an interest in the nature of these animals and how they perceive humans even if such interactions are not part of the evolved repertoire of the species. This reflects a shift from a pattern of selecting animals and procedures in order to reveal processes of psychological importance to one with more concern for the intrinsic nature and value of the animal and the human–animal relationship.

References

Alberts, J. R. (1978a). Huddling by rat pups: Multisensory control of contact behavior. *Journal of Comparative and Physiological Psychology, 92,* 220–30.

Donald A. Dewsbury

(1978b). Huddling by rat pups: Group behavioral mechanisms of temperature regulation and energy conservation. *Journal of Comparative and Physiological Psychology, 92,* 231–45.

Ashida, S. (1964). Modification by early experience of the tendency toward gregariousness in rats. *Psychonomic Science, 1,* 343–4.

Barry, H., III. (1957). Habituation to handling as a factor in retention of maze performance in rats. *Journal of Comparative and Physiological Psychology, 50,* 366–7.

Beach, F. A., & Jaynes, J. (1954). Effects of early experience upon the behavior of animals. *Psychological Bulletin, 51,* 239–63.

Bernstein, L. (1952). A note on Christie's: "Experimental naivete and experiential naivete." *Psychological Bulletin, 49,* 38–40.

Boice, R. (1973). Domestication. *Psychological Bulletin, 80,* 215–30.

Boice, R., Harmon, R., & Boice, C. (1967). Handling or mechanical removal as alternatives in an E-maze. *Psychological Record, 17,* 267–9.

Bolles, R. C., Rapp, H. M., & White, G. C. (1968). Failure of sexual activity to reinforce female rats. *Journal of Comparative and Physiological Psychology, 65,* 311–13.

Bowlby, J. (1958). The nature of the child's tie to his mother. *International Journal of Psycho-Analysis, 39,* 350–75.

Cairns, R. B., & Johnson, D. L. (1965). The development of interspecies social attachments. *Psychonomic Science, 2,* 337–8.

Candland, D. K., Faulds, B., Thomas, D. B., & Candland, M. H. (1960). The reinforcing value of gentling. *Journal of Comparative and Physiological Psychology, 53,* 55–8.

Candland, D. K., Horowitz, S. H., & Culbertson, J. L. (1962). Acquisition and retention of acquired avoidance with gentling as reinforcement. *Journal of Comparative and Physiological Psychology, 55,* 1062–4.

Daly, M. (1973). Early stimulation of rodents: A critical review of present interpretations. *British Journal of Psychology, 64,* 435–60.

Davis, H., & Pérusse, R. (1988). Human-based social interaction can reward a rat's behavior. *Animal Learning & Behavior, 16,* 89–92.

Denenberg, V. H. (1968). A consideration of the usefulness of the critical period hypothesis as applied to the stimulation of rodents in infancy. In G. Newton & S. Levine (Eds.), *Early experience and behavior* (pp. 142–67). Springfield, IL: Thomas.

Denenberg, V. H., Hudgens, G. A., & Zarrow, M. X. (1964). Mice reared with rats: Modification of behavior by early experience with another species. *Science, 143,* 380–1.

Denenberg, V. H., Schell, S. F., Karas, G. C., & Haltmeyer, G. C. (1966). Comparison of background stimulation and handling as forms of infantile stimulation. *Psychological Reports, 19,* 943–8.

Dewsbury, D. A. (1984). *Comparative psychology in the twentieth century.* New York: Van Nostrand Reinhold.

Drever, J. (1952). *A dictionary of psychology.* Harmondsworth: Penguin.

Drewett, R. F. (1973). Sexual behaviour and sexual motivation in the female rat. *Nature, 242,* 476–7.

Gallup, G. G., Jr., & Suarez, S. D. (1980). An ethological analysis of open-field behaviour in chickens. *Animal Behaviour, 28,* 368–78.

Galton, F. (1907). *Inquiries into human faculty and its development.* 2d ed. London: Dent.

40

Gertz, B. (1957). The effect of handling at various age levels on emotional behavior of adult rats. *Journal of Comparative and Physiological Psychology, 50,* 613–16.

Hale, E. B. (1969). Domestication and the evolution of behaviour. In E. S. E. Hafez (Ed.), *The behaviour of domestic animals.* 2d ed. (pp. 22–42). Baltimore: Williams & Wilkins.

Hall, C. S., & Whiteman, P. H. (1951). The effects of infantile stimulation upon later emotional stability in the mouse. *Journal of Comparative and Physiological Psychology, 44,* 61–6.

Harlow, H. F. (1932). Social facilitation of feeding in the albino rat. *Journal of Genetic Psychology, 41,* 211–21.

(1976). Monkeys, men, mice, motives, and sex. In M. H. Siegel & H. P. Zeigler (Eds.), *Psychological research: The inside story* (pp. 3–22). New York: Harper & Row.

Hebb, D. O. (1949). *The organization of behavior.* New York: Wiley.

Hebb, D. O., & Thompson, W. R. (1954). The social significance of animal studies. In G. Lindzey (Ed.), *Handbook of social psychology* (pp. 532–61). Cambridge, MA: Addison-Wesley.

Heymer, A. (1977). *Ethological dictionary.* Berlin: Paul Parey.

Hughes, C. W. (1978). Observer influence on automated open field activity. *Physiology and Behavior, 20,* 481–5.

Immelmann, K., & Beer, C. (1989). *A dictionary of ethology.* Cambridge, MA: Harvard University Press.

James, W. (1890). *The principles of psychology.* Vol. 2. New York: Holt.

Katcher, A. H., & Beck, A. M. (Eds.). (1983). *New perspectives on our lives with companion animals.* Philadelphia: University of Pennsylvania Press.

King, J. A. (1958). Parameters relevant to determining the effect of early experience upon the adult behavior of animals. *Psychological Bulletin, 55,* 46–58.

(1968). Species specificity and early experience. In G. Newton & S. Levine (Eds.), *Early experience and behavior* (pp. 1–23). Springfield, IL: Thomas.

Kintz, B. L., Delprato, D. J., Metee, D. R., Persons, C. E., & Schappe, R. H. (1965a). The experimenter as a discriminative stimulus in a T-maze. *Psychological Record, 15,* 449–54.

(1965b). The experimenter effect. *Psychological Bulletin, 63,* 223–32.

Kuo, Z. Y. (1938). Further study of the behavior of the cat toward the rat. *Journal of Comparative Psychology, 25,* 1–8.

Latané, B., & Hothersall, D. (1972). Social attraction in animals. In P. C. Dodwell (Ed.), *New horizons in psychology,* (Vol. 2, pp. 259–75). Harmondsworth: Penguin.

Lee, M. H. S., & Williams, D. I. (1974). Changes in licking behavior of rat mother following handling of young. *Animal Behaviour, 22,* 679–81.

Levine, S. (1962). Psychophysiological effects of infantile stimulation. In E. L. Bliss (Ed.), *Roots of behavior* (pp. 246–53). New York: Harper & Row.

Ligon, E. M. (1929). A comparative study of certain incentives in the learning of the white rat. *Comparative Psychology Monographs, 6*(28), 1–95.

Lindzey, G., Winston, H. D., & Roberts, L. E. (1965). Sociability, fearfulness, and genetic variation in the mouse. *Journal of Personality and Social Psychology, 1,* 642–5.

Lorenz, K. Z. (1935/1970). Companions as factors in the bird's environment. In R. Martin (Ed.), *Studies in animal and human behaviour* (Vol. 1., pp. 101–258). Cambridge, MA: Harvard University Press.

41

Maier, N. R. F. (1956). Frustration theory: Restatement and extension. *Psychological Review, 63*, 370–88.

Mason, W. A. (1983). Dogs as monkey companions. In A. H. Katcher & A. M. Beck (Eds.), *New perspectives on our lives with companion animals* (pp. 17–21). Philadelphia: University of Pennsylvania Press.

McCall, R. B., Lester, M. L., & Corter, C. M. (1969). Caretaker effect in rats. *Developmental Psychology, 1*, 771.

McCall, R. B., Lester, M. L., & Dolan, C. G. (1969). Differential rearing and the exploration of stimuli in the open field. *Developmental Psychology, 1*, 750–62.

McDougall, W. (1908). *An introduction to social psychology*. London: Methuen.

McGuigan, F. J. (1963). The experimenter: A neglected stimulus object. *Psychological Bulletin, 60*, 421–8.

McMichael, R. E. (1966). Early-experience effects as a function of infant treatment and other experimental conditions. *Journal of Comparative and Physiological Psychology, 62*, 433–6.

Morlock, G. W., McCormick, C. E., & Meyer, M. E. (1971). The effect of a stranger's presence on the exploratory behavior of rats. *Psychonomic Science, 22*, 3–4.

Newton, G., & Levine, S. (Eds.). (1968). *Early experience and behavior*. Springfield, IL: Thomas.

Porter, R. H., Wyrick, M., & Pankey, J. (1978). Sibling recognition in spiny mice (*Acomys cahirinus*). *Behavioral Ecology and Sociobiology, 3*, 61–8.

Richards, M. P. M. (1966). Infantile handling in rodents: A reassessment in the light of recent studies of maternal behaviour. *Animal Behaviour, 14*, 582.

Rosenthal, R., & Fode, K. L. (1963). The effect of experimenter bias on the performance of the albino rat. *Behavioral Science, 8*, 183–9.

Schneirla, T. C., & Rosenblatt, J. S. (1963). "Critical periods" in the development of behavior. *Science, 139*, 1110–15.

Scott, J. P. (Ed.). (1978). *Critical periods*. Stroudsburg, PA: Dowden, Hutchinson, & Ross.

Scott, J. P., & Fuller, J. L. (1965). *Genetics and the social behavior of the dog*. Chicago: University of Chicago Press.

Sheffield, F. D., Wulff, J. J., & Backer, R. (1951). Reward value of copulation without sex drive reduction. *Journal of Comparative and Physiological Psychology, 44*, 3–8.

Shelley, H. P., & Hoyenga, K. T. (1966). Rearing and display variables in sociability. *Psychonomic Science, 5*, 11–12.

Sloan, L., & Latané, B. (1974). Sex and sociability in rats. *Journal of Experimental Social Psychology, 10*, 147–58.

Sperling, S. (1988). *Animal liberators: Research and morality*. Berkeley and Los Angeles: University of California Press.

Sperling, S. E., & Valle, F. P. (1964). Handling-gentling as a positive secondary reinforcer. *Journal of Experimental Psychology, 67*, 573–6.

Suarez, S. D., & Gallup, G. G., Jr. (1982). Open-field behavior in chickens: The experimenter is a predator. *Journal of Comparative and Physiological Psychology, 96*, 432–9.

Tinbergen, N. (1953). *Social behaviour in animals*. London: Methuen.

Walsh, R. N., & Cummins, R. A. (1976). The open-field test: A critical review. *Psychological Bulletin, 83*, 482–504.

Ware, R. (1968). Development of differential reinforcing values of sexual responses in the male albino rat. *Journal of Comparative and Physiological Psychology, 65*, 461–5.

Warren, H. C. (1934). *Dictionary of psychology.* Boston: Houghton Mifflin.

Weininger, O. (1953). Mortality of albino rats under stress as a function of early handling. *Canadian Journal of Psychology, 7*, 111–14.

Weininger, O., McClelland, W. J., & Arima, R. K. (1954). Gentling and weight gain in the albino rat. *Canadian Journal of Psychology, 8*, 147–51.

Werner, C. M., & Anderson, D. F. (1976). Opportunity for interaction as reinforcement in a T-maze. *Personality and Social Psychology Bulletin, 2*, 166–9.

Werner, C., & Latané, B. (1974). Interaction motivates attraction: Rats are fond of fondling. *Journal of Personality and Social Psychology, 29*, 328–34.

West, C. H., & Michael, R. P. (1987). Handling facilitates the acquisition of lever-pressing for brain self-stimulation in the posterior hypothalamus of rats. *Physiology & Behavior, 39*, 77–81.

Wong, R. (1972). Infantile handling and sex as determinants of the negative reinforcing effects of gentling. *Psychologische Forschung, 35*, 213–17.

Young, R. D. (1965). Influence of neonatal treatment on maternal behavior: A confounding variable. *Psychonomic Science, 3*, 295–6.

Zajonc, R. B. (1965). Social facilitation. *Science, 149*, 269–74.

(Ed.). (1969). *Animal social psychology.* New York: Wiley.

—4—

The covalent animal: on bonds and their boundaries in behavioral research

John C. Fentress

... observer and object are tied together.
Delbrück (1986)

Editors' introduction

Fentress raises an important issue, echoed in subsequent contributions, about the ease with which unwarranted conclusions might be drawn about the neutral presence of a human observer. He cites the example of the changes in dominance relations that were associated with the mere presence of an observer to whom a subordinate animal was bonded. This underscores how difficult it is for the scientist to remain unobtrusive.

Yet the scientist–animal bond may provide what Fentress terms a "multiple window" from which to view animal behavior. Rather than find ways to avoid the effects of a close relationship with one's subjects, Fentress surveys some of the benefits such relationships offer behavioral research. He argues that an ideal view of animal behavior consists of a multidimensional perspective that may contain material drawn from the close relationship between scientist and animal. Many such insights, including those of an anecdotal nature, may provide enriched hypotheses for subsequent testing under more conventional "objective" conditions.

Introduction

Modern research in the physical sciences has sensitized investigators to the intimate relations between observers and their objects of investigation. It is ultimately these *relations* between observer and observed that provide scientific data. There are two primary reasons for this. First, the questions asked by investigators are necessarily selective, thus limiting the data from which we make interpretations (Fentress, 1990). Second, in the very process of measurement the observer may either intentionally or inadvertently modify the properties of the phenomena under investigation.

In the behavioral sciences, observer effects are equally likely to be important. Experimental methods, by definition, constrain animal performance and our measures of this performance. Animals may respond to the observer directly, to the observer indirectly through the constraints of apparatus, to residual cues left behind by the observer (such as olfactory traces), and so forth. Even when these responses to the observer are not dramatic, or perhaps not obvious at all, the animals may still modify their responses to other variables (see Caine, 1990, and Chapter 22, this volume). When this happens, there is an inevitable confound ("bond") between observer and observed. Even in the field, animals observe observers. The glint of sun from binoculars, novel odors, or the sound of a twig being broken may focus an animal's attention, thus modifying its response to other variables. It is not possible to predict in the abstract what events will prove to be most salient. The bottom line is that animals may be less sensitive to certain sources of experience than we had anticipated and more sensitive to other sources of information.

In the behavioral sciences, as in the physical sciences, our databases are derived from *relations* that exist between observers and their subjects. Having said this, it becomes important to manipulate observer–animal "bonds" explicitly. Only when the resulting databases are compared can we obtain a firm foundation for precise generalizations about animal performance.

Any given form of investigation can do no more than provide a single window of perception. A major emphasis of the present chapter, therefore, is that the *combination* of explicitly different forms of investigation gives us a richer perspective than that afforded by any one view. To put the matter metaphorically, multiple windows can provide a more complete picture of the landscape. Single-dimensional perspectives need to become multidimensional (as are animals).

The present chapter offers three dimensions of human–animal "bonds" taken from our ongoing study of wolves (*Canis lupus*). The primary focus is on work we have done over the past quarter-century with animals that have been hand-reared and thus socialized to humans (Fentress, 1967). Here "bonds" between animals and humans are quite explicit. The approach allows one to observe animals closely, across their lifetimes, and in multiple contexts. Furthermore, the interactions between animals and people can be used to assay directly the cues to which animals respond, their capacities of performance, and their sensitivities during development. The richness of this picture is accompanied by a need to attain formal data on wolves in more natural contexts.

The second and third dimensions of our work thus involve long-term studies of wolves that are not socialized to people and are housed within a large (3.8-hectare) enclosure at the Dalhousie Animal Behavior Field Station (DABFS) and preliminary observations my students have recently made in the field. These approaches afford opportunities to examine animals closely throughout their lifetimes under reasonably controlled conditions (DABFS) and in natural contexts. While potential observer influences are still present, their consequences may be very different. For example, wolves that are not socialized to humans tend to avoid rather than approach observers. It is also possible to vary the degree and quality of observer–animal contacts.

I have used the metaphor of the covalent animal in the title of this chapter to stress the reciprocity between organism and environment (which includes the observer). One of the classic, if not altogether useful, dichotomies in animal biology and behavior is the extent to which an animal is "doing its own thing" or being molded by the external world. This reflects an attempt to break the world into simple "either/or" dichotomies. Much of our literature in the behavioral sciences continues to pit inside versus outside, nature versus nurture, central factors versus peripheral factors, specific factors versus general factors, higher levels versus lower levels of organization, and so forth. The relations between animals and their worlds are more sophisticated than these simple dichotomies imply (e.g., Eigen & Winkler, 1981).

I have employed the two terms "bonds" and "boundaries" to refer to *relative* connections and separations between animals and certain properties of their environments (which can include observers). At a more analytic level, it is also important to attempt evaluations of relative bonds and boundaries for different properties of behavior. It is also important to consider the broader contexts of expression within which any given actions are observed. Changing contexts often produce changes in behavioral profiles.

My goal in this chapter is not to detail all of our observations. Rather, I shall select from them to show how "inevitable bonds" between observer and observed can be used to generate multiple perspectives. I start with

46

stories related to animals with whom I, and others in my group, have been "bonded" in the literal (social) sense.

Insights from hand rearing and their limitations

Let me start with some simple observations. Wolves (*Canis lupus*) are highly social animals, and they can be persuaded to accept (indeed, seek) close human contact. Wolves are predators. Success as a predator corresponds, in ways not yet adequately defined, to an alertness to even subtle cues provided by other (prey) animals. To catch prey, for example, it can be important to anticipate the prey's future actions. Wolves are also highly social; they are sensitive to multiple cues given by their companions. Wolves in captivity can be charming, annoying, playful, amusing, obstinate, inspiring, and aggressive. They are often surprisingly sensitive to even minor environmental changes and are easily frightened. They are elegant and delicate, even "moody." They can be aggressive. They refuse to be molded easily as "pets" and should never be thought of in that context (though, I confess, they can make normal "pets" appear remarkably placid and boring).

The hand rearing of wild animals is not a task to be taken lightly. It is time-consuming and often exhausting (both emotionally and intellectually). It can, improperly done, also lead to disastrous consequences for the animals concerned, and to silly science. However, close "give-and-take" observations can open up opportunities for asking questions the answers to which are not available through other means. By adjusting to an animal as well as recording its behavior, one can often glean facets of behavior that would otherwise remain obscured from view. The further reward is that the remarkable richness of animal interactions offers many lifetimes of questions that can then be pursued more formally.

A first case: Lupey

My own work with wolves began shortly after I had read an article by an established ethologist on his experiences with a hand-reared male wolf. When the animal reached the age of 10 months, it attacked this individual, who shot and killed it, then sought to generalize from his single unfortunate experience. Reading between the lines of his report, I got the impression that the attack said less about wolves than about the observer and his particular relationship to his animal. For example, he strove to maintain a clear line of authority from him to his animal. The observer became a blanket of oppression, which smothered an animal that deserved better.

The animal was, loosely speaking, psychologically squashed. It seemed to me that if one could establish a sensitive give-and-take relationship, without a single line of unilateral dominance, then a wolf might be worked with closely (Fentress, 1967). I have often wondered why this observer's wolf did not attack him sooner!

To establish a different type of relationship, one of "mutual sensitivity" and "respect," was my original goal; I was simply going to demonstrate that attacks are not inevitable. (It was also, I confess, simply a desire to work closely with an individual of a species that had long fascinated me. Much ethology has its roots in such desires, whether we admit it or not [Tinbergen, 1951; Lorenz, 1981].) At 1 year of age the animal I hoped to obtain was to be returned to the zoo. My first wolf, however, was not returned to the zoo, but remained with me in Cambridge, New York State, Oregon, and finally Nova Scotia – for his entire life.

I was perhaps fortunate that Lupey, as I cleverly named him, easily fit into this arrangement of give and take. My original plan, to return Lupey to the Whipsnade Zoo in England after about a year, evaporated through the joy of working closely with him. One danger of working closely with wild animals is that you become truly "bonded" with them. They demand an extra effort, and give much in return.

Close relatives of "man's best friend," wolves have occupied a major role in human culture and myth for generations. It is not bad science to recognize that we can (and often do) establish close affinities with other species. What is bad is that we have not pursued in any systematic way what these affinities may be telling us about either ourselves or the animals (cf. Griffin, 1982).

The first night after I removed Lupey from the zoo, at the age of 4 weeks, and brought him into my rented home in Cambridge (England) I felt I might have made a terrible error. The pup approached people and was even playful, but during his first night he began to howl in a way that would melt all but the hardest hearts. Finally I brought him into my bed, where he settled on the pillow. By the end of his second day Lupey followed me about and romped in the back garden. I spent as many hours with him, day and night, as my schedule would allow.

From this point forward, nearly every observation I had made of Lupey (e.g., Fentress, 1967) reflects the potential of observer bias and influence. An animal that becomes a friend is treated as a friend, with a necessary loss in "objectivity." This is not to say, of course, that friends fail to give us important insights into behavior. We can also seek to distill more formal questions from close observations of their behavior as it is expressed under often rich and varied contexts. We can, for example, ask, what they do when we are, or are not, obviously intruding on their lives. We can also try to share with them multiple facets of our own behavior. Their responses to our behavior offer rich material for subsequent study. For example, which cues that we or other companions give are they particularly responsive to?

I recall one vivid example of the potential sophistication of wolf sociality. My landlady on the farm outside of Cambridge heard Lupey howl in a way that was novel and that concerned her. He often howled, but the qualities of this sound seemed different. She could not explicate her observations other than to say that the sound seemed "mournful." Several minutes later her son came into the house. He had just hit with a tractor one of the family dogs, who vocalized his pain.

This particular dog was a close companion of Lupey. Even though we had no formal records, the circumstantial evidence is strong that Lupey responded to his companion's injury with a howl that differed in quality from other howls he often gave. My landlady was able to discern this. Lupey, it appears, was sensitive to signals from his bonded companion (who, by the way, turned out not to have been injured seriously). Such observations have led us into more formal analyses of wolf vocalization properties (e.g., Tooze, Harrington & Fentress, 1990; Coscia, Phillips, & Fentress, in press). The richness of hand rearing opens opportunities to address more formal, limited, and controlled questions.

Some perspectives. It is useful to recognize that research on animal behavior can flow in two quite different directions. One direction is to expand the frameworks within which we ask subsequent questions. The other direction is to clarify details within preexisting frameworks. Hand-rearing studies certainly contribute to the former. They can also contribute to the latter, but with the clear need for supportive studies that do not depend on close human–animal interactions.

In hand rearing we can observe animals, closely, in a number of contexts and across protracted periods of time. We can then begin to forge a picture of the animals' responses to these different contexts. We may even provide these contexts at will. In the process of doing so, we can begin to develop a framework that captures both the regularities and varieties of patterns in behavior that we observe. We can thus begin to establish a view that transcends particular contexts.

It is not possible to synthesize volumes of diaries, photographs, and video records into an easy "objective" list of tightly ordered themes. Rather, I have chosen to highlight but five major categories of wolf behavior I first became sensitive to through my interactions with Lupey, later reinforced by our group's efforts with nearly a score of other hand-reared wolves. While the list is somewhat arbitrary, it provides a useful gestalt that has indeed guided much of our subsequent and more formal work.

A categorization of impressions

1. *Wolves are sensitive to their surroundings.* For the purposes of the present volume I want to make this point in its broadest sense. One of the main problems of dealing with feral animals is that they often respond dramatically to small changes in their environments. As an illustration, Lupey would often struggle to escape from rooms in which furniture had been rearranged. Hebb (1946) argued that animals often show strong "fear" when confronted with a combination of familiar (e.g., room) and unfamiliar (e.g., rearranged furniture) environmental properties. This avoidance behavior is often stronger in contexts that combine familiar and unfamiliar elements than in contexts that are "totally" unfamiliar (e.g., new room). We have related observations suggesting that the presence of new observers in familiar rooms often leads to stronger avoidance than does the presence of new observers in unfamiliar rooms.

2. *Wolves often give compartmentalized responses.* They do not necessarily generalize in ways that we would expect them to. To cite one example, Lupey once confronted me with an unexpected explosion of growls when I entered his yard outside of Cambridge. As soon as I stepped back, he wagged his tail. When I stepped forward again he growled. The process could be repeated many times. Only later did I realize I had stepped on Lupey's cache of buried chicken. A friend does not step on someone else's chicken! Step back, and the friendship is renewed. Step forward, and hostility resumes.

This is an example of compartmentalized action. Compartmentalized actions are those in which rules of generalization from other action tendencies are less apparent than we might otherwise expect. ("Modularity" is a term employed in cognitive science; Posner, 1989.) The observation has many special forms, all of which lead to the same conclusion (compartmentalization). A horse, previously ignored, that rolls on the ground elicits strong tendencies from the wolf to attack. A child that falls in front of a wolf is no longer a companion, but a potential prey. A person who mistreats a wolf becomes, selectively, an object for aggression well into the future. I shall elaborate briefly on three examples.

Lupey stayed for several years in a dog kennel in New York. His cage was cleaned daily by one of the kennel workers. One day, for reasons that are not clear, this worker tapped the wolf on the head with a shovel to get him to move into the outer dog run. Lupey left. The next day, *and for every day thereafter, over a period of several weeks*, the wolf challenged this worker through an impressive, and frightening, expression of growls, standing on his hind legs, and so on. What we found equally remarkable is that this persistent aggression was not generalized to other humans. Lupey would greet them with friendly enthusiasm. Finally we had to ask the one animal caretaker never to enter Lupey's pen again.

There are many other examples I could cite. While we were at Cambridge, Lupey became accustomed to taking walks, uneventfully, in a field that contained horses. When one of these horses rolled on the ground, however, Lupey pulled violently and persistently at his lead. I am certain he would have attacked the horse given any opportunity to do so. The same horse was basically ignored when it stood.

Lupey did not generalize among people of different ages. He soon became facile at dealing with adults, whom he would either approach in a friendly manner or avoid. His responses to human infants could be strikingly different. He would often lunge at them, and more than once I praised the manufacturers of solid chains. We soon gave this game up. The potential for sudden shifts among compartmentalized actions is one reason wolves do not make good pets.

3. *Wolves are persistent.* The case recounted earlier about the animal caretaker is illustrative. As I indicated, never again could this worker enter Lupey's cage (although the rest of us could, with impunity). Persistence is an ill-defined concept, but it clearly relates to wolf behavior. As an illustration, one can appreciate the role of persistence in chasing prey. To cite a quite different example, Lupey persisted (over several hours and after discouragement on my part) in digging through a wall in our lodgings outside of Cambridge so that he could reach another room where he and I had romped the previous evening. I simply could not stop him with reasonable methods of persuasion. There were many other occasions when, to speak colloquially, he made up his mind to do something and could not be dissuaded. Now that we are alerted by such casual observations, we have often seen this single-mindedness of action in our other animals. There is a puzzle here in that wolves at other times are easily distracted by momentary events. The behavioral science field has not yet developed the tools necessary to deal adequately with such shifts between persistence and change (cf. Warren & Shaw, 1985).

4. *Wolves are sophisticated play partners.* Lupey and I initiated many games over the years we had together. Our most successful games were those that involved some form of predator–prey analog, such as rugby. It took little time to teach Lupey the twin arts of running with a ball to avoid being tackled and "tackling" his less athletic human companion. What impressed me most was Lupey's ability to adjust the pace of our games so that our contests were more or less even. This demanded a considerable reduction in his own potential performance, rather like that seen in two

dogs of very different sizes "contesting" a bone. Occasionally, and to me unpredictably, he would also "attack" or "avoid" with the full capacities he had available to him, as if there were a second message that, in spite of the games, we should be clear who *could be* the winner whenever he wished. It is hard to think of how one would design more formal experiments around such observations of sophisticated social behavior, although I am certain the attempt would be worthwhile.

5. *Wolves are crafty, but also limited.* It is difficult to summarize in "objective" terms how many times I have been taken with the apparent "intelligence" of Lupey or other wolves with whom I have had the privilege of being closely associated. They can remove locks that are not securely fastened and get into kinds of mischief that remain a source of surprise, indeed inspiration. Lupey occasionally caught unaware chickens on our farm in England, for example, after lying still with his eyes nearly closed for many minutes (except for an occasional blink). Chickens do not, apparently, avoid sleeping wolves. In ways that I do not at all understand, he succeeded in slipping his chain and *removing* a basement window to gain access to our farm family's Christmas goose (already prepared). Both Lupey and I were fortunate that our collective landlady could see nothing but good in his actions. The point is that wild animals, and certainly domestic animals as well, can often exhibit a remarkable creativity if we allow them to do so (e.g., outside the confines of standard rat mazes).

To balance the equation, they can also often show surprising inabilities to modify their behavior. To cite a trivial example, it took me only minutes to teach Lupey to "shake hands" on request. In contrast, it proved very difficult to teach him to "sit" during these play sessions. Wolves often paw one another during play; they rarely sit. To put the matter simply, they are predisposed to make certain actions and predisposed *not* to make others. We have seen wolves fail to push a barrier horizontally, even though they often employ powerful horizontally directed paw movements during the construction of their dens (Ryon, 1977). Contexts often constrain actions and flexibilities in action that are readily available to the animals in other contexts. One might wonder whether the recent flurry of enthusiasm for "constraints" in animal learning (e.g., Shettleworth, 1973; Domjan & Galef, 1983) would have been necessary if researchers had allowed themselves to observe animals more informally and in more diversified contexts.

6. *Wolves are (within limits) creative.* This is one reason they are so enjoyable to work with. Just when one begins to figure them out, they may adopt a different tact. Creativity is an exceptionally difficult concept to forge into systematic research. Both animal traditions and insights by animals have led to considerable confusion in the literature (Galef, 1990). Hand-rearing

studies have the potential to add considerable information, such as shown in the recent research by Pepperberg (1983) on imitation in parrots. We have been especially impressed by what might be called "predator games" in our wolves. Lupey and other animals would often practice complex layers of "deception" to catch us off guard. ("I'm asleep"; "I'm engaged in something else, which means I'm not watching you"; "I just went behind this tree [where you cannot see me] to rest.")

7. *Wolves benefit from their special biological heritage.* Predator games reflect an astonishing sensitivity these animals seem to have for observing behavior, both in us and in other animals. They appear to anticipate our actions and often to disguise their actions. It is a humbling experience to observe wolves watch students who are observing wolves. The students take cursory notes; the wolves stop breathing as they turn their ears to focus in on a slight sound, test the air after a student moves, and bob their heads to align the human note takers from different angles. We have not yet performed good experiments on such subtleties of wolf behavior, and they would not be easy to design. (However, we have with some students had considerable success in suggesting that if the students want to learn how to watch wolves they should monitor how the wolves are watching them.)

We are often surprised, perhaps I should say "awed," by the sophisticated performances our hand-reared wolves give the first time they have the opportunity. Although he had been an adult for several years, Lupey had not bred when we brought two pups from Alaska to him and to his two adult female cage mates. He almost immediately dug a den with remarkable architectural skill (see Ryon [1977] on the complex structure of wolf dens) and soon began regurgitating food to his "adopted" offspring. These are, of course, the sorts of observations from which ethological folklore has sprung. They indicate that animals need not be "taught" all that they can do and can often perform in remarkably sophisticated ways even when their experiences are limited (see Fentress, 1989). There are no good models of how such remarkable skills develop. (To give the reader a tantalizing tidbit, the year *after* Lupey had helped rear the pups he did breed. There are some interesting hypotheses not far beneath the surface here that could be tested by experiments [e.g., Fentress, Field, & Parr, 1978; Fentress, 1978, 1983; Fentress & McLeod, 1986]).

8. *Wolves can be surprisingly trusting and adaptable.* Several times I had to take Lupey into the office of a veterinarian. We have subsequently done this with many of our wolves. We have even had to perform manipulations (such as treatment of wounds) that must have been painful. Although the animals were alert, they accepted what must have appeared to be bizarre manipulations. These fascinating observations suggest that the animals could at some level "tune into" our intentions. (The alternative explanation,

which I doubt, is that they were too terrified to respond at all.) Obviously experiments could be designed to test such things, but I doubt whether they would add much useful information (or even be appropriate to conduct). We are, as in many cases, left with impressions. I personally am happy to have a behavioral science that allows such impressions. (It goes without saying that impressions, however rich they may be, must be followed by more formal analyses.)

When I returned home after my graduate studies at Cambridge, Lupey was placed in a kennel on the *Queen Elizabeth*. I stayed with him for much of the voyage, and he settled into this bizarre form of confined relationship with surprising dignity. He remained playful and, by all measures, trusting. When I moved from Rochester, New York, to Oregon I placed Lupey into the back of a LandRover. Although he had ridden in this vehicle before, I was terribly concerned that he would "freak out." He did not. Rather, he developed routines over the 10–day trip that informed me of his state and needs (such as banging on a suspended bucket with his paws when he needed to take a walk to relieve himself. He would then jump back into the car).

9. *Wolves are multidimensional, even contradictory.* Hand-reared wolves can live rich lives if we allow them to. The range of their actions, their flexibilities, and their constraints continue to fascinate us. They refuse, when given the opportunity to refuse, to confine their expressions to simple taxonomic "boxes." We have found that our taxonomies must be flexible as well as multidimensional. Nor do wolf actions generalize across behavioral taxa in ways that we might anticipate from laboratory studies designed in part to isolate and constrain these taxa. Typology is a product of human concept. It can buy the surface impression of objectivity and logic, but often at the expense of distorting biological fact (see Mayr, 1982). Wolves are *bio*logic, and we should seek to understand them, insofar as it is possible, in *their* terms (see Fentress, 1990).

10. *Wolves are individuals. Each* wolf with whom we have worked is, along certain dimensions, demonstrably unique. Some are "shy"; some are "bold." Some are "gentle"; some are "rough." Some appear more "wise" than others. Some are better "athletes" than are others. And so on. Each *relationship* in a series of hand-rearing studies thus also becomes unique, no matter how much one seeks to employ common methods. We have not found any simple, universal rules.

This should caution anyone searching for simple generalization. The simplicity of generalization often reflects a simplicity in measurement rather than a simplicity of that which has been measured. Both the behavioral and biological sciences need to find better tools to cope with *patterns in diversity* rather than seek to exclude diversity from our models.

The explicit use of multiple perspectives in animal behavior research can be an essential strategy in pursuing this goal.

Comment. The point of these observations is this: When we allow ourselves to work closely with an animal, we open up vistas of experience that we could not design in advance. In doing so we begin to appreciate the remarkable richness and often *contradictory* profiles of animal behavior. We begin to see the animals through several broad picture windows rather than simply through the tunnel of a preset microscope. We begin to realize that both picture windows and microscopes have their place. We use our "bonds" with animals in the literal sense of the term to help us recognize the limited visions that other forms of inquiry give us.

Human observers need sensitivity if they are to understand the nature of wolves. The point I have attempted to make thus far is that sensitive observers can often pick up relevant cues in an animal's behavior that only later become part of a formalized system of knowledge. There is a still broader problem with which we, as animal behavioral scientists, have not dealt. Wolves are contradictory. They do one thing in one context and then do something quite different in another context. That in itself is an important lesson.

Other hand-reared wolves

I will not attempt to document the full range of observations we have made on our hand-reared wolves. All of our wolves, however, have provided a common challenge. This challenge is to combine flexibility in our interactions with a search for possible insights into special features and regularities in their behavior. Occasionally we take these insights and test their validity in more formal situations.

A potential window of insight. I credit my wife Heather with one insightful approach to this problem of combining impressions and more formal methods of inquiry. In our early years at Dalhousie we hand-reared a number of wolves. These animals were also used in several more formal studies, such as documenting the responses of wolf pups to different patterns of vocalization playback (Shalter, Fentress, & Young, 1977). (It would have been interesting to test their responses to vocal playbacks with human observers both present and absent. We did make some pilot observations which suggest that the presence of a human could activate the animals' own social interactions and enhance responses to the vocalizations, but we unfortunately did not pursue formal analyses of these possibilities.) As the reader might well anticipate, many colleagues and students wished to see our animals and to interact with them. Heather's insight was to videotape the individuals and to correlate the responses of hand-reared pups to these individuals with the persons' styles of attempted interaction. We have several follow-up studies in progress.

A point that deserves emphasis is that wolf pups, like most feral animals, are (by domestic or laboratory animal standards) surprisingly sensitive to potential dangers. The wrong movement at the wrong time can send wolf pups into a frenzy of avoidance. It turns out that Heather's observations ("in progress") strongly indicate that a human observer's movement "style" is extremely important. Persons who, on our videotape records, moved either too quickly or too stealthily failed to make contact with the wolf pups. Those individuals who appeared on an independent analysis of video records to move slowly, and in a relaxed and flexible manner, were most successful in eliciting pup approach. The most successful individuals were also those who, from our video records, shifted their attention back and forth between the wolves and other foci in the room. Our impression is that wolves are especially sensitive to temporal cues and also to attentional focus. They appear to be easily "cornered" and respond by avoidance of such potential dangers. Even subtle cues on the part of human interactants, such as muscle tension and breathing patterns, may play a role, although admittedly it is difficult to obtain unconfounded measures of these multiple variables. It would be of great interest to formalize similar observations for *any* species. To my knowledge, this has not yet been done.

Our observations have convinced me that successful human–animal interactions demand a degree of sensitivity that is often difficult to capture in formal "arm's-length" experiments. Future application of explicit "interaction studies" may open the way to a richness in behavior that is otherwise difficult to capture. We are dealing here with "emotional" rather than more easily cataloged "cognitive" logic. The subsequent, and difficult, task is to work toward objective characterization.

Comment on emotional windows. Darwin (1872) admonished us to look at the "emotions" of animals, and ourselves. Emotions are not an easy scientific target. They betray and confound us at every turn. This is not to say, however, that they should be omitted from scientific (systematic) investigation.

I think that it is a basic mistake to treat animals as if they are only intellectual machines. Hand rearing, whatever its imperfections, can move us away from this simplistic perspective. Animals are emotional creatures, and their emotions deserve investigation. Wild animals can indeed be surprisingly fragile. They have not only strengths that we can admire, but also sensitivities that we should try to understand. I suspect that anyone who has worked closely with any wild vertebrate will share this conclusion.

Of course, there are dangers in subjective "bonds" that we establish with animals. Frequently through our close interactions we "tune into" events that are difficult to articulate in formal (cognitive) terms. We need more satisfactory cognitive measures of the cues we use to make our interpretations of animal behavior.

Let me give an example of the processing layers that we, and I suspect animals, employ. One of my early students at the University of Oregon (Becky Field) came to me with the strong impression that the social hierarchy in our small captive group of wolves was undergoing an important transition. I asked her to explain, in intellectual terms, why she thought this. She was initially unable to do so. I then asked her to "act out" the behavior of different animals in the group. Soon she found herself mimicking the slinky movements of certain animals and the stiff postures of others. She had processed these cues, but not initially in clear cognitive terms. By acting out the animals' movements she was *then* able to code important parameters in their behavior. Subsequent objective measurements supported her impressions.

Special observations. Hand rearing provides a number of special opportunities for investigation that reach into many of our preconceptions. Here I tell but one story that, I believe, makes the point well. The story starts with walks that Heather and I often took with a hand-reared wolf, Qochi, an animal that we kept in our Nova Scotia home. We hand-reared Qochi with the companionship of our dog Blu. Blu had nursed the young wolf pup and provided almost constant companionship for him. These animals were bonded (Figure 4.1).

Our own bonds to the wolf allowed us to perform a variety of activities, such as taking him (along with Blu) for walks near the fishing village where we then lived. Occasionally we would turn the two animals loose. Blu would return when we called; Qochi would follow. There was one situation in which this rule of Blu returning on request and Qochi following was broken.

One day Qochi and Blu ran over a small hill and did not return when we expected. We called. They still failed to return. Finally Blu returned, her face and body heavily laden with porcupine quills. Qochi still did not return. When we mounted the hill, we found the wolf inspecting an injured porcupine. Qochi had quills in his mouth as well as on his chest and legs. We put him on his leash and dragged him from the porcupine. He soon pulled us toward the ocean waters and stood submerged up to his chest. He began frothing at the mouth, dipping his head, grabbing the quills in his legs and chest with his incisors, and spitting out the quills with a quick shake of his head. He then came to shore, reached into his mouth with his forepaws, but failed to dislodge the quills in the roof of his mouth. He next

Figure 4.1. Qochi and companions in Nova Scotia. The wolf pup was adopted by our dog Blu, who nursed him and participated in many common adventures with him. These included rides in our boat to a remote island, where the animals could be let free for short periods.

grabbed a branch with his jaws and used the branch to pry the quills in his mouth. Some were dislodged, others were not. Blu, during this time, stood by rather miserably. Luckily for us, and for Qochi, he finally worked the remaining quills out. Blu was less fortunate; we had to remove her collection of quills for her.

The remarkable part of this story came several weeks later. We again took the two animals for a walk in the vicinity where they had previously met the porcupine. Without warning Qochi again began frothing at the mouth and shaking his head. It was then that we discovered that we had just passed a porcupine carcass. Blu seemed oblivious. When we took the wolf, still on his lead, several meters away, he stopped frothing. When we returned, he frothed. After several trials we decided, perhaps foolishly, to release the two animals again. Again they did not return when we expected. Over the hill we went, to find Blu once more laden with quills and Qochi starring at the porcupine from a distance. He would not approach it.

One can, of course, make too much of an anecdote. However, I remain fascinated by the fact that this one "porcupine-naive" wolf (1) exhibited such a remarkable capacity for quill removal, (2) showed frothing and head shaking several weeks later, but only in the vicinity of a porcupine carcass, and (3) showed no further inclination to attack a porcupine again. This is the type of observation that draws one to wonder at the wisdom animals carry with them in nature. Few experiments even begin to tap the processes that underlie such skills. There are two messages here. One concerns the obvious difference in the way the dog and the wolf dealt with a common situation. The other is that such an observation could be made only in the context of a close relationship with the animals.

There is a complementary lesson one can draw from "companion studies," which I shall illustrate with my final story. It is that close observations tell us not only about animals, but what we should do in the presence of animals. While our extended family (people and wolves, together) was still at Oregon, Heather crawled into a den of our university research facility that housed a mother and her pups. The entrance was tight, so we tied a rope on Heather's leg to pull her out just in case of an emergency. We had also previously lowered a microphone into the den from above, to monitor any sounds that might prove important. There was a sudden and rather loud growl, from the wolf mother. This was followed by a sigh, then silence. Moments later Heather worked her way out of the den and told this story. The female wolf had indeed growled, at which point Heather averted her gaze and emitted a deep sigh. The reason she did this was that she had often observed wolves in disadvantageous social circumstances do the same thing. Whether coincidence or not, the female wolf relaxed, and Heather was able to extricate herself safely from the den. It is reasonable to speculate that she had manipulated the female wolf's behavior, and in a most useful manner!

Summary lessons. Interactive observations have two major benefits. The first is that the observer is often in a position to observe a much wider variety of activities under close quarters than under other conditions. The second is that interactions allow the observer to experiment with different behavioral styles and to monitor the consequences of these differences on the animal's behavior. This can be of enormous value in studies of social communication. We are presently engaged in such analyses of social and individual behavior, in ways that we hope reflect the more informal insights that we have gained from our hand-reared animals. Interactive observations and experiments by definition change the behavior of animal participants. These changes can not only affect the animal's ongoing performance, but also modify the responses the animal would otherwise give to other experimental manipulations. We do not yet understand the routes over which consequences of animal–human bonds generalize to other circumstances (Fentress, 1990). For these reasons it is important to supplement the method of interactive observations with less intrusive methods. It is only when such multiple databases are accumulated that one can obtain a clear picture of relative behavioral change versus stability.

Hand rearing both alters the contexts within which animals express their behavior and affects trajectories over which capacities and constraints in performance develop. These effects are difficult to disentangle. One might view hand-reared animals as being very different from their kin that are not socialized to humans, simply because people are accepted as companions in one case and not in another. Conversely, hand-reared animals manifest many of the profiles of social communication seen in captive pack-reared and feral animals. Constraining the contexts within which an animal performs and develops can certainly alter the relative probabilities, and targets, of different actions, but the *form* of these actions may be less affected (Fentress & McLeod, 1986). It becomes important to dissect behavior in multiple ways, such as by its probabilities, targets, and morphology of movement. It is equally important to examine behavior at complementary levels (e.g., individual coordination vs. socially coordinated movements) and time frames (e.g., momentary performance and developmental history).

We have made use of our captive pack at DABFS to formalize our study of many of these issues. Since the station was established in 1974, we have obtained 18 years of uninterrupted data on individual animals (many from birth), their social interactions, relationships, and overall pack structure (see Hinde, 1982). We have combined unobtrusive and automated recording methods with simple experimental procedures that involve manipulations by observers. A major advantage of our work at this facility is that we can observe animals closely and more or less continuously. Although the animals are not socialized to humans, they tolerate routine human intrusions well.

Studies on nonsocialized animals

I shall first summarize some of our less formal observations. One of the most valuable aspects of our work with captive animals at DABFS is that we are able to observe individual animals across their entire life spans. We have thus been able to utilize a number of observational and experimental approaches that would be less available in the field. We have also attempted to combine our work on hand-reared animals with these less intrusive observations – for example, by bringing formally hand-reared animals into social groups at DABFS. The latter procedure, of course, brings with it many other avenues of scientific inquiry, concerning, for example, what happens to animals that have been transferred from their social interactions with humans to other contexts.

While our data in this regard are fascinating, they often are also incomplete, with small samples. We have seen, for example, hand-reared animals fail to enter successfully into previously established groups of wolves for reasons that appear to reflect a lack of "understanding" of the animals' new social roles (e.g., they fight older animals in the group over food sources). Here, of course, we are dealing with the potential long-term consequences of previous animal–human social bonds. There are no good data on this issue, and it is not easy to see how one might construct humane experiments to test such consequences further.

Again, our data tend toward the anecdotal. We have seen animals previously socialized ("bonded") to people relinquish their contacts with people, only to restore them during periods of stress and overt conflict with other animals. I do not think that further experiments with respect to these parameters of changed social context can be easily constructed or properly justified.

Rather, we have concentrated on (1) factors that generate obvious responses in our wolves and (2) unobtrusive observations of their ongoing behavior. With respect to the former case, we have often been surprised to find that what we, as human observers, would call major intrusions have little effect. If we drive trucks into the compound, for example, the wolves often show less avoidance than if we simply walk into the compound. Analogous observations have been made by many researchers (e.g., D. Balfour on coyotes, personal communication). W. Danilchuk, in my laboratory, has related that certain individual wolves approach her *only if* she uses a grass cutter in our compound. E. Coscia, a graduate student in my group, has told me about young wolves in the field that leap toward low-flying helicopters, even though the same wolves avoid humans. And so on. To me each of these observations suggests that animals' responses may often reflect their special sensitivities to animate as opposed to inanimate "objects." I am certain that the story is more complicated than this; its point is that there are remarkably few studies that are devoted to careful dissections of what feral animals, either in captivity or in the field, become "bonded" to. We simply do not know what they are responsive to.

We have, in part on the basis of our work with hand-reared animals, become very sensitive to routes of possible influence on wolf behavior. If we approach previously socialized animals, they may, for example, not retreat but squabble among themselves as if competing for human attention. These positive attractions to people can be as disruptive as are negative reactions.

From such observations we begin to see a source of possible future experiments, in which human observers become explicit variables. The simple tack is to record obvious responses to human observers. The more subtle tack is to use human observers to "modulate" the responses of animals to other controlled variables (as in the cognitive psychology studies of "priming" variables; e.g, Posner, 1989).

We have made a number of observations on previously hand-reared animals, now placed together into large (0.2- to 0.3-hectare) pens, to examine some of the consequences of earlier socialization to particular persons. Animals hand-reared by one individual may not generalize in terms of approach to other individuals. Thus, if a pen contains animals hand-reared by two or more persons, it is possible to ask what the consequences are of having each of these persons enter the pen. It is even more valuable to ask, "How does the presence of different observers affect the behavior of these animals to other factors?" Suppose, for example, that we play standard vocalizations to these animals with different combinations of observers present.

While this particular experiment has not been performed, for a variety of technical reasons, we *have* recorded how different observers affect priorities in access to (1) food and (2) young pups brought in for the purposes of observation. Even when an observer remains as quiet as possible, his or her presence may strongly bias access in favor of those pups with whom the person was previously socialized. Yearling (or older) animal A gains first access to food or pups in the presence of previous human partner A′; yearling (or older) animal B gains first access to food or pups in the presence of previous human partner B′.

Social interactions between the pups can also shift. This represents an interesting general problem in behavioral communication, which is the influence of a third animal on interactions within a dyad (see Cheney & Seyfarth, 1990). Unfortunately, our own data relevant to this important issue remain preliminary. However, even our limited information to date suggests how animal–human "bonds" might be used to attain future insights into the complex social structure of wolves.

Patterns in ontogeny and integration of behavior

A valuable feature of our research station is that it allows us to make close observations of the socialization process and development within a group of undisturbed animals. Several animals have now been documented on a regular basis throughout their lives. Both individual styles of behavior and socially coordinated actions can in this way be traced in a systematic manner.

Routine observations and videotape records are made from a research trailer just outside the main compound. As in the situations described earlier, these observations can be made either with or without additional (externally visible) observers present. The majority of observations have been made during feeding, which brings pack members to a clearing near the trailer. Other records are made from huts' within the enclosure, from an adjacent research building, or (most recently) from within excavated dens. To date we have more than 5,000 hours of videotape and several hundred hours of high-quality audio recordings. In addition, standardized diary records are made on a regular basis.

Notes on ontogeny. Much of our recent effort has been devoted to tracing early social ontogeny. Peter McLeod (1987; cf. Fentress & McLeod, 1988; McLeod & Fentress, 1991) was able to use combined observations of hand-reared animals and animals within the compound to document the development of social displays. Systematic video records along with formal methods for quantitative analysis allowed him to chronicle changes of coordination both in terms of behavioral sequences generated by a single animal and in terms of cross-correlations among actions within a social pair. For example, McLeod was able to demonstrate that display sequences are articulated in wolf pups before the time they exhibit obvious responses to these sequences in other animals. Such data are of particular interest to the evaluation of constraints on the role of experience during ontogeny (see Fentress & McLeod, 1986). McLeod also devised a method for unobtrusive video and audio monitoring of pups within a den, a method perfected by Elizabeth Coscia (1989). This enables us to extend our developmental analyses to the act of parturition itself.

Galvanized steel culverts 3.5 m long and 90 cm in diameter were attached to the back walls of two dens dug by the wolves (Figure 4.2). This permitted the placement of cameras, microphones, and recorders that could be activated during and outside the times human observers were present. In Coscia's study (see Coscia, Phillips, & Fentress, in press) special care was taken to measure possible indirect interference from the apparatus. For example, vocal recordings were made both with and without the video lamp being turned on (with no measurable differences in the data obtained). The ontogeny of 10 basic classes of pup vocalization could thus be traced with minimal human interference. Complementary studies are in progress by Jill Goldman at Dalhousie on mother responses to these pup vocalizations.

An extremely interesting feature of these studies is that they stress the multidimensional nature of even early wolf behavior patterns. Thus, McLeod found that displays could vary over several postural and movement combinations, and Coscia found that major categories of early vocal behavior could be segregated on the basis of at least two acoustic properties (e.g., spectral type, duration, frequency modulation, and frequency

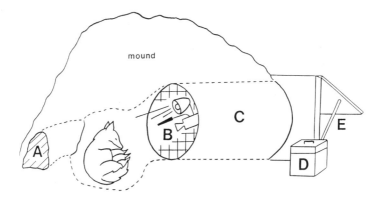

Figure 4.2. Schematic of modified wolf den. A, Wolf entrance; B, video–audio setup; C, back entrance tunnel; D, battery and recorders; E, observer entrance. (From Coscia, 1989.)

range). Furthermore, Goldman's ongoing study supports the hypothesis that a mother's reaction to vocal behavior in her pups is modulated by the presence or absence of other forms of pup activity. These multidimensional issues are ones that we became sensitive to in part through our years of close observations of hand-reared animals. It is rewarding to see the fruits of this work extend into a new phase of more formalized investigations.

Notes on movement integration. We see studies of behavioral ontogeny and integration as two temporal perspectives that provide important complementary foci for our research. One of our interests throughout has been on what might be labeled the *relational* properties of movement. In our hand-rearing studies, for example, we became very sensitive to the need for human observers to adjust and synchronize their movements to the behavioral states of the animals.

As a single illustration of the importance of interactive movements I offer my last "wolf story." A number of years ago my wife and I obtained two tundra wolf pups from Alaska. These animals were approximately 10 weeks old when we obtained them, and they remained shy for a long period of time. One trick Heather devised was to tug rhythmically on their feet, an action that mimicked what we had often seen wolves do in play. The animals would get caught up in "the game" and for various periods of time simply forgot their "shyness" (measured by tendencies to avoid rather than approach us).

Greg Moran (1978; see also Moran, Fentress, & Golani, 1981) extended work by Ilan Golani (1976) involving the application of human choreography to individually and socially coordinated movements in wolves. The basic idea is that flow of movements measured both within and across animals involves a number of mutually constrained trajectories, rather as in a dance. Animals often show relative invariances among properties of movement that, as individual properties, vary widely. The properties become "co-ordered" (Fentress, 1986). Put in other terms, only certain *mutual trajectories* are allowed.

These co-ordered movements indicate how closely individual animals monitor one another, and adjust to one another, during social behavior. When the normal co-ordering begins to break down, one often has a hint that potentially severe forms of social upheaval are imminent. This returns us to less formal suggestions I made earlier that successful human–animal bonding requires a sensitivity to even subtle cues exhibited by the animals. What started as subjective impression became the basis for solid fact.

Havkin (1981; see also Havkin & Fentress, 1985) extended Moran's study into the realm of social development. Havkin's emphasis was on the biomechanical substrates of these coordinated action patterns. For example, young wolves become increasingly sensitive to attempts by their partners to topple them over, while at the same time the wolf pups master strategies to cause their partners to lose balance. One can see how future investigations of animal play, including ones based on the impressions gained from work with hand-reared animals, might be designed to test specific hypotheses about the development of alternative tactics and strategies in the pursuit of important biological goals.

In all of this work we have sought to combine the quest for common principles of organization with observations of individual differences in performance. To illustrate the latter point, Zena Tooze (1987; see also Tooze, Harrington, & Fentress, 1990) has recently documented the existence of individual vocal signatures in wolf howling. She has also combined observations of animals within our station with field work (unpublished) to test for differences in responsiveness to individually distinctive howls.

As a final example of the advantages of close observation and simple experiments afforded by our compound, Dennis Phillips and others have detailed food-caching patterns in wolves (Phillips, Danilchuk, Ryon, & Fentress, 1990) and coyotes (Phillips, Ryon, Danilchuk, & Fentress, 1991). An interesting feature of these caching patterns is that they involve a variety of physically distinctive movements that are combined into relatively invariant and protracted sequences. For example, they are little affected by terrain or seasonal conditions. Our data also suggest that such stereotyped sequences are less sensitive to environmental perturbations than are other, more flexible behavioral sequences (e.g., Berridge & Fentress, 1987; Berridge, Fentress, & Parr, 1987).

This returns us full circle to some of the issues I raised earlier in this

65

chapter. It is important to recognize not only that animals may be more sensitive to certain sources of environmental influence (such as observer presence) than we might expect, but also that the *relative* sensitivities and buffers in behavior are often dynamically ordered (Fentress, 1976). There are times during which a manipulation might severely disrupt an animal's actions; there are other times when the same manipulation might have little effect. We have seen such phenomena often in our work with hand-reared animals. They represent important subtleties in behavior that can then be investigated under other conditions, and even across species.

Brief note on group structure and field observations. I have chosen to emphasize certain properties of wolf behavior in the context of individual animal profiles and responses. It is, of course, equally important to recognize that wolves are social animals that live within a larger group (pack) structure. We have therefore complemented our analyses of individual behavior with measures that include patterns of feeding, pup care, mating, fighting, and so forth (e.g., Fentress & Ryon, 1982; Fentress, Ryon, & McLeod, Havkin, 1986; Fentress, Ryon, & McLeod, 1987). By monitoring our animals at regular intervals and in an unobtrusive manner we can make closer observations than are normally possible in the field. We can also trace individual life histories in the context of the social group. Although our pack animals are not socialized to humans, they are habituated to our procedures. Supplementary data in the field (Tooze, Coscia) should help set the boundaries across which these captive observations can, or cannot, be generalized.

Concluding comments

In this chapter I have tried to show how different approaches to the study of animal behavior can offer complementary views. It is clear that through different approaches one becomes aware of the "bonds" that exist between human observers and animal subjects. Sometimes the bonding process is quite explicit, such as when one works closely with hand-reared feral animals. At other times these bonds are more abstract, such as when one formalizes and limits the environment within which an animal behaves. At still other times observer effects on animals are minimal, such as when automated recording methods are used in the field. In every case, however, it is important to be aware of how we might, either explicitly or inadvertently, be affecting an animal's behavior.

I have chosen to stress the more explicit bonds between observer and animals. It is my conviction that these can be used profitably to open our eyes to the richness of animal performance. The data are in that way a caution to those who believe that simplified studies can provide a complete picture. Science, of course, must deal with simplified perspectives, but it is also important to keep that fact in perspective.

One strong impression I have gained from our work with wolves is that their behavior does not map easily on a formal logic system, at least as traditionally defined in mechanistic terms. Systems of animal behavior are richly interconnected and dynamically ordered. Even in the physical sciences there is a growing dissatisfaction with earlier mechanical models. For example, the field of dynamics is seeking new ways to capture the interlocking flow of multileveled processes in nature (Abraham & Shaw, 1982), and chaos theory seeks ways to combine sudden shifts between order and unpredictability (Gleick, 1987). Each of these fields stresses the need for investigators to question simple linear models of nature.

Focus on the *relations* between observers and their animals may provide an especially rich database for similar revolutions in our thinking about animals (West & King, 1990). Of course, this is a projection into an unknown future. At the present time we can, however, take advantage of those multiple windows that an appreciation of animal–human bonds affords. The views may not be easily synthesized into a coherent picture, but the vistas they provide offer exciting opportunities for exploration.

I conclude with a few thoughts on "context." Throughout this chapter I have stressed that observers form part of the context within which an observed animal's behavior is ordered. The effects of the observer can range from severe to benign, but it is critical that we remain sensitive to the importance of contextual effects in behavior.

In many branches of natural science contextual effects are either ignored or "controlled" into a steady (invariant) state. This assists the investigator in simplifying research procedures. It often leads to reproducible ("objective") results. The difficulty arises if controlled contexts are considered irrelevant to the data obtained. Too often it has proved to be the case that when contexts are altered data patterns also change. As an illustration, the sensitive period for song learning in male white-crowned sparrows, and even the range of songs that individual white-crowned males are willing to imitate, can be expanded dramatically if visual "tutors" are employed in conjunction with vocal models (rather than the latter alone, as had been common in earlier studies; Baptista & Petrinovich, 1984).

Contextual variations, of course, come in many forms. The most fundamental point is that, in animal behavior, actions are expressed under different circumstances. These circumstances can affect performance in often dramatic ways. To cite a single laboratory illustration, Berridge and Fentress (1986) found that the salience of somatosensory input for grooming and ingestive actions in rats varied systematically as a function of behavioral phase. In neuroscience the failure to acknowledge contextual factors has often led to models that in themselves cannot be evaluated outside the limited experimental contexts within which they were developed. For example, the notion of autonomous "central pattern generators" proved to be an oversimplification once investigators included more background variation in their experiments than is normal in neurophysiological laboratories (Fentress, 1990). In animal learning experiments, failure to measure a range of activities under different combinations of contextual constraint in itself constrained the utility of many earlier models (e.g., Shettleworth, 1973). Similarly, in studies of behavioral development the restriction of variables to "obvious channels" of influence has often led to neglect of the unobvious (but critically important) processes (see Meltzoff, 1988).

To return to the main issues of this text, it is clear that animals are in some sense *both* separable from *and* richly connected to their often variable worlds. Our sensitivity to possible observer effects, as part of these variable worlds, may also sensitize us to other factors that can influence an animal's behavior. I thus see the lessons of this volume as a starting challenge rather than as a solution. We need to know what animals are sensitive to, when, and under what circumstances. The acknowledgment and future study of observer effects may be important in helping us move toward that goal.

Acknowledgments

I thank Hank Davis and Dianne Balfour for their invitation to contribute to this volume, for their constructive comments on earlier drafts, and for their saintly patience as I struggled to find something to say. I was also helped enormously by my colleagues, who participated in a variety of discussions. In particular, I thank Elizabeth Coscia, Wanda Danilchuk, Heather Parr, and Jenny Ryon. The research reported here was supported in part by infrastructure and operating grants from the Natural Sciences and Engineering Research Council, the Medical Research Council of Canada, the Dalhousie Faculty of Graduate Studies, and the IAMS Animal Food Company. Mary Racine of Cambridge University Press provided a constructive polish to my earlier draft efforts. Wanda Danilchuk provided invaluable assistance with the final preparation of this manuscript.

Finally, my thanks to the animals.

References

Abraham, R. H., & Shaw, C. D. (1982). *Dynamics: The Geometry of Behavior*. Santa Cruz, CA: Aerial Press.

Baptista, L. F., & Petrinovich, L. (1984). Social interaction, sensitive periods, and the song template hypothesis in the white-crowned sparrow. *Animal Behaviour, 36*, 1753–64.

Berridge, K. C., & Fentress, J. C. (1986). Deterministic versus probabilistic models of animal behaviour: Taste-elicited actions in rats as a case study. *Animal Behaviour, 34*, 871–80.

(1987). Deafferentation does not disrupt natural rules of action syntax. *Behavioural Brain Research, 23*, 69–76.

Berridge, K. C., Fentress, J. C., & Parr, H. (1987). Natural syntax rules control action sequence of rats. *Behavioural Brain Research, 23*, 59–68.

Caine, N. G. (1990). Unrecognized anti-predator behavior can bias observational data. *Animal Behaviour, 39*, 195–6.

Cheney, D., & Seyfarth, R. (1990). Attending to behaviour versus attending to knowledge: Examining monkeys' attribution of mental states. *Animal Behaviour, 40*, 742–53.

Coscia, E. M. (1989). Development of vocalizations in timber wolves (*Canis lupus*). Unpublished master's thesis, Dalhousie University.

Coscia, E. M., Phillips, D. P., & Fentress, J. C. (In press). Spectral analysis of neonatal wolf (*Canis lupus*) vocalizations. *Bioacoustics*.

Darwin, C. (1872). *The Expression of the Emotions in Man and the Animals*. London: John Murray.

Delbruck, M. (1986). *Mind from Matter? An Essay on Evolutionary Epistemology*. London: Blackwell Scientific.

Domjan, M., & Galef, B. G., Jr. (1983). Biological constraints on instrumental and classical conditioning: Retrospect and prospect. *Animal Learning and Behavior, 11*(2), 151–61.

Eigen, M., & Winkler, R. (1981). *Laws of the Game*. London: Allen Lane. (First published in 1975 as *Das Spiel: Naturgesetze steuern den Zufall*. Munich: R. Piper).

Fentress, J. C. (1967). Observations on the behavioural development of a hand-reared male timber wolf. *American Zoologist, 7*, 339–51.

(1976). Dynamic boundaries of patterned behaviour: Interaction and self-organization. In *Growing Points in Ethology*, P. P. G. Bateson & R. A. Hinde (Eds.), 135–69. Cambridge University Press.

(1978). Conflict and context in sexual behaviour. In *Biological Determinants of Sexual Behavior*, J. Hutchison (Ed.), 579–614. New York: Wiley.

(1983). Ethological models of hierarchy and patterning of species-specific behaviour. In *Handbook of Neurobiology: Motivation*, E. Satinoff & P. Teitelbaum (Eds.), 185–234. New York: Plenum.

(1986). Development of coordinated movement: Dynamic, relational and multi-leveled perspectives. In *Motor Development in Children: Aspects of Coordination and Control*, H. T. A. Whiting & M. C. Wade (Eds.), 77–105. Dordrecht: Martinus Nijhoff (published in coordination with NATO Scientific Affairs Division).

(1989). Comparative coordination (a story of three little *p*'s in behaviour). In *Perspectives on the Coordination of Movement*, S. A. Wallace (Ed.), 185–219. Amsterdam: Elsevier.

(1990). The categorization of behaviour. In *Interpretation and Explanation in the Study of Animal Behavior, Vol. 1: Interpretation, Intentionality, and Communication*, M. Bekoff & D. Jamieson (Eds.), 7–34. Boulder, CO: Westview Press.

Fentress, J. C., Field, R., & Parr, H. (1978). Social dynamics and communication. In *Behavior of Captive Wild Animals*, H. Markowitz & V. Stevens (Eds.), 67–106. Chicago: Nelson-Hall.

Fentress, J. C., & McLeod, P. (1986). Motor patterns in development. In *Handbook of Behavioural Neurobiology: Developmental Processes in Psychobiology and Neurobiology*, E. M. Blass (Ed.), 35–97. New York: Plenum.

(1988). Pattern construction in behaviour. In *Behavior of the Fetus*, W. P. Smotherman & S. R. Robinson (Eds.), 63–76. Caldwell, NJ: Telford Press.

Fentress, J. C., & Ryon, J. (1982). A long-term study of distributed pup feeding and associated behavior in captive wolves. In *Wolves of the World*, F. Harrington & P. Paquet (Eds.), 238–61. Park Ridge, NJ: Noyes Press.

Fentress, J. C., Ryon, J., & McLeod, P. J. (1987). Coyote adult–pup interactions in the first 3 months. *Canadian Journal of Zoology, 65*, 760–3.

Fentress, J. C., Ryon, J., McLeod, P. J., & Havkin, G. Z. (1986). A multidimensional approach to agonistic behavior in wolves. In *Man and Wolf: Advances, Issues, and Problems in Captive Wolf Research*, H. Frank (Ed.), 253–74. Dordrecht: Junk Publishers.

Galef, B. G., Jr. (1990). Tradition in animals: Field observations and laboratory analyses. In *Interpretation and Explanation in the Study of Animal Behavior, Vol. 1: Interpretation, Intentionality, and Communication*, M. Bekoff & D. Jamieson (Eds.), 74–95. Boulder, CO: Westview Press.

Gleick, J. (1987). *Chaos: Making a New Science*. New York: Viking Penguin.

Golani, I. (1976). Homeostatic motor processes in mammalian interactions: A choreography of display. In *Perspectives in Ethology*, P. P. G. Bateson & P. H. Klopfer (Eds.), 2: 69–134. New York: Plenum Press.

Griffin, D. R. (1982). *Animal Mind–Human Mind*. Berlin: Springer.

(1982). *Ethology: Its Nature and Relations with Other Sciences*. Oxford: Oxford University Press.

Havkin, G. Z. (1981). Form and strategy of combative interactions between wolf pups (*Canis lupus*). Unpublished Ph.D. thesis, Dalhousie University.

Havkin, G. Z., & Fentress, J. C. (1985). The form of combative strategy in interactions among wolf pups (*Canis lupus*). *Zeitshrift für Tierpsychologie, 68*, 117–200.

Hebb, D. O. (1946). On the nature of fear. *Psychological Review, 53*, 259–76.

Hinde, R. A. (1970). *Animal behaviour: A synthesis of ethology and comparative psychology* (2d ed.). New York: McGraw-Hill.

Lorenz, K. Z. (1981). *The Foundations of Ethology*. New York: Springer.

Mayr, E. (1982). *The Growth of Biological Thought*. Cambridge, MA: Belknap Press.

McLeod, P. J. (1987). Aspects of the early social development of timber wolves (*Canis lupus*). Unpublished Ph.D. thesis, Dalhousie University.

McLeod, P. J., & Fentress, J. C. (1991). Aggressiveness as a possible determinate of developmental changes in the responsiveness of timber wolf pups to their partners during social interactions. *Aggressive Behavior* (abstract), *17*(2), 107.

Meltzoff, A. N. (1988). Imitation, objects, tools, and the rudiments of language in human ontogeny. *Human Evolution, 3*, 45–64.

Moran, G. (1978). The structure of movement in supplanting interactions in the wolf. Unpublished Ph.D. thesis, Dalhousie University.

Moran, G., Fentress, J. C., & Golani, I. (1981). A description of relational patterns during "ritualized fighting" in wolves. *Animal Behaviour, 29*, 1146–65.

Pepperberg, I. (1983). Cognition in the African Grey parrot: Preliminary evidence for auditory/vocal comprehension of the class concept. *Animal Learning and Behavior, 11*(2), 179–85.

Phillips, D. P., Danilchuk, W., Ryon, J., & Fentress, J. C. (1990). Food caching in timber wolves, and the question of rules of action syntax. *Behavioural Brain Research, 38,* 1–6.

Phillips, D. P., Ryon, J., Danilchuk, W., & Fentress, J. C. (1991). Food caching in captive coyotes: Stereotypy of action sequence and spatial distribution of cache sites. *Canadian Journal of Psychology, 45*(1), 83–91.

Posner, M. I. (Ed.). (1989). *Foundations of Cognitive Science.* Cambridge, MA: MIT Press.

Ryon, J. (1977). Den digging and related behavior in a captive timber wolf pack. *Journal of Mammalogy, 58*(1), 87–9.

Shalter, M. D., Fentress, J. C., & Young, G. W. (1977). Determinants of response of wolf pups to auditory signals. *Behaviour, 60,* 98–114.

Shettleworth, S. J. (1973). Food reinforcement and the organization of behavior in golden hamsters. In *Constraints on Learning,* R. A. Hinde, & J. Stevenson-Hinde (Eds.) 243–63. London: Academic Press.

Tinbergen, N. (1951). *The Study of Instinct.* Oxford: Clarendon Press.

Tooze, Z. J. (1987). Some aspects of the structure and function of long-distance vocalizations of timber wolves (*Canis lupus*). Unpublished master's thesis, Dalhousie University.

Tooze, Z. J., Harrington, F. H., & Fentress, J. C. (1990). Individually distinct vocalizations in timber wolves, *Canis lupus. Animal Behaviour, 40*(4), 723–30.

Warren, W. H., Jr., & Shaw, R. E. (Eds.). (1985). *Persistence and Change.* Hillsdale, NJ: Erlbaum.

West, M. J., & King, A. P. (1990). Mozart's starling. *American Scientist, 78,* 106–14.

—5—
The phenomenon of attachment in human–nonhuman relationships

John Paul Scott

Editors' introduction

Scott's pioneering work on attachment in dogs is the basis for his survey of the attachment process in a variety of species. Scott argues that any species used for research cannot be intelligently manipulated unless the nature of its attachment process is understood. Scott elucidates the factors that underlie attachment such as the age at first contact, intensity and duration of contact, and sociality of the species.

Scott notes that the dog is unique insofar as it normally becomes part of human rather than canid society. Like Boccia et al. in Chapter 9, Scott observes that even when systematic attempts are made to standardize the research protocol, it is virtually impossible to avoid the formation of differential attachments between scientists and canid subjects.

Introduction

The phenomenon that is the subject of this book has usually been described by one of two terms, "attachment" and "bonding." It is unfortunate that this terminology is based on words in common usage that are difficult to define exactly and are loaded with surplus meaning. It is easy, however, to point out an essential difference between the two. In a social situation, attachment can be mutual and so involve two or more individuals. The word can also be applied to an asocial situation – for example, site attachment, in which only one individual is involved. Bonding, in contrast, usually implies the creation of a bond between two individuals who become, metaphorically, tied to each other. Ordinarily, it does not imply connections between more than two individuals. In the case of the interaction between humans and experimental animals, a subject that is extensively discussed in this book, there are several possibilities: mutual attachment or bonding, mutual attachment between a human and several animals, and one-way attachment on the part of either the human or the nonhuman animal.

The fact that humans became attached to animals and vice versa had been noted in popular culture long before the scientific study of the phenomenon of attachment began during the last part of the nineteenth century. A familiar example is the nursery song "Mary Had a Little Lamb," attributed to Sarah Josephine Hale (1830) by the *Oxford Dictionary of Nursery Rhymes*. Anyone who has bottle-reared a lamb can vouch for the literal truth of its words.

The first scientific reports on social attachment began with the naturalistic study of animal behavior. The work of Spalding (1873) is often cited as the first study of social attachment, and I recently have found another set of observations on such attachments in the *American Naturalist* (Caton, 1883). But it was not until the work of Heinroth (1910), who attempted to hand-rear all species of European birds, and Lorenz (1937), who systematically experimented with hand-reared birds, that social attachment was recognized as an important general phenomenon. Lorenz called the attachment process *Pragung*, which has been translated as "imprinting," implying that birds are passively acted on by their environment. Only later was it recognized that the process is an internal one and that the main function of the environment is to provide an object, social or nonsocial, to which young animals become attached. Among mammals, the most extensive studies of attachment have been done with domestic dogs, with more limited studies on domestic sheep, nonhuman primates, rodents, and humans themselves. Because my own work has been concerned largely with the dog, I shall start with that animal.

The dog

The dog is the oldest domestic animal. Its skeletal remains have been found with those of humans in burials dating back thousands of years, the oldest authentic ones having been discovered in a cave in Idaho. These remains have been dated as approximately 10,500 to 11,500 years B.P. (see Scott, 1968). As a domestic animal, the dog is unique in the degree to which an individual does not remain within the social organization peculiar to its species but normally becomes a part of human society through a process of adoption at an early age.

The process of attachment is impossible to overlook in a pet dog. It begins around 3 weeks of age and soon reaches a maximum rate that is maintained until at least 8 weeks of age. During this period, a puppy may become attached to humans through no more than a few minutes of eye contact once per day or as little as two 20-minute periods per week (Scott & Fuller, 1965). During this period a puppy may become attached to any animal that is associated with it, even including prey animals such as rabbits (Cairns & Werboff, 1967). The decline of the process of attachment is associated with a developing fear response to the strange, which is first noticeable around 7 weeks of age and reaches a maximum around 12 to 14 weeks. If an opportunity for attachment is postponed until the fear response is thoroughly established, attachment is difficult to induce without using drastic methods of restraint and forced contact over long periods. The result is never the same as if attachment had been established at an earlier period (Scott, 1987b). Individual timing is always somewhat variable, in part because of random developmental events and in part because of genetic variation. Genetic variation in dogs is particularly important with respect to the development of fear responses (Scott & Fuller, 1965).

Considered theoretically, attachment is a process that organizes the puppy's behavior in relation to those individuals, canine or human, with which it usually comes into contact during a critical or sensitive period (Scott, 1978, 1986). The extreme differences regarding developmental theory attributed to J. Schneirla and myself have been somewhat exaggerated. For example, Bateson (1981) has stated that there is a schism between those who see critical or sensitive periods as the "expression of inflexible endogenous rules" (Scott, 1962) and those who regard development as a "continuous interaction between the animal and its environment" (Schneirla & Rosenblatt, 1963).

I completely agree with the second statement and have never stated that there were "inflexible" endogenous rules. Nor do I believe that Schneirla thought that there was no regularity about development; otherwise, the egg of a cat might develop into a dog. I regard development as the interaction between genetic systems present in the fertilized egg with the various biological and social systems with which it subsequently comes into contact. All of these systems are variable, but by analyzing the effects of their respective variation we should be able to predict the outcome of development.

At the time of the initial disagreement I was unable to understand what it was based upon, and when I attempted a personal reconciliation of views I was prevented by Schneirla's illness and subsequent untimely death in

1968. It was not until I read papers by his associates on the subject of epigenesis (Gottlieb, 1970) that I began to understand his viewpoint. As a biologist trained in embryology it never occurred to me that anyone would not accept the epigenetic viewpoint, which had replaced the theory of preformation some two centuries ago. The second statement above is a general statement of this theory, with which I, of course, agree.

Schneirla's primary disagreement (Schneirla & Rosenblatt, 1963) was that he did not wish to be associated with the critical period phenomenon, with which I had attempted to link his work (Scott, 1962). This was an error on my part. Otherwise, his viewpoint was essentially no different from my own, but with some differences in emphasis, especially with respect to genetics and evolution. Another difference was emphasis on time as a variable. His viewpoint led to the idea that timing was secondary to the nature of events that occurred, whereas I believed that in studying development the major variable is time. Thus an organizational process such as attachment can vary in a number of major ways: (1) the time of inception, (2) the time of cessation, (3) the rate, (4) the degree to which it is affected by other organizational processes, (5) the degree to which variation in systems external to the organism can affect it, (6) the degree to which genetic variation affects it, and so on. This differs very little from the "probabalistic" viewpoint expressed by Gottlieb (1970).

The set of questions raised by this viewpoint led me and my associates to undertake detailed studies of the process of social attachment in dogs on the one hand, and the development of a general theory of critical periods, on the other. Briefly summarized, the time of inception in the puppy turned out to be around 28 days, rather than 3 weeks, as I had originally supposed (Gurski, Davis, & Scott, 1980). The decline in the rate of attachment proved to be associated with another organizational process, the development of the fear response to the strange (Freedman, King, & Elliot, 1961), which is first apparent at about 7 weeks and becomes increasingly important after 9 weeks.

I developed a general theory of critical periods in a series of papers (Scott, Stewart, & DeGhett, 1974; Scott, 1978; Scott, 1986). Briefly stated, an organizational process is most easily modified at the time or times in development when it is proceeding at a maximum rate. If the rate is zero, either before the process begins or after it is complete, there can be no alterations except by destroying the physical or neurological basis of organization. Such processes organize the function of a system, whether it be an organism or a social relationship. Once system organization is well established, it becomes increasingly difficult to alter without impairing or destroying the function of the system.

Thus, in the case of the attachment process in the dog, exposure to a potential object of attachment in the first 3 weeks of postnatal life produces no effect, whereas exposure during the weeks following 4 weeks will produce evidence of attachment within as little as 20 minutes (Cairns & Werboff, 1967). I concluded that the capacity for attachment never really ceases but is greatly slowed by the developing fear response to the strange, which begins to be important after 9 weeks. Previous attachments may also slow the process by preventing or limiting contact with new potential objects. I also concluded on the basis of various experiments such as that of Fisher (1955) that strong emotional stimulation may speed up the process of attachment. Finally, there is the problem of discovering the physiological and neurological basis of attachment. The work of Panksepp, Conner, Forster, Bishop, and Scoff (1983) suggests that attachment is related to the brain opioid system, but much remains to be done.

The applied aspects of this theory are obvious. Effects that can be produced with a minimum of effort at the time of maximum speed of an organizational process may take weeks, months, or years after a system has been strongly organized, if indeed they are possible at all. Then there is the problem of doing away with an unwanted or undesirable attachment, which may be as serious as that of establishing desirable ones.

In sum, the most important aspects of any science are phenomena. Theories become important only to the extent to which they enable us to understand, predict, and manipulate phenomena. The central phenomenon with which this book deals is that of social attachment. The theory that modifiability of organizational processes is related to the time at which such processes reach maximum speed enables us to manipulate easily any organizational process whose speed changes over time, including human social organization (Scott, 1987a). But the important things are the phenomena themselves – the variety of ways in which living organisms become organized.

Contrary to the situation in precocial birds such as ducks, where the following response is the best indicator of attachment, attachment in dogs is most strikingly manifested by distress vocalization, or separation distress. If a puppy is isolated from the object to which it has become attached, it begins to vocalize within a minute or so after separation and will continue to do this indefinitely. Presumably this is an adaptive response that would attract the attention of a mother or other caretaker, who would then find the separated puppy. The response is easily quantified as the number of discrete vocalizations in a given time period. Studies that vary the situation in which the puppy is isolated bring out the fact that the puppy is simultaneously becoming attached to the physical site in which it lives. Thus, there are simultaneous processes of site attachment and social attachment, and the two appear to be basically the same process. Consequently, the puppy's experience during the critical or sensitive period will determine the nature of the persons, animals, places, and objects to which it becomes attached.

If a puppy is raised in a restricted environment, it may develop a very narrow basis of attachment. More desirably from the human point of view, broader experience with a variety of humans, other animals, and places will produce a more widely adaptable dog. Restricting early experience drastically, as when a puppy is brought up in a kennel and not given outside experience or more than casual human contact, will produce a set of maladaptive symptoms that I have named the kennel dog or separation syndrome (Scott, 1970).

Genetic variation, in contrast, is not a major variable with respect to this phenomenon. In the course of testing several hundred puppies of several different breeds, I never found a puppy that did not show attachment and the resulting separation distress when it was isolated. The only differences were in the rate and amount of distress vocalizations. One breed, the African basenji, did not appear to become as strongly attached to humans as the others, but even this breed, if subjected to severe separation distress, would also become strongly attached to humans who relieved this distress. I concluded that the process of attachment is so essential to existence that little or no genetic variation is possible. This would constitute what Maynard Smith (1982) has called an "evolutionarily stable strategy," one that must be based on a stable underlying genetic organization. I also concluded that the attachment process is an integral part of any observational or experimental study of dog behavior.

Sheep and goats

Both the wild and domestic varieties of sheep and goats form cohesive flocks within which individuals spend their entire lives. Such flocks are based on an attachment process that functions from birth onward. The critical or sensitive period for attachment in young lambs begins soon after birth and persists for approximately 10 days (Scott, 1945). There is a more precisely timed attachment process in the mothers, who will become attached to any young lamb during the first 2 or 3 hours after birth (Collias, 1956). If a mother is separated from her newborn offspring and prevented from butting another penned with her, she eventually adopts the strange lamb. Thus, the critical period in an unrestrained mother is brought to an end by the maternal rejection response (Hersher, Richmond, & Moore, 1963). Thereafter, any lamb that comes near is rejected. Normally, maternal–offspring relationships are limited so that there is no competition

77

for nursing among lambs, as there might be if any lamb were free to nurse on any mother.

Since adult males are normally apart from the females during the season of parturition, the young lamb has access to, and hence can develop attachments with, only its own mother and with such other young lambs that may be present in the flock. Because of seasonal birth, all lambs in a flock are approximately the same age. The process of early attachment results in the leadership organization of sheep flocks, as each lamb tends to follow its own mother and to continue this behavior even as an adult, with the result that the oldest female with the largest number of descendants leads the flock. Among the separate male flocks that are formed in wild species, the organization is less precise, with age mates tending to stay together. When more than one age group is present, the younger males tend to follow the older. During the mating season, the male flocks join the females, at which time males actively compete for mates. The attachments formed by mating are strictly temporary, lasting only as long as any particular female happens to be in estrus (Geist, 1971).

Among goats, the attachment phenomenon is similar with respect to the existence of a limited time of attachment by mothers. Goats, however, share a behavior with many other ruminants such as deer, in that a newborn kid does not follow its mother but is left in a secluded spot where it stays without moving while the mother goes off to graze. This behavior is correlated with a less firmly organized leadership system among adult goats as compared with sheep (Hafez & Scott, 1962).

Sheep and goats form excellent models for studying maternal attachment because it is so precisely timed and is so strong. They should also be good models for studying the breaking off or weakening of the maternal bond as new infants appear during the following breeding season. Comparing the attachment process in sheep and goats with that in dogs and wolves leads to the conclusion that variation in the process of attachment is strongly related to the kinds of social organization seen in various species.

Mice and other rodents

In contrast to dogs and sheep, which normally live for many years, house mice are quite short-lived, becoming sexually mature around 60 days of age and seldom living beyond 2 years. They are also quite small in size and nocturnal in habit. They are not highly social animals. While they may on occasion live in dense populations, these are not highly organized groups, their association amounting to little more than mutual tolerance in the presence of an abundant food supply that is usually accumulated as a result of human effort. Under more natural conditions, mice usually develop a social organization based on comparatively imprecisely defined territories, in each of which a male and one or more breeding females live together with their litters. These territories rarely last long. As Bronson (1979) has said, the mouse is an opportunistic species.

Social attachment in mice has been little studied. Mice exhibit neither a following response nor a humanly audible distress vocalization in response to separation. They are capable of ultrasonic vocalization from birth on, but this has not been studied with respect to attachment. Infant mice that are picked up once per day and placed on a human hand show no escape responses as they normally would if handling were begun at a later date. But this may simply be a matter of controlling the fear response by desensitization rather than any true attachment (L. E. Roberts, personal communication).

From the experimental work on aggression, it is known that wild adult mice, either males or females, will respond to separation from a familiar cage site by thoroughly investigating the new site, and that they will defend the new site as a territory within 24 hours or less. This is often called isolation-induced aggression and is an experimental situation that has been used in a large number of experiments (Valzelli, 1965).

Territorial males in seminatural situations distinguish between familiar and unfamiliar females with respect to infanticide of their respective young (Parmigiani, 1989). This may be an indication that attachments develop between mated pairs, although such attachments have not been directly demonstrated. The evidence thus indicates that attachments in mice, at least site attachments, are readily developed at any age and are easily broken off, again fitting the concept of an opportunistic species. The data also suggest that a high degree of social organization among mammals is dependent on longevity. Short-lived animals do not have the time to develop complex organization, and frequent deaths destabilize any organization.

Like house mice, Norway rats are a highly adaptable and successful species. They are much larger and somewhat longer-lived than mice. But they are nevertheless much smaller and shorter-lived than either sheep or dogs. As Calhoun's (1963) classical study showed, rats living under seminatural conditions develop loose territorial systems associated with burrows, each territory containing one or sometimes two males and one or more breeding females with their litters. Blanchard and Blanchard (1984) have developed a laboratory colony model of aggression, each pen containing three males and three females and their offspring corresponding roughly to the groups developed by wild rats.

Separation has been intensively studied in rats using Hall's (1941) "open-field" test, which is really a test of placing a rat in a large, strange cage, thus separating it from a familiar site to which it has probably become attached and also from any cage mates to which attachments may have been made. Like mice, the rats respond by investigation and also by the emotional response of defecation. This very reliable response indicates that rats show separation distress from a familiar site, but little is known about the effects of social separation. Like mice, rats can be made into pets by hand rearing, which implies that humans become attached to them. H. Davis (personal communication) states that laboratory rats given the opportunity of living in a human home seek contact with their human associates and show strong indications of attachment.

The most highly social species of rodent so far studied is the prairie dog. These animals formerly lived in highly organized colonies of thousands of individuals on the western plains of the United States and have been studied extensively by King (1955) and Hoagland (1983). Prairie dogs form dominance relationships, establish territories, and have evolved a set of social signals that convey the presence of potential predators. They are much larger than either rats or mice, can live for several years, and are largely diurnal rather than nocturnal, spending much of their time above ground in the daylight as well as in their darkened holes. Attachment both to sites (the burrows) and to familiar individuals can be inferred. King also hand-reared prairie dogs. Such animals sought human contact and followed him closely. This evidence again supports the conclusion that attachment processes vary with the species concerned and are strongly related to the social organization developed by the species.

Studies of less social ground squirrels such as the 13-lined species (Vestal & McCarley, 1984) and Richardson's ground squirrel (Davis, 1984) show that uterine-related animals are more tolerant of one another and tend to share the same space. This again implies both social and site attachment.

With respect to attachment, Norway rats appear to be intermediate between house mice and the highly social prairie dogs. Like mice, they are predominantly nocturnal, but are larger and somewhat longer-lived, and their social groups somewhat more stable. Unlike mice, young rats are quite playful.

Davis and Pérusse (1988) have reviewed the experimental literature dealing with human–rat contact, most of which concerned the modification of emotional behavior in the so-called open-field situation and tests of the critical-period hypothesis rather than attachment per se. Davis and Pérusse found that about 50% of their subjects could be trained to bar-press for a reward of human contact, which strongly suggested that motivation was based on attachment.

Compared with the dog–human relationship, where both species develop considerable skill in interpreting each other's social signals, the human–rodent interaction presents greater difficulties in communication, especially in the nocturnal species, whose auditory signals are largely in the ultrasound range for humans, whose facial expressions are invariant, and who among themselves depend on olfactory signals that are mostly beyond human capacities for detection. The human experimenter probably misses a great deal of what is going on among his rodent subjects, and vice versa.

Primates

This group comprises a large number of species, most of which are of medium size or larger and many of which live in large social groups. They do not become sexually mature until they are several years old and usually produce only one offspring at a time. Because of these restrictions, and because nonhuman primates have never been domesticated, they are not readily available for experimental purposes and relatively little work has been done with the process of attachment per se, although primate societies have been intensively and generally studied (Smuts, Cheney, Seyfarth, Wrangham, & Struhsaker, 1988). The studies of Harlow and Harlow (1965) and Mason (1963) with rhesus infants are still among the most relevant. This species of macaque is native to India, where it lives in large social groups. Rhesus monkeys are relatively long lived and take 3 or 4 years to reach sexual maturity. Infants are born in a more mature state than human newborns. All sense organs are functional at birth, and the baby is able to move actively and cling to the mother without support. Harlow and Harlow found that newborns would cling to any inanimate object as long as it was in a vertical position and that they would become attached to this so-called surrogate mother. If raised in this fashion without normal social experience with other monkeys, the adults so reared showed markedly deviant behavior, especially with respect to mating and maternal care. In addition, Mason found that monkeys reared in this fashion and provided companionship with dogs would climb on the dogs and so become attached to them, both literally and figuratively. Thus, there is a critical or sensitive period for primary attachment that begins at birth and probably extends for some months at least, but attachments formed at later periods have not been studied in detail. Reared in seminatural situations, rhesus monkeys become attached to the members of their native troop, including the mother. The females remain in their natal troops throughout their lives, but there is a tendency for juvenile males to leave and join other troops, to which they presumably become attached (Vessey, 1968). We can infer that the attachment process goes on thoughout life, but probably at a diminished speed and intensity.

81

Humans are born with less well developed motor capacities than other primates, being unable to stand or crawl at birth. They can briefly support their own weight by grasping, but would be unable to cling to a mother for any length of time. They must be carried during the first weeks and months of life. All the sense organs are functional, however. Responses to separation (crying) indicate that the critical or sensitive period for attachment begins about 5 weeks of age and ends at approximately 7 months, when fear responses to strangers appear (Scott, 1963, 1981). Such fear responses are quite variable, and these individual differences probably have a basis in genetic variation. Children can, however, form attachments after the sensitive period and indeed throughout life, but at a much slower rate.

Most of the experimental work with infant attachment has been done between the ages of 12 and 18 months, partly because subjects are easier to obtain at this time and partly because most children begin to walk and thus move independently and so indicate their attachments by avoidance or following movements, as well as by crying (Bretherton & Waters, 1985).

Ainsworth, Blehar, Waters, and Wall (1978) developed a standardized test for attachment in infants that involves separation from the mother and reunion with her as well as exposure to strangers. They classified the quality of attachment so observed into three categories: (A) The infant snubs or avoids the mother on reunion; (B) the infant greets and seeks interaction (normal); and (C) the infant shows angry and resistant behavior. Responses A and C can be interpreted as behavior that has the effect of punishing the mother for leaving (see also Bowlby, 1973). It is obvious that this grading scale measures the relationship developed between the mother and infant, not merely the infant's attachment.

While considerable variation in responses occurs and some of this variation may be attributed to genetics (Miyake, Chen, & Campos, 1985), the results show the attachment process to be a universal one, with variation taking place primarily in the behavioral outcome rather than within the process itself.

That humans also form attachments after infancy and throughout adult life is obvious, but it is difficult to perform relevant experimental studies. Klaus and Kennel (1976) explored the possibility that human mothers responded with attachment behavior like sheep and goats with respect to their newborns. Their results and some supporting data indicate that mothers who have sensory contact with their infants immediately after birth, rather than being anesthetized or separated immediately after birth, as once was standard obstetrical practice, do subsequently behave more positively toward their infants, at least for a few weeks. But there is obviously no sharp time limitation of the maternal attachment process, as there is in sheep and goat mothers.

Because of the lack of solid data on the human attachment process at ages other than the early-infancy research described above, our best understanding of human attachment behavior comes from inferences from animal models.

Inferences from animal models

From this brief review of research on attachment, several general conclusions can be drawn. The most important is that the process of attachment in all vertebrates is essentially similar. Further, the processes of site attachment and social attachment are so similar that the most probable evolutionary explanation is that social attachment, in those species in which it exists, has evolved from the more general phenomenon of site attachment. The fact that the process of attachment has undergone relatively little evolutionary modification in diverse species suggests that it is a process so fundamental to existence that a stable genetic and physiological process ("evolutionarily stable strategy") has evolved. If this is true, it follows that we are relatively safe in generalizing findings from one species to another.

A second general conclusion is that the process of social attachment may be extended from the parent species to a nonrelated species, as in the cases of imprinting in birds and in hand-rearing experiments with mammals. This means that social attachment is always a factor in any experiment, treatment, or observation that involves prolonged contact. We can also draw several more specific conclusions, as follows:

1. Rapid social attachment among wild species is normally confined to a brief period early in life (the critical or sensitive period). Together with the normal behavioral patterns of the species, this ensures that an infant animal will become attached primarily to its own species and to close genetic and social relatives within that species. In addition, most highly social species are capable of forming attachments in later life, but they do so at much slower rates and are limited by earlier attachments that may either restrict or enhance contacts with other individuals.

2. The process of attachment is an internal one and not dependent on rewards and punishments. All that is necessary is that the object of attachment, whether animate or inanimate, be noticed by the individual concerned and that the contact be maintained long enough for the attachment process to take place. The necessary contact time will vary from a few minutes during the critical period to much longer periods during adulthood. In any case, the strength of attachment should increase with prolonged contact. This should apply to any extended psychotherapeutic process (Scott, 1987b).

3. External rewards and punishments are not essential to the attachment process, but may modify it. For example, a reward may serve to keep the individual in contact until the attachment process occurs. Among humans in a work situation, individuals previously unknown to each other may be kept in contact for eight hours per day for indefinite periods, allowing both site and social attachments to take place.

In contrast, punishment, including painful or other noxious stimulation, may shorten contact time and so prevent the occurrence of attachment if the individual can escape. But if the individual is unable to escape, as with young children and prisoners, the attachment process is speeded up and intensified by the emotions generated by punishment. This accounts for the numerous instances of seemingly maladaptive attachments – between slaves and masters, jailers and prisoners, children and abusive parents, wives and abusive husbands, husbands and abusive wives. A general hypothesis may be stated: *The occurrence of any strong emotion, whether pleasant or noxious, will speed up and intensify the process of attachment.*

Under natural conditions the emotion of fear, to which the usual response is to escape or retreat, largely governs the expression of the attachment process. Thus, a fearful response is the usual primary reaction to being placed in a strange situation, causing the animal to try to find its way back to the home area to which it is attached, and in most species that have been studied, a developing fear response to the strange is the process that brings the critical or sensitive period to an end.

4. The neurophysiological and emotional bases of the attachment process are not well known, resting largely on inference and a few experiments. Panksepp et al. (1983) suggested that the brain opiate system may be involved. Certainly, some dramatic changes can be produced by drugs that affect the opiate system. For example, dogs raised in a restricted environment exhibit a typical kennel-dog syndrome when placed in a novel situation: extreme fear, attempts to escape, and complete avoidance of a strange human experimenter. Given a dose of naloxone, a morphine antagonist, the same dogs will make constant contact with a strange experimenter, nosing, licking, and tail wagging, showing the behavior of an attached dog in an exaggerated form. It would be expected that the altered behavior would persist only after prolonged and repeated human contact accompanying drug treatment.

Also, kennel dogs placed in a strange kennel become hyperactive and cannot be trained to sit still. But if treated with amphetamine sulfate, they become trainable, presumably because the amphetamine inhibits the fear response (Corson et al., 1980). Such separation may be one cause of hyperactivity in children and may interfere with learning as well.

5. The function of the attachment process varies from one species to another, being related to the usual social organization of the species concerned. Social attachment is essential to the maintenance of permanent or semipermanent social groups. Without it, only temporary aggregations can occur. It is also related to the longevity of the species, being more prominent in long-lived animals such as dogs or humans. This means that any species that is used for experimental purposes cannot be intelligently manipulated unless the nature of its attachment process, both to sites and to social objects, is known and allowed for.

6. A major factor in the expression of any social behavior is the nature of the relationship in which it occurs. This generalization can be extended to nonsocial behavior (e.g., interaction with nonliving objects) that occurs or is elicited within a social relationship. This means that experimenters should always be conscious of the kinds of relationships that are developing between them and their experimental animals and keep these under control so that they do not become a random variable that can distort the experiment.

Experimental control and observations of the attachment process in dogs

I shall now describe the way in which we attempted to control these processes in one particular experiment on genetics and social behavior in dogs (Scott & Fuller, 1965) for which we designed an artificial cultural and environmental system that we called the "School for Dogs."

1. *Canine contacts.* Puppies remained in social contact with their own species throughout the 1-year period in which each was included in the experiment. Each pregnant mother was assigned to a large nursery room, in which she bore and fed her litter until she was removed at 10 weeks. The litter remained together continuously except when individuals were briefly removed for testing. It was moved to a large outdoor run when the puppies were 16 weeks of age, where it remained until the puppies were 1 year of age, at which time they left the experiment.

2. *Routine human contact.* Male animal caretakers came in once per day to feed and water the animals, pick up feces, and add to absorbent shavings as needed. Every 4 weeks, the mother and litter were removed to a holding cage while a thorough steam cleaning was done and fresh bedding added. The daily feeding and cleaning (without special cleaning) were continued in the outside runs. The caretakers were instructed to make only necessary contact with the puppies and not to make pets of them.

3. *Experimenter–puppy contacts.* Except on Sundays, experimenters made daily contact with the puppies in order to take measurements and perform tests on them. Some tests involved removal from the home pen (separation from a familiar site); almost all experiments involved individual testing (separation from familiar individuals).

4. *Sex of experimenters.* In order to control for human sex differences, all experiments were done by male–female pairs, making possible attachment to individuals of opposite sexes. The puppies had no contacts with human children or adolescents.

5. *Behavior of experimenters.* Experimenters were instructed to treat every puppy alike, in order to maximize the effect of genetic variation among the breed populations and individuals within breeds. They were told not to make special pets and (because this would inevitably be a variable) not to talk to the puppies or to talk while an experimental observation was being made. Finally, they were told never to punish the animals unless this was a part of the test.

This regimen was rigorously adhered to with all puppies reared during a period of some 13 years. When the data were analyzed, almost every test and observation indicated some degree of genetic variation, as had been expected. In addition, some unexpected results were obtained from the tests that involved the dog–human relationship.

First, humans became strongly dominant over the puppies, even though no punishment had been employed in their interaction, and the puppies were correspondingly subordinate. This resulted from the fact that the puppies had been picked up and handled from the first day of existence and almost daily thereafter for a period of 1 year. The handling was begun before the puppies had developed any capacity for escape or aggressive responses, and they consequently developed a strong habit of passivity while being held by the experimenter. The result was that the puppies could be easily controlled simply by being picked up, even as adults. One could even stop a dog fight in this way. We concluded that the most desirable method of developing dominance over experimental animals, an essential requirement in any dog–human relationship, could best be achieved not by punishment but by restraint.

Second, the puppies did not seem to develop their maximum capacities as observed among dogs reared in homes or trained as work dogs. This could be explained by our standardizing of human responses to the dogs. A normal social relationship involves reciprocal interaction. In our dealing with the puppies, we did not respond differentially to what they did, as would be the case in any fully developed relationship. In short, we eliminated feedback. In some cases the puppies appeared to be trying to figure out what the experimenter wanted them to do and of course, were receiving no signals other than those that were built into the test involved. Furthermore, the four breeds that are used for hunting and hence whose work involved some sort of food rewards, responded very well to the standard test situations that employed food rewards. In contrast, the Shetland sheep dogs, which are related to working dogs that are trained not with food rewards but with praise and punishment by the trainers, did relatively badly on the tests that involved food rewards. Therefore, these experimental animals did not develop the behavioral capacities that one might have seen in a training situation involving dog–human interaction.

Third, in spite of our efforts, there were indications that the puppies responded differentially to male and female experimenters. The Handling Test was developed to measure individual differences in the responses of puppies to standardized human handling. It was administered every 2 weeks from 5 through 15 weeks of age. Responses were rated on four levels according to their vigor, and several scores were developed. These turned out to be good measures of the process of attachment to humans, which we called primary socialization because it involved developing the first social relationships with humans.

With every litter, male and female handlers were alternated from one test to the next. The male handler in almost every case was myself, and the female was one of several research assistants, but the same team ran through the test from beginning to end on any given litter. Thus, a given litter might be handled by a male at 5 weeks and by a female at 7 weeks, and vice versa.

Table 5.1 compares the mean scores of Scale 2, which we called playful aggression. This scored positive contacts with a human handler such as approaching, pawing, and nosing, usually playfully aggressive in nature. Individual scores ranged from 0 to 22 at 5 weeks and 0 to 35 at 7 weeks. Biting almost never occurred. The score was therefore a good indicator of attachment.

In every breed population at 5 and 7 weeks, except for fox terriers at 5 weeks, the mean scores for female handlers were higher than those for the male. This, of course, held true for the total scores including all breeds. There was also an increase in mean scores from 5 to 7 weeks in every breed, indicating the progress of the attachment process.

It follows that in some way the puppies were distinguishing between the experimenters, probably because of subtle differences in movements. It is also possible that the puppies were actually becoming more attached to the female experimenters who had somewhat more contact with the puppies than did the male in these earlier weeks of development. We may conclude that one of the variables that enters into any experiment involving contact

Table 5.1. *Mean scores of playful aggressiveness toward male and female handlers at 5 and 7 weeks of age in three breeds of dog*

Breed	Sex of handler	5 weeks		7 weeks	
		Score	*n*	Score	*n*
Basenji	M	1.7	26	9.1	17
	F	6.6*	17	11.8**	26
Beagle	M	2.6**	32	9.6	18
	F	4.7	18	12.3**	32
Cocker spaniel	M	3.0	22	5.7	36
	F	4.2**	36	8.5**	22

Note: The score reflects the number of times that a puppy reaches response thresholds for a given set of behavior patterns. The result is essentially a rank-order score and may be analyzed accordingly. All three breeds exhibited higher response scores to female handlers at both ages.

The most striking result was shown by the basenji puppies handled by males, which received 17 zero scores out of 26 animals at 5 weeks. Because of this distribution, the χ^2 test was used. Puppies of this breed tend to be fearful of human handlers at this age and apparently discriminated between males and females. Results in two breeds other than the above could not be analyzed because of the small number of litters tested by a male handler, but were essentially similar.
* $p < .001, \chi^2$. ** $p < .002$, Wilcoxon–Mann–Whitney.

between experimenter and subject is the possibility of differential attachment, as well as differences in responses based on differences in experimenter behavior.

Final comments on human–nonhuman attachment

Experimenters can and do become attached to the animals they study. For example, Jane Goodall (1965) obviously became attached to the members of the chimpanzee troop that she studied. Her study method was to establish herself in an area where the troop members could see her and remain quiet until they began to approach, first cautiously and eventually somewhat indifferently, as they would a new harmless animal. As she began to know them ("knowing" is perhaps an euphemism for attachment) and to recognize individuals, she gave them familiar human names as well as descriptive ones, never identifying them by abstract numbers or letters. It is very possible that the chimpanzees became reciprocally attached to her.

Hers was not an unusual case, and it may not be confined to students of animal behavior. It was once the fashion among experimental zoologists to concentrate one's research on a single species and to learn as much as possible about it from every aspect. I once observed that after a lifetime of associating with one species, an experimenter might come to resemble his favorite subjects, not only in behavior but sometimes even in appearance. Among psychologists also there is a tendency to become attached to a particular species, if not to the individuals within the species. Once a "rat runner," always a "rat runner." There are other reasons for such persistence, but experimenters seldom switch.

Whether or not experimenters become socially attached to their subjects, they definitely become attached to their laboratories, an obvious case of site attachment. Field-workers also become attached to their study sites, and this may limit their work.

The degree to which the human capacity for social attachment is extended to other animals depends on several factors. One is the length of time one remains in contact with individual experimental subjects, which might range from a few minutes in certain rodent experiments to several years in some primate studies. A second factor is the number of individuals involved. It would be difficult to become attached to an individual inbred mouse in one brief experiment, although in the course of many experiments one might become attached to house mice in general. Still another factor is the nature of the species concerned. It should be easier for a human to become attached to another mammal than to an insect. Although I have no proof that this is true, it should be easier to extend human–human relationships to similar species than to distantly related ones.

Similar considerations apply to the attachments of nonhuman animals to experimenters. Vertebrate animals are normally protected from forming easy attachments to humans, partly by those behaviors that delay or inhibit attachment beyond the critical period, such as the developing fear of the strange, which in most species is a normal reaction to anything new and unusual, and hence potentially dangerous. But this response may in itself interfere with an experiment and must always be controlled or eliminated in some way.

These negative responses to experimenters can be overcome in time, as happened with the chimpanzees in Goodall's case. Likewise, captive wild wolves are extremely fearful and therefore not amenable to training. But Ginsburg and his colleagues (Woolpy & Ginsburg, 1967) found that if they enforced visual contact by daily entering a wolf's cage for half an hour or so, they could bring about attachment as shown by positive approaches after approximately 6 weeks of training.

These limitations and difficulties do not apply during the critical period

89

in young animals, when limited contact will often activate the attachment process within a few minutes.

The effects of the attachment process on experimental procedures and results in any species will therefore depend on the sociality of the species concerned, the age at which contact is brought about, the degree of contact, and its duration. In addition, the nature of the individual experimenter may be a factor, as we found in our puppy experiments. In no case can an experimenter assume that he or she is invisible, unheard, or even unsmelled. Nor can an experimenter assume that he or she is immune to the attachment process.

References

Ainsworth, M. D. S., Blehar, M. C., Waters, E., & Wall, S. (1978). *Patterns of attachment: A psychological study of the strange situation.* Hillsdale, NJ: Erlbaum.

Bateson, P. (1981). Control of sensitivity to the environment during development. In K. Immelmann, G. W. Barlow, L. Petrinovich, & M. Main (Eds.), *Behavioral development: The Bielefeld Interdisciplinary Project.* Cambridge University Press, pp. 432–53.

Blanchard, D., & Blanchard, R. J. (1984). Affect and aggression: An animal model applied to human behavior. In R. J. Blanchard & D. C. Blanchard (Eds.), *Advances in the study of aggression.* New York: Academic Press, vol. 1, pp. 1–62.

Bowlby, J. (1973). *Attachment and loss, Vol. II: Separation: Anxiety and anger.* New York: Basic.

Bretherton, I. , & Waters, E. (Eds.) (1985). *Growing points of attachment theory and research. Monographs of the Society for Research in Child Development, 50*(1, 2, Serial no. 209).

Bronson, F. G. (1979) The reproductive ecology of the house mouse. *Quarterly Review of Biology, 54,* 265–79.

Cairns, R. B. & Werboff, J. (1967). Behavior development in the dog, an interspecific analysis. *Science, 159,* 1070–2.

Calhoun, J. B. (1963). *The ecology and sociology of the Norway rat.* Bethesda, MD: U. S. Department to Health, Education, and Welfare.

Caton, J. D. (1883). Unnatural attachments among animals. *American Naturalist, 17,* 359–63.

Collias, N. E. (1956). The analysis of socialization in sheep and goats. *Ecology, 37,* 228–39.

Corson, S. A., Corson, E. O'L., Becker, R. E., Ginsberg, B. E., Trattner, A., Conner, R. L., Lucas, L. A., Panksepp, J., & Scott, J. P. (1980). Interaction of genetics and separation in canine hyperkinesis and differential responses to amphetamine. *Pavlovian Journal of Biological Science, 15,* 5–11.

Davis, H., & Pérusse, R. (1988). Human-based social interaction can reward a rat's behavior. *Animal Learning and Behavior, 16,* 89–92.

Davis, L. (1984). Behavior interactions of Richardson's ground squirrels: Asymmetries based on kinship. In J. O. Murie & G. R. Michener (Eds.), *The* biology of ground-dwelling squirrels: Annual cycles. behavior ecology and sociality. Lincoln: University of Nebraska Press, pp. 424–43.

Fisher, A. E. (1955). *The effects of differential early treatment on the social and exploratory behavior of puppies.* Unpublished doctoral thesis, Pennsylvania State University.

Freedman, D. G., King, J. A., & Elliot, O. (1961). Critical period in the social development of dogs. *Science, 135,* 1016–17.

Geist, V. (1971). *Mountain sheep: A study of behavior and evolution.* Chicago: University of Chicago Press.

Goodall, J. (1965). Chimpanzees of the Gombe Stream Reserve. In I. DeVore (Ed.), *Primate behavior; Field studies of monkeys and apes.* New York: Holt, Rinehart, & Winston, pp. 425–73

Gottlieb, G. (1970). Conceptions of prenatal behavior. In L. R. Aronson, E. Tobach, D. S. Lehrman & J. S. Rosenblatt (Eds.), *Developmental and evolution of behavior: Essays in memory of T. C. Schneirla.* San Francisco: Freeman.

Gurski, J. C., Davis, K., & Scott, J. P. (1980). Interaction of separation discomfort with contact, comfort and discomfort in the dog. *Developmental Psychobiology, 13,* 463–7.

Hafez, E. S. E. & Scott, J. P. (1962). The behavior of sheep and goats. In E. S. E. Hafez (Ed.), The behavior of domestic animals. London; Balliere, Tindall, & Cox, pp. 3–20.

Hall, C. S. (1941) Temperament: A survey of animal studies. *Psychological Bulletin, 38,* 909–43.

Harlow, H. F., & Harlow, M. K. (1965). The affectional systems. In A. M. Schrier, H. F. Harlow, & F. Stollnitz (Eds.), *Behavior of nonhuman primates.* New York: Academic Press, pp. 287–334.

Heinroth, O. (1910). Beitrage zur Biologie, namentlich Ethologie und Psychologie der Anatiden. *Verhandlung des V: Internationalen Ornithologen-Kongresses, 5,* 589.

Hersher, L., Richmond, J. B., & Moore, A. U. (1963). Modifiability of the critical period for the development of maternal behavior in sheep and goats. *Behavior, 20,* 311–19.

Hoagland, J. L. (1983). Nepotism and alarm calling in the black-tailed prairie dog (*Cynomys ludovicianus*). *Animal Behavior, 31,* 472–9.

King, J. A. (1955). *Social behavior, organization and population dynamism in a black-tailed prairie dog town in the Black Hills of South Dakota.* Contributions from the Laboratory of Vertebrate Biology, No. 67. Ann Arbor: University of Michigan Press.

Klaus, M. H., & Kennell, J. H. (1976). *Maternal–infant bonding.* St. Louis, MO: Mosby.

Lorenz, K. (1937). The companion in the bird's world. *Auk, 54,* 245–71.

Mason, W. A. (1963). The effects of environmental relocation in the social development of rhesus monkeys. In C. H. Southwick (Ed.), *Primate social behavior.* New Yort: Van Nostrand, pp. 161–3.

Miyake, K., Chen, S., & Campous, J. J. (1985). Infant temperament, mother's mode of interaction, and attachment in Japan: An interim report. In I. Bretherton & E. Waters (Eds.), *Growing points of attachment theory and research. Monographs of the Society for Research in Child Development. 50*(1, 2, serial no 209), 276–97.

Panksepp, J., Conner, R. L., Forster, P. K., Bishop, P., & Scott, J. P. (1983). Opioid effects on social behavior of kennel dogs. *Journal of Applied Ethology, 10,* 63–74.

Parmigiani, S. (1989). Maternal aggression and infanticide in the house mouse: Consequences on social dynamics. In P. F. Brain, D. Mainardi, & S. Parmigiani (Eds.), *House mouse aggression: A model for understanding the evolution of social behavior.* London: Harwood, pp. 161–78.

Schneirla, T. C., & Rosenblatt, J. S. (1963). "Critical periods" in the development of behavior. *Science, 139,* 1110–14.

Scott, J. P. (1945). Social behavior, organization, and leadership in a small flock of domestic sheep. *Comparative Psychology Monograph, 18*(4), 1–29.

(1962). Critical periods in behavior development. *Science, 138,* 949–58.

(1963). The process of primary socialization in canine and human infants. *Monographs of the Society for Research in Child Development, 28*(1, serial no. 85), 1–47.

(1968). Evolution and domestication of the dog. In T. Dobzhansky (Ed.), *Evolutionary biology.* New York: Appleton-Century-Crofts, Vol. 2, pp. 243–75.

(1970). Critical periods for the development of social behavior in dogs. In S. Kazda & V. G. Denenberg (Eds.), *The postnatal development of phenotype.* Prague: Academia, pp. 21–32.

(Ed.) (1978). *Critical periods.* New York: Academic Press.

(1981). Biological and psychological bases of social attachment. In H. Kellerman (Ed.), *Group cohesion: Biological, psychological and sociological reflections.* New York: Grune & Stratton, pp. 206–24.

(1986). Critical periods in organizational processes. In F. Falkner & J. M. Tanner (Eds.), *Human growth,* 2d ed. New York: Plenum, Vol. 3, pp. 223–41.

(1987a). Critical periods in processes of social organization. In M. H. Bornstein (Ed.), *Sensitive periods in development.* Hillsdale, NJ: Erlbaum, pp. 247–68.

(1987b). The emotional basis of attachment and separation. In D. P. Schwartz, J. L. Sacksteder, & Y. Akabane (Eds.), *Attachment and the therapeutic process.* Madison, CT: International Universities Press, pp. 43–61.

Scott, J. P., & Fuller, J. L. (1965). *Genetics and the social behavior of the dog.* Chicago: University of Chicago Press.

Scott, J. P., Stewart, J. M., & DeGhett, V. J. (1974). Critical periods in the organization of systems. *Developmental Psychobiology, 7,* 489–513.

Smith, J. M. (1982). *Evolution and the theory of games.* Cambridge University Press.

Smuts, B. B., Cheney, D. L., Seyfarth, R. M., Wrangham, R. W., & Struhsaker, T. T. (Eds.). (1988). *Primate societies.* Chicago: University of Chicago Press.

Spalding, D. A. (1873). Instinct, with original observations in young animals. *MacMillan's Magazine, 27,* 282–93.

Valzelli, L. (1965). Aggressive behavior induced by isolation. In S. Garattini & E. B. Sigg (Eds.), *Aggressive behavior.* New York: Wiley, pp. 70–6.

Vessey, S. H. (1968). Interactions between free-ranging groups of rhesus monkeys. *Folia Primatologica, 8,* 228–39.

Vestal, B. M., & McCarley, I. T. (1984). Spatial and social relations of kin in thirteen-lined and other ground squirrels. In J. O. Murie & G. R. Midremer (Eds.), *The biology of ground-dwelling squirrels: Annual cycles, behavioral ecology, and sociality.* Lincoln: University of Nebraska Press, pp. 404–23.

Woolpy, J. H., & Ginsburg, B. F. (1967). Wolf socialization: A study of temperament in a wild social species. *American Zoologist 7,* 357–63.

—6—

Humanity's "best friend": the origins of our inevitable bond with dogs

Benson E. Ginsburg and Laurie Hiestand

Editors' introduction

Like Scott, Ginsburg and Hiestand are concerned with the intense bond that exists between scientists and canid subjects. Their focus is on the evolutionary factors that account for the ease with which humans and canids interact. Because of this emphasis, Ginsburg and Hiestand stress the critical nature of knowing the species, breed, and individual histories of canid subjects. These authors also underscore the sensitivity of wolves and dogs to minimal human cues and discuss how such discriminations might affect differential responding by canids to human observers.

It will surprise no one to find the dog included in this volume. After all, when one thinks of the bonds we form with nonhuman species, our relationships with dogs inevitably come to mind. Humans and dogs have led a mutually cooperative existence for some 15,000 years (Wayne, Benveniste, Janczewski, & O'Brien, 1989). As a result of this close association, and probably also because many people enjoy working with them, dogs have often been the subject of scientific scrutiny. The dramatic variations in their size and other physical and physiological attributes, as well as the variability in their temperament and behavior, attract the attention of both amateurs and scientists. It is because dogs frequently serve as experimental subjects that their degree of sensitivity to the presence of humans should be examined. In order to understand more fully the nature of the dog–human bond, it is thus important to consider certain aspects of the behavior of the dog's progenitor, the wolf.

The sensitivity of the domestic canid to human behaviors and even moods has been well documented (Buytendijk, 1936; Woodbury, 1943; Fox, 1971). Johnson (1913; cited in Rosenthal, 1966) showed that allegedly successful auditory discriminations made by dogs were actually the result of a reliance on subtle cues that emanated from the experimenter, such as posture, respiration, and tensing and relaxing of muscles. When the tests were repeated with the experimenter hidden from the dogs, and the person who was collecting the data was ignorant of the desired results, the animals were no longer able to make the discriminations.

This fine communicative sensitivity between humans and dogs has been both useful and problematic. In work done at Guide Dogs for the Blind in San Rafael, California (Pfaffenberger, Scott, Fuller, Ginsburg, & Bielfelt, 1976), it was found that some trainers who were highly successful with German shepherd dogs were less so with breeds of a different temperament (e.g., retrievers). The opposite was also true. The same principle of matching human–dog affinities applied as well to the blind person receiving the dog. Generally a class at the school consisted of eight students. The trainer had 12 dogs available to him whose personalities and working habits were known. In evaluating each blind candidate before assigning a dog, the objective was to find the best match, not only for physical characteristics (a larger dog for a larger person) but also for personality. A less confident person, for example, required a dog that worked more deliberately. The manner in which a student gave commands and signals was also a factor in the evaluation. The proper matching of the person with the dog played a significant role in the speed of bonding and progress in training.

The gender of a human handler can also differentially affect dog behavior. Lore and Eisenberg (1986) found that while female dogs approached unfamiliar humans of either sex readily, male dogs tended to avoid unfamiliar male humans. They reported that most dog owners, and many veterinarians, felt that dogs were less reactive toward women. J. P. Scott (Chapter 5, this volume) similarly found an effect of the sex of the handler on a measure of playful aggression in three breeds of dog.

In the long and varied history of dogs' service to people, an area where they have proved especially useful is the military. There, too, the potential effects of humans on dog performance have been observed. During World War II, for example, some dogs showed an uncanny ability during training to detect mines, even when these had been buried under tide flats, thereby obscuring olfactory and visual cues. One hypothesis was that the dogs were responding to subtle cues from persons in their vicinity who knew where the mines were buried. A team of parapsychologists was asked to design an evaluative program in which such cues could be either identified or eliminated. However, no conclusive evidence for reliance on social cuing could be verified.

Another example of the dogs' reliance on subtle behavioral cues obtained from wartime was their success in detecting a variety of booby traps, for which they could not have been specifically trained since the types of traps that would be encountered were unknown. There was wide variation in the performance of the dogs, which was correlated in part with the reactions of their handlers when under fire. Thus, the state of mind of the handler was communicated to the dog, a phenomenon known to anyone who has taught a dog-obedience class and one that cannot be ignored by scientists involved in behavioral research with this species.

One example of the presence of such effects in the laboratory occurred at the Jackson Laboratory at Bar Harbor, where the behavioral reactions of dogs on a Pavlov conditioning stand were evaluated. Researchers used a series of standard tests, with the objective of attempting to alter extreme behaviors pharmacologically. It was discovered that the ratings of the dogs varied between testers. Two different investigators elicited differential behavior from the dogs, even though they administered the same tests in (presumably) the same way to the same dogs. It was then ascertained that one of the raters was intimidated by the larger breeds, while the other had a dislike of "small, yappy dogs." Since the animals were being rated by an independent observer behind a one-way glass, the dogs were most likely responding to subtle cues that actually caused them to alter their behavior. The pharmacological work was therefore done using a conditioned foreleg response that was not tester dependent (James & Ginsburg, 1949).

Even the physiology of the domestic dog can be differentially affected merely by the presence of a human. This has potential implications for biomedical research. For example, Gantt (1964; cited in Rosenthal, 1966) observed that a dog's heart rate could drop from 160 to 140 beats per minute simply as a result of a particular human's presence in the room.

Dogs are able to use auditory and olfactory cues as well as visual information to detect humans, a fact that our species, with its heavy reliance on vision, may often forget. As many canid researchers can attest, it is necessary to test dogs in an undisturbed laboratory setting, since the animals are easily distracted by the presence (visual, auditory, or olfactory) of both unfamiliar and familiar people. They are often aware of such intrusions before the researcher is (Arons, 1989; L. Hiestand, personal observations).

Dogs' reactions also depend on the situation. As an example, a consistently friendly dog who knew one of our students well and paid little attention to him if he entered the professor's house while someone was at home threatened the same student viciously when he entered to leave a manuscript when no one was there. Most owners of dogs have encountered similar situational experiences. These extend as well to the laboratory setting, where a dog's behavior may be a function of several variables, such as its previous experience in that setting, its reaction to novelty, and its reaction to lab personnel, both as individuals and in terms of the differential behavioral cues they generate.

This acute sensitivity that dogs have to their social environment is not the product of domestication, although the degree of its expression can be modified by that process; instead, it has been inherited from the wild progenitor of the species. The domestic dog is most closely related to the wolf, of all the extant members of the Canidae (Scott & Fuller, 1965, pp. 54–6; Wayne & O'Brien, 1987). The wolf is well known for its complex social interactions and individuation of roles within its social group, the pack (Mech, 1970). If we assume that the modern wolf still has the capacities and characteristics of its ancestral form, what have been the major results of more than 15,000 years of artificial selection? The most obvious changes are seen in the tremendous physical and behavioral variations that characterize present breeds (Stockard, Anderson, & James, 1941; Scott & Fuller, 1965; American Kennel Club, 1989). These are so striking and diverse that Darwin (1872) was hesitant to ascribe a single origin to the domestic dog.

Lorenz (1952) reinforced the idea of a polyphyletic origin and concluded that the northern breeds were descended from wolves (*Lupus* dogs), while the others were descended from jackal stock (*Aureus* dogs). These two types, in his view, had quite different behavioral characteristics, the *Aureus* dogs being more juvenile and servile than the more independent *Lupus* dogs. The former were also considered to have a more labile affiliation to the human master, while the latter took longer to develop the human–dog alliance, although once this was established, the bond remained strong. *Lupus* dog behavior was interpreted as deriving from the social structure of a wolf pack, with the human in the role of the dominant wolf. This view has permeated much of the popular literature about dog behavior and has served as the basis for many training methods. Current evidence from the electrophoretic analysis of blood proteins does not support the idea of a multiple origin of the dog. Instead, it points to a single origin from wolf stock (Wayne & O'Brien, 1987; Wayne et al., 1989).

How, then, can we account for the diversity in morphology, physiology, and behavior of the domestic dog? Three events occurred during domestication that gave rise to the variability we see today: One is the loss of phenotypic buffering mechanisms characteristic of wild species, that is, the maintenance of fairly uniform physical characteristics by natural selection;

the second is the effect of artificial selection for characteristics of value to humans; and the third is humanity's preference for the exotic, which has favored and maintained mutations that would otherwise have been lost (Ginsburg, 1976a,b).

Current breeds of dog have been selected to express behaviors differentially that were inherited from their wolf ancestors. One obvious example of this lies in their predatory ability. The wolf has evolved a cooperative, affiliative society to allow it to exploit the large ungulate prey biomass. Predatory behavior still exists in the domestic dog, to various degrees depending on the breed. In the Siberian husky, for example, which as a breed originated from individuals that were chosen for their ability to fend for themselves, full predatory behavior remains in the behavioral repertoire. In others, such as sheep-herding dogs, some elements of the full predatory repertoire remain, such as stalking, chasing, and confronting, but without the ensuing attack (Vauk, 1953; Coppinger et al., 1987; Arons, 1989). In breeds developed to live with sheep and to guard them, characteristic behaviors are territoriality centered around the location of the sheep, as well as some degree of affiliation between dog and sheep. The sheep are not attacked. Aggressive, xenophobic behaviors are instead directed against intruders. In pointers and setters, appetitive aspects of predatory behaviors are typical, but the consummatory aspects are weakened or missing. Selection by humans has modified the predatory characteristics of the original canid repertoire so that some are retained to various degrees and in various combinations, but others are weakened or lost (Coppinger et al., 1987).

Studies of feral dogs suggest that when left to fend for themselves, domestic dogs of various breeds can revert to complete predatory behavior. Some of these observations are based on mongrels, so that we do not know what is genetically represented. Many feral dogs are scavengers. The "packs" that are formed are usually opportunistic associations with none of the organized characteristics of the wolf pack (Scott, 1950, 1973; Beck, 1973; Daniels, 1987). To what degree the full predatory patterns remain as potentially available behaviors in breeds in which selection has been for various combinations of appetitive behaviors is not clearly established. What we do know is that in order to maintain the typical genetic characteristics of a breed, whether behavioral or morphological, continued selection pressure is necessary, suggesting that sufficient genetic variability remains within the gene pool of most breeds to permit further modification, including reversion to behaviors previously selected against but not eliminated from the gene pool of the breed in question. Thus, at Guide Dogs for the Blind, where the foundation stock of German shepherd dogs were all outbred, it was possible within a relatively short span of

generations to select for a more compact body form and a milder temperament than obtained in the original stock (Ginsburg, 1976b).

Apart from the differences in predation, what other wolf behaviors have
been modified in dogs? Particularly, how may we understand dog–human
affiliation? Assuming a wolf ancestry, and assuming that the present-day
wolf reflects the characteristics of that ancestry, it may be surmised that the
ancestral progenitors of the modern wolf varied in temperament and that
interactive (i.e., more sociable) types were the ones that were taken as pets
when a den was raided or that became the camp followers which interacted
and affiliated with early humans to some degree, while the more wary,
fearful animals remained wild. In this view, there was already a genetic
differentiation of behavioral types that were predisposed to the potential
for domestication.

Wolves fall into four temperament categories from a very early age, and
these basic types persist throughout life, regardless of experiential influences (MacDonald, 1983). At the University of Connecticut, we found, as
did Zimen (1981), that when wolf pups were handled, even in our very
early interactions with them (i.e., when they were 1 or 2 weeks old) as we
attempted to weigh, measure, and sex them, they exhibited four distinct
temperaments that did not change fundamentally throughout life (see
Ginsburg, 1987). These were (1) positively interactive, (2) actively avoiding,
(3) avoiding but aggressive if pressed, or (4) aggressive.

These temperaments did not seem to have any direct associations with
eventual roles or status in the pack (Jenks & Ginsburg, 1987; Schotté &
Ginsburg, 1987). They were important, however, in predicting the ability of
an animal to withstand isolation and were directly related to later success
in socialization to human handling. Attempts to raise a wolf pup of Type 2
temperament as a pet were, at most, marginally successful. Repeating the
procedure under the same circumstances with a pup of Type 1 temperament resulted in a socialized wolf that generalized well to other persons
and situations (Slade, 1983). Indeed, the importance of social stimulation
for wolves of Type 1 temperament was discovered in a series of isolation
experiments (MacDonald & Ginsburg, 1981). A Type 1 wolf refused to eat
and was in danger of starving during the isolation phase, until given a brief,
daily period of human handling. The extent of this wolf's sensitivity to
humans was indicated by the fact that a degree of socialization could be
maintained solely by the sound of voices outside its enclosure. When total
isolation was reimposed, the wolf again refused food and was considered
to be severely "depressed." Reinstituting the handling regime resulted in
improvement.

Domestic dogs similarly seek out social stimulation and typically bond
easily with humans. However, predicting the adult temperament of a dog

Figure 6.1. Captive wolves in enclosure at the University of Connecticut. (Photo courtesy Dr. Cynthia S. Schotté.)

from a very young age, unlike predicting the temperament of a wolf, is a matter of behavioral roulette. Despite the various puppy tests that have been developed, their predictive value depends on knowing the character- istics of the breed and of the progeny produced by close relatives. Because of this, when young adult dogs are evaluated for their suitability for train- ing to lead the blind, predictions are more certain. However, the same behaviors exhibited by one particular breed may have a different predictive value than identical behaviors found in another breed (Ginsburg, 1976b). Also, it has been shown that for some dogs there is an optimal time for socialization to people (Scott, 1962). Once socialized, some breeds (not all) are prone to the development of a separation syndrome if they are re- moved from familiar situations and/or people (Davis, Gurski, & Scott, 1977).

The separation syndrome was an important factor in research conducted by several laboratories. At the University of Connecticut a replicable hybrid stock showed consistent hyperkinesia when one requirement of the experimental protocol was the ability to sustain an inhibitory task for which the dogs were trained (i.e., to sit and stay on command) (Ginsburg, Becker, Trattner, & Bareggi, 1984). As with hyperkinetic children, many of the dogs were able to perform the task when pretreated with amphetamine. These results were not replicable in the laboratory where the dogs had been bred and from which they were shipped to Connecticut. In a series of investigations carried out to discover the reason for this discrepancy, it was established that not only was it necessary for the dogs to have the genetic susceptibility to hyperactivity, but in addition this had to be triggered by the induction of a separation syndrome; this accounted for the disagreement in behavioral results between the laboratory where they had been bred and the one to which they had been sent (Ginsburg et al., 1984). In contrast, transferring socialized young adults of many breeds, as must be done when training for special tasks such as police work or leading the blind is involved, does not usually present a problem and new affiliations are readily formed at this time (Pfaffenberger et al., 1976). This phase corresponds to a time when maturing wolves bond to other pack members, and thus their attachments to individuals are more flexible. There are, then, demonstrable individual and breed differences with respect to the lability or fixity of affiliative behaviors.

Affiliative behaviors are especially important in social vertebrates, such as the wolf, that are characterized by dominance–deference hierarchies. These have been described as peck orders in domestic fowl, as nip orders in fish, and in various additional hierarchical terms for other species, all with agonistic connotations. Organized groups, whether of wolves, chimps, or humans, are typically xenophobic among groups and cooperative within their own social structures. The xenophobic tendencies result in a partitioning of the gene pool, thereby promoting a more rapid evolution of the species (Wright, 1939; Ginsburg, 1968). The within-group dominance hierarchies often determine the mating structure of the group and limit its tendency to increase while, at the same time, further restructuring the gene pool.

Although these adaptations have been richly described in the literature and will not be elaborated here, it is not sufficiently appreciated that without a countervailing social cement, that is, affiliation, the aggressive behaviors that bring about, maintain, and change dominance hierarchies would be unable to do so. Aggressive behaviors, by themselves, would destroy rather than promote social structure. It is only in the context of affiliative behaviors, which derive from affective preferences that develop and persist by means of communication of affect, that the characteristic group behaviors that we see in nature could have evolved. It is the genetic

Figure 6.2. Captive wolves at play. (Photo courtesy Dr. Cynthia S. Schotté.)

capacity for the communication of affect that constitutes the basis of social behavior, and it is because dogs have the capacity to communicate in an appropriate interspecific context that our relationships with them are possible (Ginsburg, 1975, 1976a; Buck & Ginsburg, 1991).

Within a pack, the affiliative factors that promote cooperation are seen in the ontogeny of individual and group behavior. There is recognition between mother and young (Mech, 1970, 1988). Vocalizations on the part of young pups promote care-giving behavior on the part of the mother and others in the group. All, or nearly all, of the members of a pack assist in feeding a litter by regurgitating food to them, a response that is both spontaneous and elicited by begging (Rabb, Woolpy, & Ginsburg, 1967; Fentress & Ryon, 1982; Mech, 1970, 1988).

The appropriate communicative patterns are learned by wolf pups at an early age. They seek attention and beg for food from the pack. They soon learn which members are forthcoming and which are neutral or even hostile. As juveniles, they react not only to specific signals such as play invitation or threat, but also to the affective state of the sender. This is based in part on their previous interactions with other individuals in the pack. It is also based on their ability to monitor subtle cues as well as to read autonomic signals. Some of these, such as piloerection, are obvious. Others are less so. Wolves in dyadic confrontations monitor each other head to head, especially in equivocal encounters where the outcome is not predictable (personal observations). It has been hypothesized that the typical yellow eye of the wolf is a communicative adaptation that enables a wolf to determine the affective state of another by monitoring the change in pupillary size during a confrontation, constriction signaling anger or arousal, and dilation signaling fear (Ginsburg, 1976a).

Thus, communication among the members of a wolf pack has an important visual component (Mech, 1970). Tail and facial postures are among the repertoire of signals that a wolf uses. Schenkel (1947) described 8 facial expressions and 12 tail positions of the wolf, which make it capable of fine discrimination of the mood and intent of conspecifics (Mech, 1970). Wolves also use visual as well as olfactory cues to locate their prey (Mech, 1970). Since vision is also important in human communication, this common perceptual basis facilitates the interactions of canids and humans.

In our interactions with adult wolves, particularly those not socialized to human handling, we have found the communicative signals they used to be reliable. Two components of signals are involved: One is the affective state of the animal, and the other is the specific posture, gesture, or vocalization that is used as a specific signal. Because of the reliability of wolves' communicative behavior, intragroup conflict is minimal. Fights are replaced by threats.

Interactions between canids, and between canids and humans, are facilitated by the stability and observability of communicative patterns. In interacting with a wolf, an experienced observer can rely on the communicative lexicon. A full-blown threat from a wolf is a clear signal of intent (Woolpy & Ginsburg, 1967). Selection for consistency of communication has been maintained in the wild. However, under domestication the signal has often been divorced from its meaning. This may make it difficult to predict what a threatening dog will do unless one knows the dog. The intent it is trying to communicate may be ambiguous, for example, if it is wagging its tail and growling at the same time, or its behavior may not correspond to our interpretation of the signal being emitted (Ginsburg, 1976a).

In our laboratory's attempts to socialize adult wolves to human handling, we have found that, like dogs, wolves can quickly determine our affect by monitoring the smoothness or slight incoordination in our movements. Our most successful handlers have been persons who are empathic to the wolves and are able to control their own body movements. Some have been dancers or athletes. The subjective reactions of the handlers as reported during interactive sessions do not communicate the anxiety they feel and express verbally unless this is reflected in slight hesitations and incoordination in their approach. The wolves are adept, but not perfect, in interpreting our affective states – an extension of their interpretive capacity with their own species. In the early domestication of the dog, these communicative capacities and affiliative tendencies of the interactive types facilitated cross-species communications, making it possible for dogs to become integrated into the human community. Once this occurred, dogs could be perceived as reliable in finding game, giving alarm signals, and performing other tasks that could be made more useful in an immediate sense through training and over generational time spans by means of artificial selection.

Guide dogs for the blind, hearing ear dogs for the deaf, guard dogs for persons and territories, bomb- and narcotic-detecting dogs, hunting dogs, retrievers, sled dogs, tracking dogs, search dogs, and other service dogs are the symbiotic results of long-term breeding and training programs building on the natural capacities of the dog, derived from its wolf origins. The social cement that enables dogs to become integrated into the human community depends on their communicative capacities both as senders and as interpreters and on our ability to interpret and communicate with the dog. The genetic foundations of the dog's selective affiliative behaviors are derived from its wolf ancestry, where cohesiveness within packs and territorial demarcations between packs are well developed. This translates into territorial behaviors on the part of the dog such that it comes to prefer its particular human care givers and, to various degrees, to protect them and the area in which they both live.

That many canids appear adept at discriminating among individual humans is not surprising for species that have dominance hierarchies and that must be aware of their relationships with other individuals in their social group. Wolves, for example, like dogs, commonly are more at ease around female human handlers. Coyote–beagle hybrids also react to specific individuals (personal observations); however, their reactions depend on environmental factors. When some of these animals were moved from one enclosure to another, for instance, they maintained an affiliation to some, but not all, of their previous handlers.

As with dogs, environmental variables may also affect wolves. Wolves maintained in opaque enclosures become much more wary of unfamiliar humans than wolves allowed to look beyond their cages, such as at a zoo (Ginsburg & Schotté, 1979). However, even the wary wolves in the opaque enclosures at the University of Connecticut could recognize the approach of familiar humans presumably by olfactory and auditory cues such as the sound of their footsteps, and would often not even rise to their feet from a recumbent position. In contrast, when strangers approached, the wolves would cower together in a corner or dash nervously around the enclosure.

Thus, one should take into account the past social and housing experiences of any canid subject when interpreting its behavior. Canids tend to form hierarchies and coalitions when kept as a group. Human observers might affect the behavior of individuals, potentially causing disruptions in the formation and maintenance of hierarchies. This could occur if some of the animals were hesitant to interact when observed. Differential handling of captive wolves could also interfere with the development of social relationships in the pack.

It is thus very important to know the species, breed, and individual histories of the canids one works with. A domestic dog from a pound or a former pet might have very different reactions to humans than a kennel-reared dog. There are questions of scientific validity, and also ethical considerations, when dogs of unknown genetic and experiential background are used as experimental subjects. Similar problems may arise when a dog is removed from a stable social environment and placed into isolation in a laboratory. The results obtained from such an animal would not be the same as from one that was well adjusted and behaviorally "normal." Instead it might display excessive fear or attachment to humans that would be reflected in physiology as well as behavior.

However, there is some hope of reestablishing species-typical behavior in canids that have experienced restricted rearing conditions. We have found that the normal behavioral capacities of wolves are still present in socially deprived animals and may still be expressed under appropriate circumstances (Ginsburg, 1975; MacDonald & Ginsburg, 1981). Isolates emit all or most of the communicative signals typical of the species. They do not, however, use them in a patterned fashion or understand them when exposed to normal conspecifics. Initially, when wolves that had been isolated were placed together, they were fearful and avoided interaction. This "affective opacity" had to be surmounted so that cognitive systems could be engaged and the animals could begin to learn from one another. In effect, each wolf served as a behavioral pacemaker for other wolves. Eventually, in a series of 24-hour round-robin dyadic encounters, the fundamentals of essential communicative behaviors came to be expressed, after which the isolates were placed together as a group and were left to interact within a familiar enclosure and observed via a remote video system.

By means of these procedures, normal play, threat, and other behaviors were initiated and responded to, and all of the isolates were eventually able to communicate in a normal manner both as senders and as receivers (MacDonald & Ginsburg, 1981). Their genetic capacities had not been lost and were relatively easily evoked. The final stage was to introduce a normal, adult wolf that was accustomed to interacting with unfamiliar conspecifics. This approach has not been attempted with similarly isolated dogs. Although it is possible that the same procedures would prove effective with some individuals, the dog's communication signals are not as stereotyped as the wolf's, which might make it harder to find effective models and tutors.

The period during a canid's life history at which behavioral observations are conducted should also be taken into account, because of the effect of age-dependent variables. For example coyote–beagle hybrids that have been subjected to postpubertal social stress over an extended period of time may reach a threshold and begin to display the coyote threat pattern of gaping and hissing (which is not within the genetic repertoire of the domestic dog), rather than the typical dog threat behaviors of snarling and growling that they used initially. This is associated with a post-pubertal rise in cortisol levels attendant upon social stress (Moon & Ginsburg, 1985). Hiestand (1989) found that although wolf pups resembled adult dogs in their lack of ability to solve complex spatial tasks, at about 6 to 7 months of age, when wolves begin attending to vertically oriented objects in their environment in ways not typical of dogs, they simultaneously become capable of accomplishing difficult spatial manipulations. As mentioned previously, puberty can also alter the behavior of dogs, thus making predictive puppy tests potentially unreliable.

Because of the degree of sensitivity to their social environment that dogs have inherited from wolves, it is necessary for canid researchers to take these factors into account when analyzing the behavior of their subjects. Obviously, the use of video cameras and remote sensing devices is one way to circumvent the problem. If this is not feasible, other steps that might be taken include having one researcher test all of the dogs, so that any influences are expressed uniformly across the subject population. If possible, the use of a screen or blind, or staying behind the dogs as they work, can reduce the possibility of cuing.

Although it is important to be cautious, differential bonding with canid subjects does not always affect the data collected. In one study of the ability of dogs to perform sequential manipulations, no correlation was found between the researcher's affinity for, or bond with, a particular dog and its success at solving the problems (Hiestand, 1989). However, subtle cuing no doubt enabled the poodle of Sir John Lubbock (1888; cited in Williams, 1926) to discriminate, after only 1 month of training, between cards with the words "food," "bone," "out," "water," and so on printed on them. Therefore, the data collected from canids should always be examined for potential influences of this kind.

Because humans are a very salient aspect of a domestic dog's social environment, thanks to its wolf heritage, it is impossible for us to be an insignificant feature of the laboratory. Socialized animals will anticipate interaction with us, while unsocialized dogs will exhibit fear responses. This does not imply that dogs are unsuitable subjects for behavioral and physiological research. It only means that we should be aware of our potential influence on the results, as well as of the dog's social history. We can even benefit from this "inevitable bond": Petting has been shown to be an effective reinforcer for training dogs, although verbal praise alone was not (McIntire & Colley, 1967), and interaction with even a passive human can serve as an effective reward (Bacon & Stanley, 1963).

In conclusion, our species cannot take the credit for creating humanity's best friend from raw material. According to Gould (1986), the domestic dog is our closest nonhuman companion because "its ancestor, the wolf *Canis lupus*, had evolved behaviors that, by a fortunate accident of history, predisposed it for human companionship."

References

American Kennel Club. (1989). *The Complete Dog Book*, 17th ed. Howell Book House, New York.

Arons, C. D. (1989). Genetic variability within a species: Differences in behavior, development, and neurochemistry among three types of domestic dogs and their F_1 hybrids. Unpublished Ph.D. dissertation, University of Connecticut.

Bacon, W. E., & Stanley, W. C. (1963). Effect of deprivation level in puppies on performance maintained by a passive person reinforcer. *Journal of Comparative and Physiological Psychology, 56*, 783–5.

Beck, A. M. (1973). *The Ecology of Stray Dogs: A Study of Free-Ranging Urban Animals*. York Press, Baltimore.

Buck, R., & Ginsburg, B. E. (1991). Spontaneous communication and altruism: The communicative gene hypothesis. In M. S. Clark (Ed.), *Review of Personality and Social Psychology*. Sage, London, pp. 149–75.

Buytendijk, F. J. J. (1936). *The Mind of the Dog*. Houghton Mifflin, Boston.

Coppinger, R., Glendinning, J., Torop, E., Matthay, C., Sutherland, M., & Smith, C. (1987). Degree of behavioral neoteny differentiates canid polymorphs. *Ethology, 75*, 89–108.

Daniels, T. J. (1987). The social ecology and behavior of free-ranging dogs. Unpublished Ph.D. dissertation, University of Colorado.

Darwin, C. (1872). *The Expression of the Emotions in Man and Animals* (reprinted in 1965). University of Chicago Press, Chicago.

Davis, K. L., Gurski, J. C., & Scott, J. P. (1977). Interaction of separation distress with fear in infant dogs. *Developmental Psychobiology, 10*, 203–12.

Fentress, J. C., & Ryon, J. (1982). A long-term study of distributed pup feeding in captive wolves. In F. H. Harrington & P. C. Paquet (Eds.), *Wolves of the World*. Noyes, Park Ridge, NJ, pp. 238–61.

Fox, M. W. (1971). *Behaviour of Wolves, Dogs and Related Canids*. Harper Row, New York.

Ginsburg, B. E. (1968). Breeding structure and social behavior of mammals: A servo-mechanism for the avoidance of panmixia. In D. G. Glass (Ed.), *Genetics: Biology and Behavior Series*. Rockefeller University Press and Russell Sage Foundation, New York, pp. 117–28.

(1975). Nonverbal communication: The effect of affect on individual and group behavior. In P. Pliner, L. Krames, & T. Alloway (Eds.), *Nonverbal Communication of Aggression: Vol. 2, Advances in the Study of Communication and Affect*. Plenum, New York, pp. 161–73.

(1976a). Evolution of communication patterns in animals. In M. E. Hahn & E. C. Simmel (Eds.), *Communicative Behavior and Evolution*. Academic Press, New York, pp. 59–79.

(1976b). Genetics of guide dog production. In C. J. Pfaffenberger, J. P. Scott, J. L. Fuller, B. E. Ginsburg, & S. W. Bielfelt, *Guide Dogs for the Blind: Their Selection, Development, and Training*, Volume 1 of Developments in Animal and Veterinary Sciences Series. Elsevier, Amsterdam, pp. 161–87.

(1987). The wolf pack as a socio-genetic unit. In H. Frank (Ed.), *Man and Wolf.* Dr. W. Junk Publishers, Amsterdam, pp. 401–13.

Ginsburg, B. E., Becker, R. E., Trattner, A., & Bareggi, S. R. (1984). A genetic taxonomy of hyperkinesia in the dog. *International Journal of Developmental Neuroscience*, 2, 313–22.

Ginsburg, B. E., & Schotté, C. S. (1979). The effect of visual isolation in a captive wolf pack. *Animal Behavior Society Abstracts*, No. 68.

Gould, S. J. (1986). The egg-a-day barrier. *Natural History*, 95, 16–24.

Hiestand, N. L. (1989). A comparison of problem-solving and spatial orientation in the wolf (*Canis lupus*) and dog (*Canis familiaris*). Unpublished Ph.D. dissertation, University of Connecticut.

James, W. T., & Ginsburg, B. E. (1949). The effect of prostigmine on the conditioned response of "inhibited dogs." *Journal of Comparative and Physiological Psychology*, 42, 6–11.

Jenks, S. M., & Ginsburg, B. E. (1987). Socio-sexual dynamics in a captive wolf pack. In H. Frank (Ed.), *Man and Wolf.* Dr. W. Junk Publishers, Amsterdam, pp. 375–99.

Lore, R. K., & Eisenberg, F. B. (1986). Avoidance reactions of domestic dogs to unfamiliar male and female humans in a kennel setting. *Applied Animal Behaviour Science*, 15, 261–6.

Lorenz, K. Z. (1952). *King Solomon's Ring*. Apollo, New York.

MacDonald, K. (1983). Stability of individual differences in behavior in a litter of wolf cubs (*Canis lupus*). *Journal of Comparative Psychology*, 97, 99–106.

MacDonald, K. B., & Ginsburg, B. E. (1981). Induction of normal prepubertal behavior in wolves with restricted rearing. *Behavioral and Neural Biology*, 33, 133–62.

McIntire, R. W., & Colley, T. A. (1967). Social reinforcement in the dog. *Psychological Reports*, 20, 843–6.

Mech, L. D. (1970). *The Wolf: The Ecology and Behavior of an Endangered Species*. Natural History Press, New York.

(1988). *The Arctic Wolf*. Voyageur Press, Stillwater, MN.

Moon, A., & Ginsburg, B. E. (1985). Genetic factors in the selective expression of species-typical behavior of coyote × beagle hybrids. Invited paper presented at the 19th International Ethological Conference, Toulouse, August.

Pfaffenberger, C. J., Scott, J. P., Fuller, J. L., Ginsburg, B. E., & Bielfelt, S. W. (1976). *Guide Dogs for the Blind: Their Selection, Development, and Training*, Volume 1 of Developments in Animal and Veterinary Sciences Series. Elsevier, Amsterdam.

Rabb, G. B., Woolpy, J. H., & Ginsburg, B. E. (1967). Social relationships in a group of captive wolves. *American Zoologist*, 7, 305–11.

Rosenthal, R. (1966). *Experimenter effects in behavioral research*. Appleton-Century-Crofts, New York.

Schenkel, R. (1947). Expression studies of wolves. *Behaviour*, 1, 81–129.

Schotté, C. S., & Ginsburg, B. E. (1987). Development of social organization and mating in a captive wolf pack. In H. Frank (Ed.), *Man and Wolf*. Dr. W. Junk Publishers, Amsterdam, pp. 349–74.

Scott, J. P. (1950). The social behavior of dogs and wolves: An illustration of sociobiological systematics. *Annals of the New York Academy of Science, 51*, 1009–21.

(1962). Critical periods in behavioral development. *Science, 138*, 949–58.

(1973). Comparative social psychology. In D. A. Dewsbury & D. A. Rethlingshafer (Eds.), *Comparative Psychology: A Modern Survey*. McGraw-Hill, New York, pp. 124–60.

Scott, J. P., & Fuller, J. L. (1965). *Genetics and the Social Behavior of the Dog*. University of Chicago Press, Chicago.

Slade, V. R. (1983). Individual differences in social bond formation of two hand-reared prepubertal wolves. Unpublished M. S. thesis, University of Connecticut.

Stockard, C. R., Anderson, O. D., & James, W. T. (1941). *The Genetic and Endocrine Basis for Differences in Form and Behavior*. Wistar Institute of Anatomy and Biology, Philadelphia.

Vauk, G. (1953). The modification of prey capture behavior in dogs in the course of domestication (tr. N. L. Hiestand). *Deutsche Zoologische Gesellschaft, 46*, 180–4.

Wayne, R. K., Benveniste, R. E., Janczewski, D. N., & O'Brien, S. J. (1989). Molecular and biochemical evolution of the Carnivora. In J. L. Gittleman (Ed.), *Carnivore Behavior, Ecology, and Evolution*. Cornell University Press, Ithaca, NY, pp. 465–94.

Wayne, R. K., & O'Brien, S. J. (1987). Allozyme divergence within the Canidae. *Systematic Zoology, 36*, 339–55.

Williams, J. A. (1926). Experiments with form perception and learning in dogs. *Comparative Psychology Monograph, 4*, 1–70.

Woodbury, C. B. (1943). The learning of stimulus patterns by dogs. *Journal of Comparative and Physiological Psychology, 58*, 317–20.

Woolpy, J. H., & Ginsburg, B. E. (1967). Wolf socialization: A study of temperament in a wild social species. *American Zoologist, 7*, 357–63.

Wright, S. (1939). *Statistical Genetics in Relation to Evolution*. Hermann, Paris.

Zimen, E. (1981). *The Wolf: A Species in Danger* (tr. E. Mosbacher). Delacorte Press, New York.

—7—

The use of dog–human interaction as a reward in instrumental conditioning and its impact on dogs' cardiac regulation

Ewa Kostarczyk

Editors' introduction

Kostarczyk's chapter provides empirical support for an intuitively reasonable premise: Interaction with a human may have profound behavioral and physiological effects on a dog. Kostarczyk's observation that not all dogs respond in a typical manner to human contact underscores Ginsburg and Hiestand's comment regarding the importance of knowing the background of individual canid subjects.

Kostarczyk makes an important point in passing: By attempting to minimize the degree of social contact between scientist and dog we may be creating a degree of social deprivation that yields an abnormal baseline for research.

Rewarding dog–human interactions

The story told here is an unexpected result of our search for an appetitive nonalimentary reward that could be used in experimental conditioning procedures. A variety of stimuli, often independent of biological needs, can be effectively used as rewards. Sensory stimuli such as changes in light onset (Berlyne & Koenig, 1965), changes in temperature (Stellar, 1980), auditory stimulation (Barnes & Kish, 1961), and tactile stimulation (Wenzel, 1959) have all been successfully used. The opportunity to explore the environment (Montgomery, 1954) or manipulate objects (Harlow & MacLearn, 1954; Kish & Barnes, 1961) may also act as a reward.

Such rewards, however, are very labile and apt to produce satiation. Their reinforcing value usually depends on the subject's level of deprivation. Comparative behavioral studies that examine the performance of conditioned responses established under similar experimental conditions but based on different motivation are rare. Our original objective was to find an appropriate method that would allow us to (1) make behavioral comparisons among biologically different but clearly appetitive rewards in terms of their ability to motivate performance; (2) make comparisons of the physiological peripheral parameters affected by the action of these rewards.

The most suitable paradigm would be one that considered rewards of two types: those that satisfy the biological primary needs of an organism and those that belong to a category of learned motivation based on secondary forms of reinforcement (Cofer & Appley, 1964; Altman, 1966). This chapter presents the theoretical basis for, and the results from, experiments related to this secondary form of reinforcement; specifically social rewards. The reader should remember, however, that most of the experiments described here have parallels with experiments that use alimentary rewards.

The dog is an optimal subject for research in this field because its emotional expression is readily understood by humans (Thorpe, 1963) and because its attachment to humans has been demonstrated by a long history of mutual social relations. If it is true that attachment is not only due to primary reinforcement (Wickler, 1976), it should be possible to establish emotional attachment and a repertoire of instrumental responses using only social rewards. The repertoire of possible social rewards is large. Most forms of sensory cues deriving from the experimenter could play a reinforcing role. Although studies of the nature and strength of such rewards are lacking, dogs have been trained to perform a variety of tasks using social rewards (Scott & Fuller, 1965; Pfaffenberger & Scott, 1975; Zagrodzka, Korczynski, & Fonberg, 1981). Harlow (1962) demonstrated that attachment is more dependent on tactile stimulation than on alimentary reinforcement. Frank and Frank (1988) reported that timber wolf pups had shorter latencies on problem-solving tasks when reinforced by the opportunity to interact with another canid than they did with a food reward. Petting, therefore, seems more likely than other sensory stimuli to be an effective reinforcer, although this remains controversial.

Stanley and Elliot (1962) demonstrated that a passive experimenter is more attractive to some puppies (basenjis) than is one who pets them. Candland, Pack, and Matthews (1960) reported the aversion of rats to handling, but Davis and Pérusse (1988), using a less regimented form of contact, successfully trained about half of their rats to lever-press using only petting and social interaction with a human as a reward. These differ-

Figure 7.1. Use of social contact to reinforce paw extension.

ential findings may reflect species and breed differences (Freedman, 1957), the variable ages of the animals used in the experiments, different degrees of socialization or prior social exposure to a human (Davis & Pérusse, 1988), differences in the procedures used (especially the rate of petting), and the nature of the human contact (Werner & Anderson, 1976). Further research is needed to determine whether tactile stimuli, which are necessary for normal biological and emotional development in many species, can be as effective as alimentary reinforcement. Our first objective was to design a behavioral test that used social forms of reinforcement while employing training procedures similar to accepted methods of instrumental conditioning (Konorski & Miller, 1933; Skinner, 1938; Konorski, 1967).

Our second objective was to evaluate the efficacy and motivational basis of this type of training. All experiments were conducted on experimentally naïve, mongrel, male dogs, 2 to 4 years old. Their history of socialization was unknown. We performed two critical experiments that are described in detail elsewhere (Fonberg, Kostarczyk, & Prechtl, 1981). In the first, we found that dogs were able to learn and sustain the natural instrumental responses (CRs) of sitting, paw extension, and lying prostrate when reinforced only by social rewards such as stroking the back and head and vocal encouragement (the words "good dog") (Figure 7.1).

The fact that purely social forms of reinforcement could be used as potent rewards for training dogs (see also Bacon & Stanley, 1963; Frank & Frank, 1988) came as no great surprise given the history of human–dog

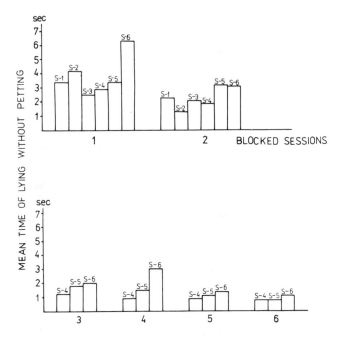

Figure 7.2. Mean time of lying without petting per 30-second trial. Each dog is represented by a different bar. The critical sessions in (A) two blocks of five sessions, (B) four blocks of overtraining. (From Fonberg, Kostarczyk, & Prechtl, 1981.)

interactions. Moreover, overtraining did not produce a deterioration of performance but, on the contrary, resulted in a continual decrease of latencies (Figure 7.2). The strength of petting as a determinate of the dog's behavior was indicated by the fact that a high proportion of the total time spent lying prostrate was associated with petting. The reciprocally low proportion of time spent in postpetting lying was due to the fact that the dogs did not maintain the prostrate position if the petting had been withdrawn (Figure 7.2).

Since the natural instrumental responses trained in the first experiment could be familiar to the dogs from prior experiences with humans, we wanted to verify these results in an artificial, experimental situation. We also wanted to determine whether the acquisition of instrumental responses, reinforced either socially or by alimentary reward, would be similar under identical experimental conditions (i.e., requiring the same type of instrumental response and using the same intertrial intervals, frequency of sessions, and number of trials in each session). Therefore, in the second

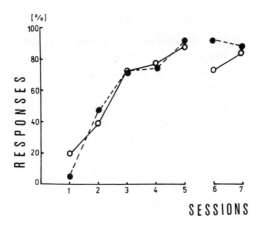

SESSIONS

Figure 7.3. Percent correct responses at successive stages of training to the criterion (5) and during two blocks of 10 sessions of overtraining (6 and 7) in the social group (solid line) and in the alimentary group (dashed line). Analysis of variance of vincentized scores (with raw scores transformed to arc sin P) showed the statistically significant influence of training stage and the lack of influence of reward type on performance. Interaction between training stage and reward type was not significant. (From Kostarczyk & Fonberg, 1982a.)

experiment we examined instrumental responses to auditory conditional stimuli in two groups of dogs. Dogs were trained to put a foreleg on the food tray in the reflex chamber in response to the conditional stimulus, a tone of 500 Hz and 20 seconds' duration. The first group (nine dogs) was reinforced by food (delivered automatically), and the second group (eight dogs) was reinforced exclusively by petting. Details of the experiment are described in Kostarczyk and Fonberg (1982a). A similar course of learning and level of performance during overtraining sessions was found in both groups (Figure 7.3), indicating that petting serves as a good reinforcement and that the rewarding value of social reinforcement is comparable to that of food reinforcement. In recent experiments (unpublished data) high rates of instrumental responding were established in a dog reinforced by petting, using a second-order schedule of presentation (1-second tone reinforcing one or more instrumental responses).

These findings support the theory of hedonistic learning, which is based primarily on alimentary instrumental learning tasks (Pfaffman, 1960; Fonberg, 1969; Wyrwicka, 1975; Trojniar & Cytawa, 1978; Cabanac, 1979). They also provided a precise method for comparing behaviors motivated by different appetitive rewards. Thus, our next step was to examine the physiological mechanisms (central and peripheral) that govern reward.

113

Ewa Kostarczyk

Cardiac responses as indicators of emotional state

Changes in the vegetative system that accompany particular behavioral performance are often used to evaluate the dynamics of emotional processes (James, 1884). Functional interactions within the vegetative system are complex, not only because complex behavioral acts involve a number of effectors with different sympathetic and parasympathetic innervation, but also due to the various functional states of particular organs. The action of the sympathetic and parasympathetic systems is essentially antagonistic: If the activity of one increases, the activity of the other is likely to decrease. This statement, however, is a gross oversimplification, and many deviations from that norm are known to exist.

There is some evidence to support the idea of stereotype of the vegetative response complexes accompanying particular behaviors (Lacey & Lacey, 1958; Hilton, 1975; Obrist, 1976; Kostarczyk, 1985, 1986a). Cardiovascular changes associated with orientation responses, aversive stimuli, and defensive behavior, for instance, have been found to be invariable in dogs, cats, and monkeys. The cardiovascular components include vasodilation in the skeletal muscles, vasoconstriction in the mesentery and renal glomeruli, elevated arterial blood pressure, and accelerated heart rate (Martin, Sutherland, & Zbrozyna, 1976). However, conflicting experimental data are common in the literature.

Nonspecificity of reactions occurs in response to a variety of emotionally different stimuli. In the dog, for instance, heart rate accelerates when either a meal (Konorski, Santibanez, & Beck, 1968) or punishment (Soltysik & Kowalska, 1960) is expected. Similar patterns are found for ECG activation, changes in blood pressure, and galvanic response. The autonomic nervous system conveys information about arousal or its absence, and arousal may, of course, be both appetitive and aversive. Although the dog's heart rate would increase when the animal was waiting for either food or electric shock, only waiting for food would also stimulate the dog to salivate. Thus, the behavior-related specificity of autonomic responses should not be neglected.

Species-specific differences and individual variability in reactions of the vegetative system are other factors complicating experimental research in this field. Anticipatory aversive stimuli elicit cardiac deceleration in humans, cats, rabbits, and occasionally rats, but acceleration in dogs and pigeons (Obrist, 1976). Cardiac responses are the most commonly used vegetative indicators, and when they accompany the orientation response (i.e., a response to a new, unexpected stimulus) they provide a good illustration of the previously described problems. In most species, the orientation response is initially accompanied by cardiac acceleration, which then apparently slows down. The response is elicited both by simple single auditory stimuli (Gantt, 1944; Soltysik, Jaworska, Kowalska, & Radom, 1961) and by complex stimuli – for example, a new environment (Candland, Pack, & Matthews, 1967). In humans, the orientation response often runs a two-phase course, with the initial acceleration of the heart rate followed by deceleration, even though the stimulus is continued (Graham & Clifton, 1966). The changes accompanying the orientation response are related to the duration of the stimulus (Soltysik et al., 1961), as well as the adaptation processes occurring during exposure (Black, Flower, & Kimbrell, 1964; Snapper et al., 1965). The question of whether the cardiac response undergoes extinction with repeated stimulus presentation remains controversial. Although extinction has been observed in rats (Snapper, Ferraro, Schoenfeld, & Locke 1965) and dogs (Soltysik et al., 1961; Martin et al., 1976), other authors have failed to confirm extinction in these species (Black, Flower, & Kimbrell, 1964; Snowden, Bell, & Henderson, 1964; Roessler, Collins, & Burch, 1969). Moreover, Martin, Sutherland, and Zbrozyna (1976) stressed that they were often unable to achieve extinction of the response. Divergent findings were most likely due to differences in the quality and complexity of the stimuli employed. As established by Robinson and Gantt (1947) and Gantt (1960), a novel stimulus does not necessarily elicit an increase in the heart rate. Rather the effect depends on stimulus type; in dogs, for example, the cardiac orientation response is elicited exclusively by auditory stimuli (Jaworska & Soltysik, 1960).

Animal personality also has an impact on cardiac response. Fox (1978) found that dominant wolf cubs have higher baseline heart rates than submissive cubs. There is evidence that the accompanying neuroendocrine arousal, which involves sympathetic adrenomedullary activation, can be associated with arousal of the pituitary adrenocortical system. Although their roles have not yet been determined, these two systems can be differentiated. Dominant animals display primarily adrenomedullary and catecholamine arousal, animals that are less dominant but still aggressive experience adrenocortical and catecholamine arousal and submissive subjects respond with the adrenocortical system only (Henry, 1982). Thus, animals use a coordinated series of responses (behavioral, cardiovascular, and endocrine) as the means to a common homeostatic end.

Although there is evidence that cardiac responses depend on motor and respiratory activity (Coote, 1975; Obrist, 1976), influences from the motor and respiratory systems do not have a direct impact on the cardiac activity; that is, a movement does not necessarily evoke cardiac acceleration. Nevertheless, these influences may act via central regulations; for example, anticipation and preparation of movement could accelerate heart rate in an immobile subject (Vanderwolf, 1971; Obrist, Sutterer, & Howard, 1972). Whether emotional factors act in connection with activation of the motor system is unclear, but there is no doubt that cardiac responses are not the simple result of motor activity. Cardiac acceleration, for example, appears *before* a conditioned instrumental response is to be performed (Perez-Cruet & Gantt, 1959; Jaworska, Kowalska, & Soltysik, 1962), curarized animals still demonstrate conditioned cardiac responses (Royer & Gantt, 1966; Yehle, Dauth, & Schneiderman, 1967), and emotional states during stress are often accompanied by extreme cardiovascular responses with only minor motor activity (Galosy & Gaebelein, 1977).

This brief review of the evidence for cardiac response as an indicator of emotional state demonstrates clearly that a multitude of independent causes can elicit the same effect. However, as our studies suggest (Kostarczyk & Fonberg, 1982a,b; Kostarczyk, 1986a), analyzing several vegetative responses before and after the event in question may help in the detection of specific emotions, especially when these data are compared with behavioral observations made at the same time.

Cardiac changes in the dog associated with rewarding interaction with humans

The methodological problems associated with identifying which peripheral autonomic responses are associated with appetitively reinforcing stimuli are substantial. Data obtained using natural appetitive rewards seem to be suitable for approaching these issues. Alterations in peripheral autonomic functions, such as heart rate, have been observed in the course of various appetitive events: food consumption in dogs (Konorski, Santibanez, & Beck, 1968; Kostarczyk & Fonberg, 1982c; Kostarczyk, 1986a, b), milk drinking in cats (Ninomiya & Yonezawa, 1979), pleasant tactile stimulation in dogs and humans (Gantt, 1944; Gelhorn, 1967), and sexual activity in dogs (Gantt, 1944). It is interesting that morphine and opioid peptides, which evoke pleasure sensations and consequently increase the danger of addiction, also evoke bradycardia when injected intravenously in rats. Tolerance to this effect develops, and the fall in heart rate can be used as an in vivo index of opioid drug sensitivity (Kiang & Wei, 1984).

The hypothesis that heart rate deceleration is associated with a state of well-being was supported by data demonstrating specific dynamic changes in the dog's heart rate and salivation during a complex alimentary situation (Kostarczyk, 1986a, b). The onset of eating was accompanied by heart rate acceleration without a similar increase in salivation, but during the course of eating heart rate decelerated to baseline and salivation increased. Dogs that were rejected from our program of instrumental training because of lack of progress did not show this typical cardiac pattern while eating (Kostarczyk, 1980). The pattern of cardiac acceleration at the beginning of eating, with subsequent cardiac deceleration during eating, was observed *only* when the food was attractive to a dog (Kostarczyk & Fonberg, 1982c). Furthermore, in hypophagic, amygdalohypothalamically lesioned dogs, which show little interest in food, the amount of cardiac deceleration toward baseline during food consumption was smaller (Kostarczyk & Fonberg, 1982b). Presumably, these different behavioral and cardiovascular responses are associated with differential activation of neuroendocrine arousal, as noted previously.

There is evidence that the decelerative cardiac response is not bound to specific stimuli, but rather to their positive or pleasant emotional contents. The petting of a dog by an experimenter, for example, generally evokes cardiac deceleration (Gantt, 1944). Such tactile stimulation from an experimenter also normalizes the cardiac acceleration caused by aversive stimuli both in dogs (Anderson & Gantt, 1966; Lynch & Gantt, 1968) and in humans (Gattozzi, 1971). In schizophrenic patients who had difficulty dealing with human relationships, however, tactile stimulation by an experimenter evoked acceleration of the heart rate (Gattozzi, 1971). Similar observations were made on dogs; when a dog was petted by an experimenter who had previously punished it, cardiac acceleration was observed (Gantt, 1944). These examples suggest that tactile stimulation evokes cardiac deceleration only in cases when its emotional content is positive.

Training-conditioned instrumental responses (CRs) in dogs with tactile stimulation, (e.g., petting) as the reinforcement has several advantages. This method provides an opportunity to record heart rate changes in dogs during a time-controlled reward period. In most techniques involving natural rewards such as food or sex, by contrast, the duration of the reward cannot be experimentally controlled. Our method also excludes artifacts of heart rate monitoring derived from motor excitation because the dog's posture during petting is a passive one. The analysis of changes in heart rate during petting and during conditioned stimuli reinforced by petting seems, therefore, to be appropriate for studying the relationships between reward induced by petting and peripheral autonomic response.

The basic experiment, as previously described, involved training dogs to put a foreleg on the food tray (CR) in the reflex chamber in response to conditioned stimuli (CSs). Correct responses were reinforced by petting the back and head of the dogs for 20 to 30 seconds. Heart rate was recorded using three electrodes fastened on a leather belt placed around the dog's chest (Figure 7.4).

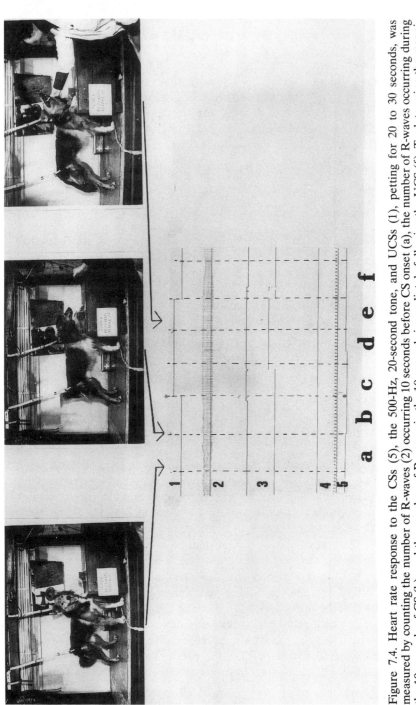

Figure 7.4. Heart rate response to the CSs (5), the 500-Hz, 20-second tone, and UCSs (1), petting for 20 to 30 seconds, was measured by counting the number of R-waves (2) occurring 10 seconds before CS onset (a), the number of R-waves occurring during the 10 seconds of CS (b), and the number of R-waves in the 10 seconds immediately following the UCS (f). To determine the magnitude of the heart rate response to the US, duration of each UCS was divided into three equal intervals (c, d, e) and the number of R-waves occurring in each interval was counted.

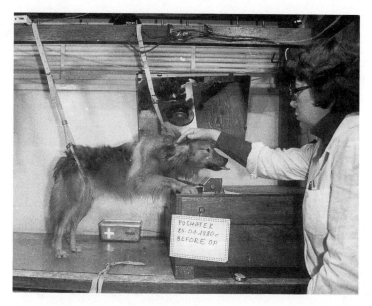

Figure 7.5. Petting of dog during simultaneous assessment of instrumental and cardiac responses.

The fact that heart rate slowed when dogs were petted (see Figures 7.5 and 7.6) suggests that even casual interactions in the laboratory environment may affect the physiology of this species. The common feature of the five dogs that demonstrated cardiac deceleration in response to petting was that they stably performed conditioned instrumental responses for a petting reward; that is, for them petting was, without any doubt, reinforcing. Cardiac deceleration usually occurs immediately upon the onset of petting, while the cessation of petting results in sudden heart rate acceleration.

Three dogs that were unsuccessful in instrumental training had cardiac responses to petting that were the opposite of those of typical, successfully trained dogs (Figure 7.7). These animals displayed cardiac acceleration instead of deceleration. The reason for their failure to be trained is unclear. Behavioral responses to petting were highly variable. One dog, for instance, wagged its tail while another did not, but neither tried to escape from the experimenter's petting hand. Although we recorded such behavioral changes as tail wagging, ear drooping, and relaxation of the body, the absence of escape from, and the positive approach to, the experimenter's hand were used as the primary criterion for a dog's acceptance of petting. In the critical stage of training, we also evaluated the level of a dog's

119

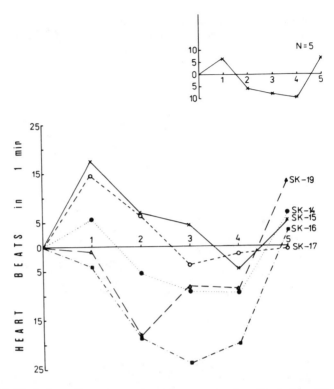

Figure 7.6. Typical cardiac patterns in response to petting: 0 = before CS; 1 = CS; 2, 3, 4 = consecutive periods of petting; 5 = after petting.

acceptance of petting by making the dog maintain a fixed body or leg position while being petted. The same experimenter routinely both restrained and petted the dogs.

The different cardiac patterns in response to petting may explain why some dogs failed in the instrumental training. The dogs that failed to train also displayed cardiac acceleration, suggesting that petting may have caused an aversive or indifferent response and therefore an ineffective reward. These data are consistent with the hypothesis that tactile stimulation evokes cardiac deceleration when its emotional content is pleasant and rewarding.

The relationship between cardiac deceleration and pleasant sensations has been stressed by Lacey (1967; Lacey & Lacey, 1980). Physiological support for this hypothesis was provided by Kumada and co-workers (1979), who demonstrated the link between cardiac deceleration and activation of sensory afferents. Since the petting reward acts mainly through

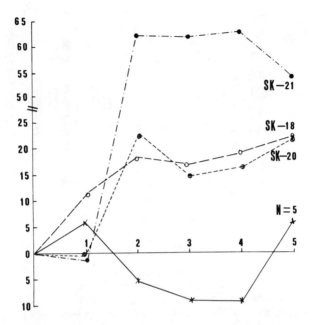

Figure 7.7. Cardiac pattern in three atypical dogs (dashed lines) compared with mean pattern from five typical dogs (solid line): 0 = before CS; 1 = CS; 2, 3, 4, = consecutive periods of petting; 5 = after petting. (From Kostarcyzk & Fonberg, 1982a.)

the activation of sensory afferents, the decelerative response during petting seems to be strongly related to sensory mechanisms.

Parasympathetic and sympathetic contributions to the cardiac responses that accompany social reward

The relative influences of the parasympathetic and sympathetic nervous systems on the heart responses that accompany various behavioral states are not well understood. According to Obrist and co-workers (1982), both sympathetic and parasympathetic innervation of the heart are involved in mediating the translation of life's events into myocardial adjustments. Sympathetic (adrenergic) effects dominate under conditions in which control is possible because they prepare or mobilize the organism for action. Vagal influences dominate, or sympathetic effects are less prominent, under conditions in which action is not possible, since there is nothing to be gained by a mobilization of those visceral processes that would expedite action.

Ewa Kostarczyk

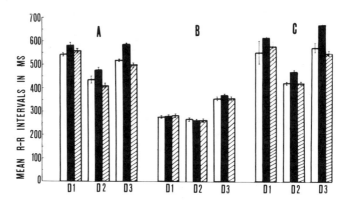

Figure 7.8. Effects of petting on heart rate interval in (A) the pretreatment period, (B) the atropine (s.c. 0.25 mg/kg) treatment period, and (C) the post-atropine period. Each bar represents the mean of 15 trials (3 days × 5 trials). White bars: before petting; black bars: during petting; diagonal bars: after petting. The test session consisted of five trials during which ECG recordings of three 30-second phases (before, during, and after petting) were made. The ECG activity was recorded on an FM magnetic recorder. Subsequently analog–digital conversion was made off-line and interbeat (R–R) intervals identified by computer.

Since many of our data, like those of other authors, point to parasympathetic activation (reflected by cardiac deceleration) as an essential characteristic of consummatory behavior, we wanted to further explore this problem. We chose cardiac deceleration in response to petting as a model of the appetitive consummatory response and examined the effects of atropine treatment, which is known to block the parasympathetic system. The passive position adopted by our dogs during petting rules out massive motor activation, making our experimental paradigm very suitable for external ECG recordings. As expected, atropine's blockage of the parasympathetic system resulted in a significant heart rate increase and abolished "petting-induced" heart rate deceleration (Figure 7.8). Under atropine treatment the dogs also appeared to be much less interested in being petted by the experimenter; instead they were excited, anxious, and tense. Avoidance of petting and aggression toward the experimenter's petting hand were also noted (Jelen & Kostarczyk, 1991). Identical doses of atropine also abolished differential heart responses to conditional stimuli of variously palatable foods (Kostarczyk, 1989).

These results are consistent with the hypothesis that changes in parasympathetic activity (either increases or decreases) could act as a "programmer" of the drive and reward components of appetitive behavior (Kostarczyk, 1986a). Atropine, in eliminating parasympathetic activation, abolishes "hedonic-like" heart rate changes and disturbs consummatory appetitive behavior (i.e., relaxation during petting). However, these findings do not provide information about whether the decreased need for social contact (petting) depends on central mechanisms or on feedback from peripheral autonomic reflexes. Further studies on the heart rate changes that accompany appetitive rewards, together with the use of neuropharmacological agents that do not cross the blood–brain barrier (i.e, methylatropine), are needed to answer this question.

In a recent experiment we examined sympathetic involvement in a hedonic process by measuring the effect of peripheral administration of propranolol (β-adrenergic blocker) on the dog's typical heart rate deceleration response to petting. The results indicated a tendency for atypical cardiac reactions to petting during propranolol treatment (i.v. 1 mg/kg), but cardiac deceleration in response to petting was still present. Analyses of autocorrelation coefficients (Kostarczyk, Jurewicz, Blinolvaska, & Kasicki, 1991) demonstrated that the dog's heart rate is autocorrelated; that is, in an untreated animal the first R–R interval has an impact on the following intervals. However, both petting and propranolol treatment induced a decrease in and even a disappearance of this autocorrelation (Figure 7.9). This suggests that petting may inhibit the sympathetic system by various mechanisms. Inhibition could occur by an increase in sympathetic tone (cardiac deceleration), which in turn produces sympathetic withdrawal and supports bradycardia. The alternative is primary sympathetic withdrawal that promotes parasympathetic prevalence or a quite independent action of both systems regulating the frequency of heart beats and cardiac rhythmicity.

These results indicate that petting has a strong impact on the general psychophysiological functions of an organism. To some extent petting also disorganizes cardiac regulation. Why, then, is it effective as a reward? There is no doubt that momentary bradycardia may provide a physiological benefit (the major factor determining a reward), but accompanying it is the potentially dangerous disappearance of cardiac rhythmicity. Recent data (Kostarczyk & Fonberg, 1988) demonstrated that while the need for petting in bradycardiac, amphetamine-treated dogs increases, their conditioned instrumental responses for a petting reward deteriorate. Petting continued for up to 25 minutes could potentially produce a more profound disorganization of cardiac regulation, which could, in turn, contribute to the deterioration of the dogs' behavior. Perhaps this or another, yet unknown physiological disorganization is the main reason for several behavioral phenomena that have previously been explained by purely psychological arguments such as the theory of learned helplessness as the result of uncontrolled rewards (Seligman, 1975) and the well-

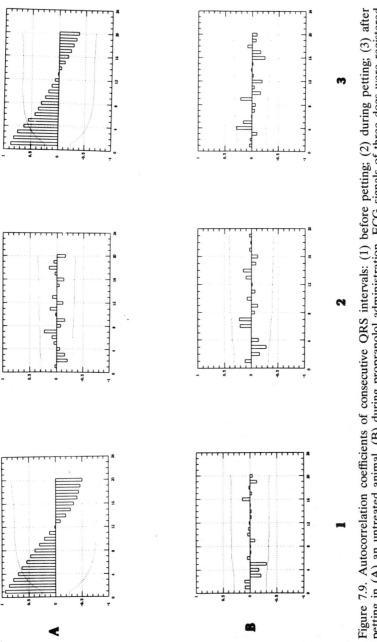

Figure 7.9. Autocorrelation coefficients of consecutive QRS intervals: (1) before petting; (2) during petting; (3) after petting in (A) an untreated animal, (B) during propranolol administration. ECG signals of three dogs were registered using an FM magnetic recorder. After analog–digital conversion, the interbeat (R–R) intervals were analyzed off-line by computer. Statistical analysis indicated the tendency for atypical cardiac reactions to petting during propranolol treatment, but cardiac deceleration still occurred in response to petting. Analysis of the serial autocorrelograms and power spectra of the heart beats revealed that dogs' heart beats are autocorrelated.

documented finding that instrumental training proceeds much better with partial reinforcement (Ferster & Skinner, 1957). The results of our pharmacological experiments support the hypothesis that parasympathetic activation is associated with the hedonic components of reinforcement. However, there is no correlation between the degree of parasympathetic activation and the apparent value of the reward. Moreover, it seems that prior sympathetic arousal, as reflected by cardiac acceleration and/or inhibition of salivation when a food reward is presented (Kostarczyk, 1986a), is due to the excitatory function of conditioned and unconditioned stimuli and is an important factor that could influence the reward's value (Kostarczyk & Fonberg, 1982b). According to Mowrer's (1960) homeostatic theory of reinforcement, appetitive arousal is "motivating in a sense, even punishing." Reinforcement is equated with the attenuation of this unpleasant arousal. Thus, the interplay between a state of arousal and its subsequent decrease is a crucial one, and the capacity to be aroused seems to be an important condition for the hedonic value of reward.

Conclusions

Behavioral, hormonal, and cardiovascular mechanisms are not separate, independent systems but are integrated at all levels of an organism, from the periphery to the brain, and function to maintain overall homeostasis. Our studies indicate the potential for selective activation of the autonomic effectors that underlie specific emotional and behavioral states. Since most of the current research in this area deals with negative emotional states (anxiety, stress, conflict), our data strengthen the concept of the importance of socioemotional stimuli in maintaining overall homeostasis. In conclusion, serveral points should be made.

First, interaction with a human can be positively reinforcing for a socialized domestic dog, as determined by both behavioral and physiological measures. Experimenters should be aware that seemingly casual contact with this species may differentially affect the behavior and physiology of individual subjects.

Second, experimental situations are not neutral for animals. Their "need" for social contact depends on their genetic heritage and life experience. If we neglect to interact socially with a socialized animal, our scientific discoveries will be made with deprived, homeostatically disturbed animals. The opposite situation will occur when the experimental paradigm includes close social interaction with humans, and the animal subjects are wild and unsocialized.

Third, social interactions, which have an important impact on physiological regulation, may be a useful tool in clinical diagnosis and therapy for psychiatric patients or people that suffer arrhythmias and tachycardias. Both cardiovascular and socioemotional responses share some pathways in the brain, with the amygdala and hypothalamus being of the greatest interest (Cornish & Hall, 1979; Lang, Innes, & Tansy, 1979; Cechetto, Ciriello, & Calaresu, 1983; Cechetto & Calaresu, 1985). Amygdalohypothalamic lesions disturb both emotional cardiovascular responses (Johansson & Jonson, 1977; Kapp, Gallagher, Frysinger, & Applegate, 1981) and socioemotional reactions (Fonberg & Kostarczyk, 1980; Thompson, 1981; Kostarczyk, Fonberg, & Prechtl, 1986). The variety and complexity of the amygdala and hypothalamic efferents and afferents, and their connections with the sensory, endocrine systems (for a review see Kostarczyk, 1986b), suggest that there might be localized specific sites that organize cardiovascular responses to social stimuli. There is a need to identify and define the neuroanatomical sites and neurotransmitters used in mediating these integrated responses. Opioids having an impact on both social behavior (Panksepp et al., 1980; Schenk, Britt, & Atalay, 1982; Schenk et al., 1987) and cardiovascular regulations (Schaz et al., 1980; Kiang & Wei, 1984) may play an important role in these processes, as they seem to be involved in the emotional cardiac responses triggered via the amygdala (Gallagher, Kapp, & Pascoe, 1982).

References

Altman, J. (1966). *Organic foundations of animal behavior* [in Polish]. New York: Holt, Rinehart, & Winston.

Anderson S., & Gantt, W. H. (1966). The effect of person on cardiac and motor responsivity to shock in dogs. *Conditioned Reflex, 1*, 181–9.

Bacon, W. E., & Stanley, W. C. (1963). Effect of deprivation level in puppies on performance maintained by a passive person reinforcer. *Journal of Comparative and Physiological Psychology, 56*, 783–5.

Barnes, G. W., & Kish, G. B. (1961). Reinforcing properties of the onset of auditory stimulation. *Journal of Experimental Psychology, 62*, 164–70.

Berlyne, D. E., & Koenig, I. D. V. (1965). Some possible parameters of photic reinforcement. *Journal of Comparative Physiological Psychology, 60*, 276–80.

Black, R. W., Fowler, R. L., & Kimbrell, G. (1964). Adaptation and habituation of heart rate to handling in the rat. *Journal of Comparative and Physiological Psychology, 57*, 422–5.

Cabanac, M. (1979). Sensory pleasure. *Quarterly Review of Biology, 54*, 1–29.

Candland, D. K., Pack, D. K., & Matthews, T. J. (1967). Heart rate and defecation frequency as measures of rodent emotionality. *Journal of Comparative and Physiological Psychology, 64*, 146–50.

Cechetto, D. F., & Calaresu, F. R. (1985). Central pathways relaying cardiovascular afferent information to amygdala. *American Journal of Physiology, 17*, R38–R45.

Cechetto, D. F., Ciriello, J., & Calaresu, F. R. (1983). Afferent connections to cardiovascular sites in the amygdala: A horseradish peroxidase study in the cat. *Journal of Autonomic Nervous System, 8*, 97–110.

Cofer, O., & Appley, M. H. (1964). *Motivation: Theory and research.* New York: Wiley.

Cornish, K. G., & Hall, R. E. (1979). Heart rate changes caused by chemical stimulation of the amygdaloid body. *Physiology and Behavior, 22*, 947–54.

Coote, J. H. (1975). Physiological significance of somatic efferent pathways from skeletal muscle and joints with reflex effects on the heart and circulation. *Brain Research, 87*, 139–44.

Davis, H., & Pérusse, R. (1988). Human-based social interaction can reward a rat's behavior. *Animal Learning and Behavior, 16*, 89–92.

Ferster, C. B., & Skinner, B. F. (1957). *Schedules of reinforcement.* New York: Appleton-Century-Crofts.

Fonberg, E. (1969). The role of hypothalamus and amygdala in food intake, alimentary motivation and emotional reactions. *Acta Biologiae Experimentalis, 29*, 335–58.

Fonberg, E., & Kostarczyk, E. (1980). Motivational role of social reinforcement in dog–man relations. *Acta Neurobiologiae Experimentalis, 40*, 117–36.

Fonberg, E., Kostarczyk, E., & Prechtl, J. (1981). Training of instrumental responses in dogs socially reinforced by humans. *Pavlovian Journal of Biological Sciences, 16*, 183–93.

Fox, M. W. (1978). *The dog: Its domestication and behavior.* New York: Garland STPM Press.

Frank, M. G., & Frank, H. (1988). Food reinforcement versus social reinforcement in timber wolf pups. *Bulletin of the Psychonomic Society, 26*, 467–8.

Freedman, D. G. (1957). Constitutional and environmental interactions in rearing of four breeds of dogs. *Science, 127*, 585–6.

Gallagher, M., Kapp, B. S., & Pascoe, J. P. (1982). Enkephalin analogue effects in the amygdala central nucleus on conditioned heart rate. *Pharmacology, Biochemistry and Behavior, 17*, 217–22.

Galosy, R. A., & Gaebelein, C. J. (1977). Cardiovascular adaptation to environmental stress: Its role in the development of hypertension, responsible mechanisms, and hypotheses. *Biobehavioral Review, 1*, 165–75.

Gantt, W. H. (1944). Experimental basis for neurotic behavior. *Psychosomatic medicine monographs* (Vol. 3, p. 211). New York: Woeber.
　(1960). Cardiovascular components of the conditional reflex to pain, food and other stimuli. *Psychological Reviews, 40*, Suppl. 4, 266–91.

Gattozzi, R. E. (1971). The effect of person on conditioned emotional responses of schizophrenic and normal subjects. *Conditioned Reflex, 6*, 190–1.

Gelhorn, E. (1967). *Principles of autonomic–somatic integrations: Physiological basis and psychological and clinical implications* (Vol. 12). Minneapolis: University of Minnesota Press.

Graham, F. K., & Clifton, R. K. (1966). Heart rate change as a component of the orienting response. *Psychological Bulletin, 65*, 305–20.

Harlow, H. F. (1962). Development of affective behavior in primates. In E. L. Bliss (Ed.), *Roots of behavior* (pp. 157–66). New York: Harper & Row.

Harlow, H. F., & MacLearn, G. E. (1954). Object discrimination learned by monkeys on the basis of manipulation motives. *Journal of Comparative and Physiological Psychology, 47*, 73–6.

127

Henry, J. P. (1982). Psychosocial stimulation of mice in complex population cages and the mechanism of cardiomyopathy. In O. A. Smith, R. A. Galosy, & S. M. Weiss (Eds.), *Developments in neuroscience* (Vol. 15, pp. 267–76). Amsterdam: Elsevier Biomedical.

Hilton, S. M. (1975). Ways of viewing the central nervous control of circulation: Old and new. *Brain Research, 87*, 213–19.

James, W. (1884). The physical basis of the emotions. *Psychological Review 1*, 516–29.

Jaworska, K., Kowalska, M., & Soltysik, S. (1962). Studies on aversive classical conditioning: 1. Acquisition and differentiation of motor and cardiac conditioned classical reflexes in the dog. *Acta Biologiae Experimentalis 22*, 23–34.

Jaworska, K., & Soltysik, S. (1960). Cardiac responses to the acoustic stimuli in dogs [in Polish]. *Acta Physiologica Polonica 11*, 740–1.

Jelen, P., & Kostarczyk, E. (1991). Atropine abolishes the heart deceleration during petting in dogs. Unpublished manuscript.

Johansson, G., & Jonsson, L. (1977). Myocardial cell damage in the porcine stress syndrome. *Journal of Comparative Pathology 87*, 67–74.

Kapp, B. S., Gallagher, M., Frysinger, R. C., & Applegate, C. D. (1981). The amygdala, emotion and cardiovascular conditioning. In Y. Ben-Ari (Ed.), *The amygdaloid complex.* Inserm Symposium 20 (pp. 355–66). Amsterdam: Elsevier.

Kiang, J. G., & Wei, E. T. (1984). Peripheral opioid receptors influencing heart rate in rats: Evidence for endogenous tolerance. *Regulatory Peptides 8*, 297–303.

Kish, G. B., & Barnes, G. W. (1961). Reinforcing effects of manipulation in mice. *Journal of Comparative and Physiological Psychology 54*, 713–15.

Konorski, J. (1967). *Integrative activity of the brain: An interdisciplinary approach.* Chicago: University of Chicago Press.

Konorski, J. & Miller, S. (1933). *Les principes fondamentaux de la théorie physiologique des mouvements acquis: Les reflexes conditionnels moteurs* [in Polish with French summary]. Warsaw: Ksiaznica Atlas TNSW.

Konorski, J., Santibanez, G., & Beck, J. (1968). Electrical hippocampal activity and heart rate in classical and instrumental conditioning. *Acta Biologiae Experimentalis 28*, 168–83.

Kostarczyk, E. (1980). The role of hedonic values of the reinforcer in the conditioning of the alimentary and social instrumental and autonomic responses. Unpublished Ph.D. thesis, Polish Academy of Science.

(1985). The role of arousal in hedonic evaluations. Commentary on R. A. Wise's article: "Neuroleptics and operant behavior: The anhedonia hypothesis." *Behavior and Brain Science 8*, 177.

(1986a). Autonomic correlates of alimentary conditioned and unconditioned reactions in the dog. *Journal of the Autonomic Nervous System 17*, 279–88.

(1986b). The amygdala and male reproductive functions: I. Anatomical and endocrine bases. *Neuroscience and Biobehavioral Reviews 10*, 67–78.

(1989). Atropine abolishes the differential heart responses to conditioned stimuli of various foods. *Pavlovian Journal of Biological Science 24*, 156–9.

Kostarczyk, E., & Fonberg, E. (1982a). Heart-rate mechanisms in instrumental conditioning reinforced by petting in dogs. *Physiology and Behavior 28*, 27–30.

(1982b). Autonomic responses accompanying conditioned and unconditioned alimentary reactions in amygdalo-hypothalamically lesioned dogs. *Acta Neurobiologiae Experimentalis 42*, 43–58.

(1982c). Characteristics of the heart rate in relation to the palatability of food in dogs. *Appetite 3*, 321–8.

(1988). Amphetamine effects on unconditional and conditional instrumental responses with alimentary and social rewards in dogs. *Pavlovian Journal of Biological Science 23*, 10–14.

Kostarczyk, E., Fonberg, E., & Prechtl, J. (1986). Changes in socio-emotional behavior under imipramine treatment in normal and amygdalo-hypothalamic dogs. *Acta Neurobiologiae Experimentalis 46*, 187–203.

Kostarczyk, E., Jurewicz, P., Blinowska, K., & Kasicka, S. (1991). Effects of propranolol upon the cardiac reactions to petting in dogs. Unpublished manuscript.

Kumada, H., Reis, D. J., Tervi, N., & Dampney, R. A. L. (1979). The trigeminal depressor response and its role in the control of cardiovascular functions. In C. Brooks, K. Koizumi, & A. Sato (Eds.), *Integrative functions of the autonomic nervous system* (pp. 319–30). Tokyo: University of Tokyo Press.

Lacey, J. I. (1967). Somatic response patterning and stress: Some revisions of activation theory. In M. H. Appley & R. Trumbull (Eds.), *Psychological stress: Issues in research* (pp. 14–44). New York: Appleton-Century-Crofts.

Lacey, B. C., & Lacey, J. I. (1958). Verification and extension of the principle of autonomic response stereotype. *American Journal of Psychology 71*, 50–73.

(1980) Cognitive modulation of time-dependent primary bradycardia. *Psychophysiology 17*, 209–21.

Lang, I. M., Innes, D. L., & Tansy, M. F. (1979). Areas in the amygdala necessary to the operation of the vagosympathetic pressor reflex. *Experientia 35*, 57–8.

Lynch, J. J., & Gantt, W. H. (1968). The heart rate component of the social reflex in dogs: The conditioned effects of petting and person. *Conditioned Reflex 3*, 69–80.

Martin, J., Sutherland, G. J., & Zbrozyna, A. W. (1976). Habituation and conditioning of the defence reactions and their cardiovascular components in cats and dogs. *Pflugers Archive 365*, 37–47.

Montgomery, K. C. (1954). The role of exploratory drive in learning. *Journal of Comparative and Physiological Psychology 47*, 60–9.

Mowrer, O. H. (1960). *Learning theory and behavior.* New York: Wiley.

Ninomiya, I., & Yonezawa, Y. (1979). Sympathetic nerve activity, aortic pressure and heart rate in response to behavioral stimuli. In C. Brooks, K., Koizumi, & A. Sato (Eds.), *Integrative functions of the autonomic nervous system* (pp. 433–42). Tokyo: University of Tokyo Press.

Obrist, P. A. (1976). The cardiovascular–behavioral interaction – as it appears today. *Psychophysiology 13*, 95–107.

Obrist, P. A., Light, K. C., Langer, A. W., Grignolo, A. G., & Koepke, J. P. (1982). Behavioral–cardiovascular interaction. In O. A. Smith, R. A. Galosy, & S. M. Weiss (Eds.), *Circulation, neurobiology and behavior* (pp. 57–76). Amsterdam: Elsevier Biomedical.

Obrist, P. A., Sutterer, J. R., & Howard, J. L. (1972). Preparatory cardiac changes: A psychobiological approach. In A. H. Black, & W. F. Prokasy (Eds.), *Classical conditioning II* (pp. 312–40). New York: Appleton-Century-Crofts.

Panksepp, J., Herman, B. H., Vilberg, T., Bishop, P., & DeEskinazi, F. G. (1980). Endogenous opioids and social behavior. *Neuroscience and Biobehavioral Reviews 4*, 437–87.

Perez-Cruet, J., & Gantt, H. W. (1959). Relation between heart rate and "spontaneous" movements. *Bulletin of the Johns Hopkins Hospital 105*, 315–21.

Pfaffenberger, C. J., & Scott, J. P. (1975). Early rearing and testing. In J. P. Scott, J. L. Fuller, R. N. Ginsburg, & S. W. Bielfelt (Eds.), *Guide dogs for the blind: Their selection, development and training* (pp. 13–37). Amsterdam: Elsevier.

Pfaffman, C. (1960). The pleasure of sensation. *Physiological Review 67*, 253.

Robinson, J., & Gantt, W. H. (1947). The orienting reflex/questioning reactions: Cardiac, respiratory, salivary and motor components. *Bulletin of the Johns Hopkins Hospital 80*, 231–53.

Roessler, R., Collins, F., & Burch, N. R. (1969). Heart rate response to sound and light. *Psychophysiology 5*, 359–69.

Royer, F. L., & Gantt, W. H. (1966). Effect of movement on the cardiac conditioned reflex. *Conditioned Reflex 1*, 190–4.

Schaz, K., Stock, G., Simon, W., Schlor, K. H., Unger, T., Rockhold, R., & Ganten, D. (1980). Enkephalin effects on blood pressure, heart rate and baroreceptor reflex. *Hypotension 2*, 395–407.

Schenk, S. M., Britt, M. D., & Atalay, J. (1982). Isolation rearing decreases opiate receptor binding in rat brain. *Pharmacology, Biochemistry and Behavior 16*, 841–52.

Schenk, S. M., Hunt, T., Klikowski, G., & Amit, Z. (1987). Isolation housing decreases the effectiveness of morphine in the conditioned taste aversion paradigm. *Psychopharmacology 92*, 49–51.

Scott, J. P., & Fuller, J. L. (1965). *Genetics and the social behavior of the dog.* Chicago: University of Chicago Press.

Seligman, M. E. P. (1975). *Helplessness: On depression, development and death.* San Francisco: Freeman.

Skinner, B. F. (1938). *The behavior of organisms: An experimental approach.* New York: Appleton-Century.

Snapper, A. G., Ferraro, D. P., Schoenfeld, W. W., & Locke, B. (1965). Adaptation of the white rat's cardiac rate to testing conditions. *Journal of Comparative and Physiological Psychology 94*, 25–35.

Snowden, C. T., Bell, D. D., & Henderson, N. D. (1964). Relationships between heart rate and open field behavior. *Journal of Comparative and Physiological Psychology 58*, 423–6.

Soltysik, S., Jaworska, K., Kowalska, M., & Radom, S. (1961). Cardiac responses to simple acoustic stimuli in dogs. *Acta Biologiae Experimentalis, 21*, 235–52.

Soltysik, S., & Kowalska, M. (1960). Studies on avoidance conditioning: Relations between cardiac (Type I) and motor (Type II) effects in the avoidance reflex. *Acta Biologiae Experimentalis 20*, 157–70.

Stanley, W. C., & Elliot, O. (1962). Differential human handling as reinforcing events and as treatment influencing later social behavior in basenji puppies. *Psychological Reports 10*, 775–88.

Stellar, E. (1980). Brain mechanisms and hedonic processes. *Acta Neurobiologiae Experimentalis 40*, 313–24.

Thompson, C. I. (1981). Long-term behavioral development of rhesus monkeys after amygdalectomy in infancy. In Y. Ben-Ari (Ed.), *The amygdaloid complex* (pp. 259–270). Inserm Symposium 20. Amsterdam: Elsevier.

Thorpe, W. C. (1963). *Learning and instinct in animals.* London: Methuen.

Trojniar, W., & Cytawa, J. (1978). Hedonestesia: The nervous process determining motivated ingestive behavior. *Acta Neurobiologiae Experimentalis 38*, 139–51.

Vanderwolf, G. H. (1971). Limbic–diencephalic mechanisms of voluntary movement. *Psychological Review 78*, 83–113.

Wenzel, B. M. (1959). Tactile stimulation as reinforcement for cats and its relation to early feeding experience. *Psychological Reports 5*, 297–300.

Werner, C., & Anderson, D. F. (1976). Opportunity for interaction as reinforcement in a T-maze. *Personality and Social Psychology Bulletin 2*, 166–9.

Wickler, W. (1976). The ethological analysis of attachment, sociometric motivations and sociophysiological aspects. *Zeitschrift fur Tierpsychologie 42*, 12–28.

Wyricka, W. (1975). The sensory nature of reward in instrumental behavior. *Pavlovian Journal of Biological Science 10*, 23–51.

Yehle, A., Dauth, G., & Schneiderman, N. (1967). Correlates of heart-rate classical conditioning in curarized rabbits. *Journal of Comparative and Physiological Psychology 64*, 98–104.

Zagrodszka, J., Korczynski, R., & Fonberg, E. (1981). The effects of imipramaine on socio-emotional and alimentary motivated behavior in dogs. *Acta Neurobiologiae Experimentalis 41*, 361–70.

—8—

Behavioral arousal and its effect on the experimental animal and the experimenter

Alastair J. S. Summerlee

Das sind die bande die mich binden.
Walküre, Act II (R. Wagner, 1852)

Editors' introduction

The point of Summerlee's chapter is basic and central to the theme of our book: The human is part of the animal's environment, and the presence of the human results in cortical arousal that can be measured physiologically. That such arousal is not general, but reflects the familiarity of particular individuals to the animal, offers direct evidence of the specificity of the scientist–animal bond.

Summerlee's caveats are aimed at those working with direct physiological measurement, but their implications for behavioral research are inescapable. Merely handling an animal yields measurable physiological changes. When such handling is performed by a researcher with whom there is no familiarity, cortical activity borders on that associated with alarm reactions.

Introduction

In this chapter I shall discuss the way that the environment can affect the spontaneous activity of cells within the cerebral cortex of unanesthetized, unrestrained animals. I shall demonstrate that there is an inevitable link between an animal's level of arousal (a composite response to the animal's environment and its internal milieu) and the activity of nerve cells within the cortex. I will then show that an animal's level of arousal is adjusted, and the activity of cortical neurons modified, to the familiarity of its environment, which includes the presence and actions of the investigator. Having recognized and established this link, I shall explain how the responses of the animal, both behavioral and neuronal, can modify the behavior and attitudes of the experimenter completing the ring of the inevitable bond between the animal, the scientist, and the animal.

Much of the initial experimental work described was derived from cortical activity recorded from two rats, Brunnehilde and Wotan, who created the first dent in my physiologist's approach to experiments, but credit should be given to the many other animals that have contributed to this work.

Patterns of activity from cortical neurons

Nerve cells of mammalian cerebral cortex discharge in a series of randomly spaced action potentials that appear, at first glance, to have no obvious relation to the animal's level of arousal or to the external environment. There are some neurons within specific areas of the cortex – for example, neurons in primary visual cortex – that have clearly defined functions (i.e., they respond to specific visual patterns), but many neurons have no function that can be related to the external environment. Yet all these neurons show spontaneous electrical activity. Even the neurons in visual cortex continue to discharge when the cells are completely deprived of external input (i.e., when the animal is placed in complete darkness) (Webb, 1976a). The pattern of firing of the cells is thought to encode information that is used in processing information (Klemm & Sherry, 1981, 1982), and there have been numerous attempts to find a consistent relationship between an animal's level of arousal and rate of discharge of cortical neurons, all without success. It is easy, with hindsight, to appreciate why these early attempts were not successful. An animal's level of arousal flickers from moment to moment, and simple measurement of the frequency of firing of a neuron, even over a short period of time, is likely to include several different levels of arousal, which could lead to inconsistent findings. Moreover, measuring frequency of discharge from a neuron is not an adequate definition of a pattern of discharge. This is explained more fully in Figure 8.1. Over a set length fo time, perhaps 10 seconds, either a neuron could discharge with a series of regularly spaced action potentials spread out over the period, or the same number of potentials might be concentrated into a brief burst of activity at the beginning of the 10 seconds, leaving the rest of the period silent. In both cases the frequency of firing over the 10-second period is identical but the pattern of activity is quite different. In practice, neurons recorded from a variety of sites in cerebral cortex show a mixture of randomly spaced single potentials interrupted by short bursts of discharge

133

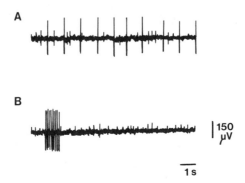

Figure 8.1. Spontaneous activity of a neuron in the parietal cerebral cortex of a conscious unrestrained rabbit. The figure illustrates that simple measurement of the rate of discharge of a cell may not describe accurately the pattern of discharge. In Part A there are 10 action potentials spread evenly over the 10-second period; in Part B the 10 potentials are grouped together at the start of the period to form a burst of discharge. In both cases the frequency of firing is the same but the pattern of discharge is completely different; effectively, the modal interspike interval is long in the first trace and short in the second. The recording of site of this neuron was later localized to Layer III of the cortex.

Figure 8.2. Neuronal activity recorded from one neuron located in Layer IV of parietal cortex to illustrate the pattern of firing over a period of time. Note that activity is composed of a series of short bursts interrupted by periods of relatively regular activity and periods of effective silence. The pattern of discharge varies according to the level of arousal of the animal.

and periods of silence (Figure 8.2). The challenge has been, therefore, to find a means of describing the temporal pattern of discharge in a statistically acceptable manner and then to find some measure of that description which can be related, in a consistent way, to an animal's level of arousal.

There have been three major steps toward these goals. The first was described as early as 1964 by Evarts. Recording from neurons in the cerebral cortex of conscious monkeys, he reported that the most common phenomenon observed in the patterned discharge of these cells when an animal fell into peaceful sleep was a shortening of the most common interval between successive action potentials (the modal interspike interval). The second step was achieved by Burns and Webb (1976), who demonstrated that interval distributions derived from short trains of action potentials from any cortical neurons tested could be described adequately by logarithmic-normal functions. Theoretically the shape of any logarithmic-normal distribution is defined by only three parameters, the modal (most common) interval, the modal count (height at the mode), and the geometric standard deviation of the distribution (effectively the scatter about the mode). Burns and Webb (1976) argued that any aspect of neuronal activity that might be influenced by the behavioral state of the animal would be extremely unstable. For this reason, they restricted their analyses to trains of 200 action potentials, which was a compromise between one so small that sampling error would be unacceptable and one so large that recording time would be too long and the animal's level of behavioral arousal might not remain constant. This description of patterned discharge holds true for cortical neurons recorded from the cat (Burns & Webb, 1976; Burns, Stean, & Webb, 1979), rabbit (Paisley & Summerlee, 1983; Slaney & Summerlee, 1983), and rat (Summerlee, 1980), provided that the rate of discharge is faster than 2.5 potentials per second. Interval distributions derived from neurons firing more slowly than 2.5 potentials per second could not always be described by logarithmic-normal functions due to an excess number of long intervals. However, intervals up to 10 times the modal interval still provided statistically acceptable fits to logarithmic-normal distributions (Burns & Webb, 1976; Paisley & Summerlee, 1984a). These data effectively confirm that attempts to correlate mean frequency of discharge with behavioral arousal are unprofitable because the mean rate of firing is dependent on two parameters, a modal interspike interval and the scatter about that mode. An alteration in either of these parameters could result in substantial shifts in the mean frequency of firing without a necessary change in the other parameter. This point is illustrated, theoretically, in Table 8.1. The final major advance has been the quest to find a universally acceptable description of different levels of arousal, which was achieved by Webb (1976a,b).

Effects of changing the levels of arousal on neuronal activity

Given the problem of trying to link behavioral arousal to the activity of cortical neurons, it became necessary to find an adequate assessment of the state of arousal of an animal. Many of the earlier studies used definitions of behavior proposed by Hobson (1969), who described the behavioral state of an animal by three parameters: (1) sensory input, the

Table 8.1. *Predicted mean frequency of discharge (action potential/second) of a central cortical neuron*

Geometric standard deviation	Modal interval (milliseconds)					
	10	20	30	40	50	60
1.5	79.0	39.5	26.3	19.6	15.8	13.1
2.0	50.9	25.4	17.0	12.7	10.2	8.5
2.5	31.7	15.9	10.5	7.9	6.3	<1
3.0	20.2	10.1	6.7	5.0	4.0	<1
3.5	13.2	6.6	4.4	3.3	2.6	<1
4.0	9.0	4.5	3.0	2.3	1.8	<1

Note: This table shows that mean frequency of discharge may be affected by a shift in either the modal interspike interval or scatter about the mode (geometric standard deviation). The table was constructed by arithmetic integration of theoretical normal distributions with given modal intervals and geometric standard deviations. *Source*: Burns and Webb (1976).

degree to which an animal is receptive to its environment; (2) motor output, the degree to which an animal moves in its environment; and (3) the electroencephalogram (EEG). In practice, however, this definition is difficult to implement. It is not possible to measure the total sensory input to an animal; the motor output could best be measured by describing the behavioral reactions and responses of the animal to its environment, and value of the EEG for defining level of arousal is questionable. Jouvet (1967) stated, "In no way does the state of the corticogram (EEG) allow us to presume that an animal is asleep or awake." These early experiments were typical of the genre: an attempt by the experimenter to describe levels of arousal in an animal by seeking to control all the experimental variables including the behavior of the animal. The inevitable consequence of such intervention was an upset in the animal's level of arousal! There was no universally acceptable measure of level of arousal. Webb used an alternative approach; she argued that there were two natural extremes of arousal, peaceful sleep and alarm, which could be adequately defined by behavioral criteria. This classification was less affected by errors of interpretation and certainly less invasive than other definitions affording the animal opportunity for complete freedom of movement and behavior. Moreover, any behavioral state between these two extremes should show intermediate values in parameters measured, thereby allowing Webb to make a systemic study of the effects of level of arousal on the temporal pattern of discharge of cortical neurons.

If Webb's descriptions of the animal's behavioral responses (Hobson's motor output) were sufficiently strict, the EEG would become an extra parameter that could be monitored, but not an absolute measure of arousal. Subsequently, Summerlee and Paisley (1982; Paisley & Summerlee, 1984a,b) confirmed that Webb's behavioral criteria for quiet sleep, rapid eye movement (REM) sleep, and alarm were always associated with particular patterns of EEG activity. The reverse was not always true. The behavioral descriptions were critical to the experiments: Webb (1976a) defined a sleeping cat as an animal that lay with its head supported by some part of the recording box, showing slow, deep, regular respiratory movements, its eyes shut, and ears unresponsive to low-level laboratory sound. REM was identified by jerky movements of the eyes beneath closed lids, rapid shallow breathing, and twitching of whiskers, facial muscles, and tail in a sleeping cat.

Throughout the experiments the only intervention practiced was the recording of neuronal activity, and Burns, Stean, and Webb (1974) had already described a technique for making long-term recordings (continuous recordings for several days) from single cortical neurons in completely unanesthetized, unrestrained cats. There were no signs that the animals were concerned by the recording apparatus, and it has subsequently been shown that this technique can be used to take recordings from rabbits in labor or when lactating (Paisley & Summerlee, 1984b; O'Byrne, Ring, & Summerlee, 1986). Both these behaviors are extremely sensitive to stress; for example, a female rabbit may even stop nursing her young if the technical staff responsible for daily cleaning of the cages is changed.

Webb (1976a) found that the transition from wakefulness to quiet sleep was always accompanied by a shortening of the modal interspike interval. The mode remained short for the duration of the sleeping period, even if the cat started to show periods of REM sleep. On average the modal interval in sleep was about one-third of the value observed for the same neuron in an awake animal, and interval distributions with modes of less than 20 milliseconds were typical of neuronal activity recorded from a sleeping cat (Webb, 1976a). This general principle also holds for cortical neurons recorded from the rat (Summerlee, 1981) and rabbit (Summerlee & Paisley, 1982; Paisley & Summerlee, 1984a,b) that is, the modal interval decreases by approximately one-third of the value detected in the waking animal, although, in both these species, sleep was associated with modal intervals of less than 50 milliseconds. Peaceful sleep was generally accompanied by an increase in the geometric standard deviation from the mode. The average geometric standard deviation was 42% greater in sleeping cat (Webb,

1976a) or 46% greater in sleeping rats and rabbits (Paisley & Summerlee, 1983, 1984a) compared with the scatter in awake animals. In contrast, REM sleep was associated with a decrease in the scatter about the shortened mode, which means that neurons tended to fire faster in animals during REM sleep, a finding that is entirely consistent with several reports recording changes in firing rate only (Evarts, 1962; Hobson & McCarley, 1971; Noda & Adey, 1973; Steriade, Deschenes, & Oakson, 1974).

Alarm, the opposite extreme of arousal, was induced by a hiss of compressed gas introduced into the experimental box when the animals were awake and relaxed (Webb, 1976b). Since in some of the experiments, recordings were taken from neurons in primary visual cortex, the experiments were carried out in the dark (the cats were under infrared camera observation) to eliminate contamination by visual responses. In these experiments, as in the subsequent experiments on rats and rabbits (Paisley & Summerlee, 1983, 1984a), the responses obtained from all the cortical neurons recorded were the opposite of those obtained when the animal fell asleep: The modal interval increased and the geometric standard deviation decreased. Table 8.1 shows that, theoretically, there need not be a change in firing frequency of these cells as the temporal pattern of discharge changes, and published data support this hypothesis (Webb, 1976b; Summerlee, 1980; Summerlee & Paisley, 1982): There is no consistent change in the firing frequency of cortical neurons as an animal is alarmed from a state of quiet rest. When a whole series of behavioral states defined by simple behavioral criteria (i.e., quietly awake, eating, drinking) is studied in the same neuron (Figure 8.3), it can be seen that there is a progression of logarithmic-normal curves from sleep through alarm with a gradual increase in the modal interspike interval and reduction in the scatter about the mode. These data imply that there is a systematic relationship between the level of arousal of an animal and the temporal pattern of discharge of cortical neurons.

Perhaps the findings are most remarkable for REM sleep. There is no significant change in the modal intervals between peaceful sleep and REM sleep despite the substantial differences in EEG activity. There is, however, a decrease in the scatter about the mode, which was observed in cats (Webb, 1976a) and in rats and rabbits (Paisley & Summerlee, 1984a). From the theoretical table of the basis of temporal patterns of firing it is possible to predict that during REM sleep there should be an overall increase in the mean firing rate of the cortical cells compared with peaceful sleep, a finding that is entirely consistent with several reports in the literature (Noda & Adey, 1970, 1973; Hobson & McCarley, 1971). Moreover, these results indicate that the most important factor governing the temporal pattern of activity of cortical neurons is the lack of consciousness

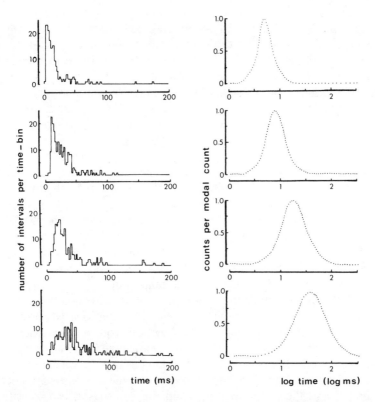

Figure 8.3. The effect of four states of behavior (quiet sleep, REM sleep, wakefulness, and alarm) on the temporal pattern of discharge of one neuron in association cortex of a conscious rabbit. On the left, the graphs show the change in shape of the interval distributions for the four behavioral states. Each histogram has been standarized and replotted on a logarithmic time scale on the right. The mode becomes shorter progressively as the level of arousal falls toward sleep and remains short even during REM sleep. In this case the scatter about the mode decreases slightly as the level arousal falls toward sleep, which is not the more common finding. (From Paisley & Summerlee, 1984a.)

and not the activity of the EEG. The intriguing relationship between the temporal pattern of activity of cortical neurons and the EEG is discussed further elsewhere (Paisley & Summerlee, 1984a).

Effects of a continuum of arousal

The data presented on the changes in temporal pattern of discharge recorded from neurons in a wide variety of cortical sites and in different species show characteristic and consistent changes in activity between sleep and wakefulness, and between wakefulness and alarm. The direction of these changes implies that there is a continuum of effects of arousal. Personal experience of levels of attention would lead one to predict that this would be the case. But wakefulness encompasses an enormous range of behavioral states, from sleep at one extreme to alarm at the other. There is no independent, and accepted, measure of level of arousal; consequently, it is not easy to demonstrate a continuous relationship between level of arousal and the temporal pattern of discharge from cortical neurons. Trulson and Jacobs (1979), using a combination of behavioral, electro-encephalographic, and electromyographic monitors, divided behavior into 12 distinct categories. The definitions of these states are not easy to implement and analyze, and the data are still subject to subjective interpretation. Furthermore, the results leave us with only 12 discrete points on a continually flickering scale of arousal. We need to know the sequence of events that occurs to the temporal pattern of discharge of a cortical neuron when an animal is alarmed, recovers, relaxes, and finally falls asleep. We would predict that the modal interval should lengthen in alarm, gradually shorten as the animal relaxes, and finally become shortest when the animal falls into peaceful sleep. But when one is working with conscious, unrestrained animals, this behavioral sequence practically never happens; an alarmed animal relaxes quickly after the initial arousing stimulus, then may take hours before finally falling asleep. In many experiments it was difficult not to believe that it almost became a competition between the animal and the experimenter; could the animal stay awake until the experimenter fell asleep in the quiet, warm, semidark laboratory? Despite the soporific nature of the experiments, we managed to obtain several complete records whereby the activity of a cortical neuron in a rabbit was tracked through a progression from wakefulness to alarm, relaxation, and finally sleep. An example of a such an experiment is shown in Figure 8.4. These data support the hypothesis that there is a continuum of arousal that is paralleled by changes in the temporal pattern of discharge of cortical neurons.

It may be misleading, however, to consider that there is a continuum of effect on the temporal pattern of discharge of cortical neurons from sleep at one extreme to alarm at the other, because the stimulus used to alarm might have initiated a classic "fear, fight, flight" response by the body and sympathetic activation might have distorted the observations. Was there an alternative behavioral state that could be studied such that it would be possible to investigate the effects on the temporal pattern of discharge of cortical cells? We had been recording the activity of cortical neurons in suckling rabbits and were intrigued by the findings. There are several

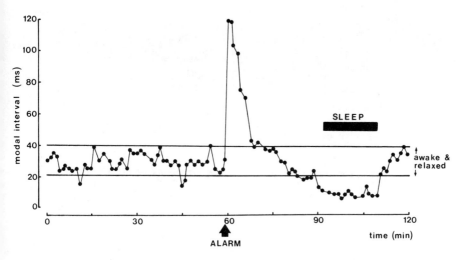

Figure 8.4. The relationship between level of arousal and the temporal pattern of activity in a neuron in association cortex of a conscious rat. Each point represents average data from 10 trains of 200 consecutively recorded action potentials. After 1 hour of control recording, when the rat was peacefully awake or moving undisturbed around the run, the animal was alarmed with a 5-second hiss of compressed gas. The range of modal intervals during the control period when the animal was awake and relaxed is shown. The most consistent change observed when the animal was alarmed was a profound lengthening of the modal interspike interval, which remained significantly elevated for about 10 minutes. In this case the rat settled down quickly and fell asleep. As the rat's level of arousal fell, so the modal interspike interval became shorter until it was significantly shorter than in the control period. These data suggest that there is a continuum of arousal and the response of the cortical neurons.

reports in the literature that suckling is soporific (Newton & Newton, 1948; Lincoln, 1974; Lincoln et al., 1980), yet the most consistent finding in the rabbit cortex was that suckling was associated with a lengthening of the modal interspike interval (Paisley & Summerlee, 1983, 1984a), which is characteristic of alarm. Recordings of EEG activity confirm that suckling in rabbits and also in miniature pigs is associated with EEG arousal (Poulain, Rodriguez, & Ellendorff, 1981; Neve, Paisley, & Summerlee, 1982). In some cases during suckling the neurons stopped firing almost completely, which might be equivalent to a substantial increase in the modal interval leaving too few potentials to calculate a meaningful mode (Figure 8.5). Webb (1976b) also observed some neurons in visual cortex that stopped firing when the cats were alarmed by the hiss of compressed gas, and other reports in the literature confirm that very high levels of arousal are

141

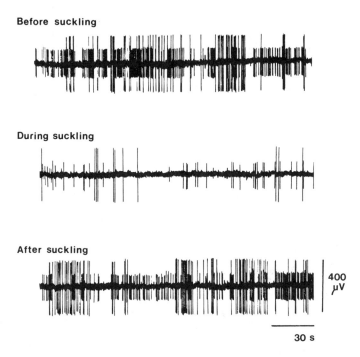

Figure 8.5. Extracellular activity recorded from two neurons in association cortex of a conscious rabbit in association with suckling. The rate of discharge of both cells decreased dramatically in suckling (<0.05 Hz) and the interspike interval lengthened. (From Paisley & Summerlee, 1983.)

associated with a hiatus in the firing of cortical neurons. Among these reports the most dramatic, and ethically disturbing, is that of Lomo and Mollica (1962), who reported that many cells in visual cortex of the unanesthetized, restrained, rabbit stopped firing when the animal struggled against the restraint of the recording apparatus. These animals were bolted, without anesthetic, analgesic, or even sedative to the stereotaxic apparatus, the skull trephined, and electrodes sunk into exposed areas of visual cortex. The disturbing interpretation of their observations might be that the escape behavior the rabbits were trying to manifest by struggling would have been associated with extreme forms of alarm and hence a dramatic lengthening of the modal interspike interval.

Further circumstantial evidence to support the hypothesis that the shifting continuum of arousal has consistent effects on the temporal pattern of discharge of cortical neurons can be obtained from reports on the effects of anesthetics on the temporal pattern of discharge. The prediction would

142

be that anesthetizing an animal should result in changes in the interval distributions akin to peaceful sleep. Both Noda and Adey (1973), working with lightly anesthetized, semirestrained cats, and Foster (1984; Foster, Paisley, & Summerlee, 1982), working with halothane-anesthetized Japanese quail, reported that records obtained from these mammals and birds were virtually indistinguishable from records obtained during quiet sleep.

Changing levels of arousal and recognition

If there is an inevitable, consistent relationship that binds level of arousal and the patterned discharge of cortical neurons, it should be possible to carry out the reverse experiment and use the moment-to-moment changes in neuronal activity to identify changes in level of arousal. This could be used to investigate the effects of an animal encountering a novel experience on cortical neuron activity and arousal and study how these features adjust to repeated exposure to either familiar and/or unfamiliar circumstances.

Rats and rabbits, close to skeletal maturity, were fitted with platinum microwire recording electrodes implanted into various regions of association and parietal areas of cerebral cortex during sterile surgical operations. The technical details are reported elsewhere (Summerlee & Paisley, 1982; Paisley & Summerlee, 1983). Briefly, the wires were inserted to a depth of 0.5 to 1.5 mm from the surface of the cortex through a small hole drilled in the skull, and each wire was soldered to a miniature electrical socket that was attached to the parieto-occipital region of the skull. The hole in the skull was sealed and the animals allowed to recover from anesthesia. A daily routine was established to test the electrical activity present on the implanted electrodes. Since the animals were implanted when they were skeletally immature, continued growth caused a natural form of micromanipulation, and the exposed recording face at the tip of wires was pulled very slowly backward through the cortex. The extremely slow movement dragged the tip of the wires past a succession of nerve cells. In consequence, each wire yielded neuronal activity that could be recorded from an average of five cells a month. The average length of recording time for each neuron was 1.5 days (range 5 hours to 24 days), so the long-term stability of these recordings afforded the possibility of making repeated tests on the same cell to confirm previous results. As alluded to earlier, this technique for recording neuronal activity does not appear to compromise the animals in any way: Rabbits that are sensitive to the stress of changes of routine in the animal house will continue to reproduce successfully over several months of recording (Paisley & Summerlee, 1984b; O'Byrne et al., 1986).

The activity of individual neurons was relayed by a small telemetry device worn on the animal's collar to a radio receiver and the data recorded onto magnetic tape accompanied by a verbal commentary on the behavior of the animal. The animals were allowed to move freely about a run constructed in the laboratory. During recordings, extraneous noise was kept to a minimum and the lighting was subdued. Initial analysis was carried out on-line, but the data were always analyzed fully at a later stage. First, the response of each neuron to a number of standard procedures was tested. Neuronal activity was recorded for at least 10 minutes while the animal was sitting quietly but not asleep to establish the mean and range in fluctuation of the modal interspike interval for successive trains of 200 action potentials. A further period of recording was taken while the animal was walking about the run or eating or drinking. Where possible, the activity was also recorded when the animal was peacefully asleep and showing REM sleep (behavioral definitions for sleep and REM sleep were given earlier), but there were some occasions when the animals declined to fall asleep despite very long periods of observation. The animals were then alarmed by a 5-second hiss of compressed gas released from a cylinder and neuronal data recorded continuously until the modal interspike interval had returned to within 95% confidence limits of the average value when the animals were quietly awake. In all cases the temporal pattern of discharge of the neurons changed in the anticipated manner; compared with the resting state, the modal interval always decreased when the animal started to doze or fell asleep. In contrast, the modal interval increased when the animal was awake, eating, or drinking and increased further when the animal was alarmed. An example of one experiment is shown in Figure 8.6. The time taken for the modal interval of successive trains of potentials to return to within 95% confidence limits of the range observed when the animal was awake and undisturbed was 7 ± 2.7 minutes (mean ± SEM) for all neurons recorded. The geometric standard deviation of the distributions was not as reliable a guide to the level of arousal since the mean time for return to control values was 9 minutes (range 3 to 18 minutes).

Once these standard responses for each neuron had been characterized, the effects of a number of procedures on the temporal discharge pattern were tested:

1. lifting the animal out of the run,
2. repeatedly lifting the animal out of the run,
3. introducing a new person into the laboratory,
4. the new person lifting the animal out of the run,

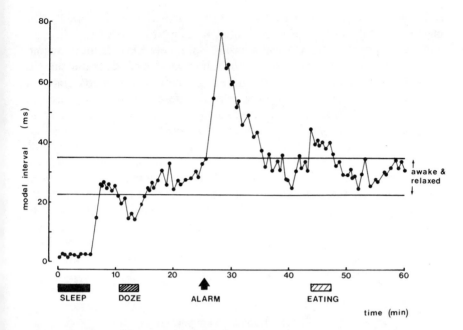

Figure 8.6. The effect of changing levels of arousal on the temporal pattern of discharge of a cortical neuron in a conscious, unrestrained rabbit. Neuronal activity was recorded continuously and analyzed in relation to the behavior of the animal. The relationship between the modal interspike interval (each point is the average of 5 trains of 200 action potentials) and various aspects of behavior is shown. The range of modal intervals when the animal was awake and relaxed is shown. Note that as level of arousal increases when the animal wakes-up, the modal interval becomes progressively longer. It is longest when the rabbit is frightened by a hiss of compressed gas. As the animal relaxes after being alarmed, the modal interval decreases.

5. the new person repeatedly taking the animal from its run and stroking it.

The changes in the temporal pattern of discharge of the neuron were determined, again, by computing the changes in the modal interval and the geometric standard deviation about the modes and the length of time taken for the modal interval and the standard deviation to return to within the range experienced in the quiet, awake state. The changes observed are shown in Table 8.2, and an example of the patterned discharge from one neuron is shown in Figure 8.7. In summary, lifting the animal out of the run produced a transient (less than 2 minutes) elevation in the modal interval and usually a decrease in the scatter about the mode that was similarly

145

short-lived, but these responses depended on experience. Repeatedly lifting the animal out of the run resulted in a habituation of the arousal-induced response of the cortical neurons. After several trials there was no longer a consistent increase in the modal interval detected. As the animals became accustomed to the procedure, the shift in the modal interval became less marked but individual cells in different animals appeared to habituate to the experience at different rates. Most animals appeared to become accustomed to the experimental protocol (neuronal response no longer observed) within five trials. Neuronal activity recorded from rats seemed to habituate to the experience more quickly than rabbits. This may be related to an apparent appreciation of physical contact demonstrated by the rats but not the rabbits. Most of the rats learned that they would be picked up, and once the experiments were in progress the rats started to solicit attention. Several of the rats began climbing out of the cage to meet the experimenter. Moreover, their attempts to climb out of the run were often associated with a lengthening of the modal interspike interval, which confounded interpretation of the results.

Recognition of a new person

Introducing a new person into the laboratory had a surprisingly rapid and profound effect on the temporal pattern of discharge of cortical neurons in both rats and rabbits. Within milliseconds there was a substantial increase in the length of the modal interspike interval that was usually associated with whisker twitching and sniffing the air. Most of the animals, however, seemed to accept the intruder and stopped displaying behavioral traits of concern, and the modal interval slid toward the normal range for relaxation within a relatively short period of time (less than 2 minutes). The speed with which the modal interval declined was not consistent for neurons from different animals but tended to be more consistent within the same animal. An intruder that lifted either a rat or rabbit out of the run produced substantial and prolonged shifts in both the modal interval and scatter about the mode. In many cases the neurons stopped firing (less than one potential every 10 seconds) for at least 60 seconds. Where it was possible to record a sufficient number of potentials to create interval distributions, the modal interval was always greater than two and a half times the length of the mode at rest and the scatter 56% smaller than control values. These data suggest that such a procedure caused considerable angst in the animals. Repeated exposure to the same intruder and repeated lifting out of the run became a less traumatic experience as judged by the activity of the cells in the cortex. The leap in the length of the modal interval became

Table 8.2. *Effects of changing environmental conditions on neuronal activity*

Time after treatment	1 minute		5 minutes		10 minutes	
	Mode	Scatter	Mode	Scatter	Mode	Scatter
Control	32.4 ± 6.8	2.5 ± 0.6	34.6 ± 7.1	2.4 ± 0.7	32.5 ± 6.9	2.5 ± 0.6
Lifting animal out	68.1 ± 11.1↑	1.4 ± 0.5↓	38.9 ± 9.2	2.2 ± 1.4	34.8 ± 7.0	2.4 ± 0.7
Repeated lift out (fifth time)	38.4 ± 8.4	1.8 ± 0.7	36.4 ± 3.7	2.4 ± 0.6	33.7 ± 9.2	2.5 ± 1.1
Intruder in laboratory	75.9 ± 9.6↑	1.4 ± 0.3↓	49.4 ± 7.6↑	2.5 ± 0.9	37.4 ± 8.3	2.3 ± 0.9
Intruder lifts animal out	94.4 ± 12.8↑	1.3 ± 0.1↓	76.3 ± 8.4↑	2.4 ± 1.1	41.3 ± 7.3	2.3 ± 0.8
Intruder lifts animal out (tenth time)	51.4 ± 6.9↑	2.6 ± 0.7	41.7 ± 7.2	2.5 ± 0.9	38.8 ± 6.3	2.5 ± 0.9

Note: The data show the effects of various changes in an animal's environment on the temporal-discharge pattern of cortical neurons in a group of 10 rats. Successive trains of 200 action potetials were analyzed for at least 10 minutes after one of five experimental paradigms and the effects on modal interspike interval and scatter about the mode calculated. The data were compared by analysis of variance, and significant ($p < .05$) increases or decreases compared with control period are indicated by the arrows.

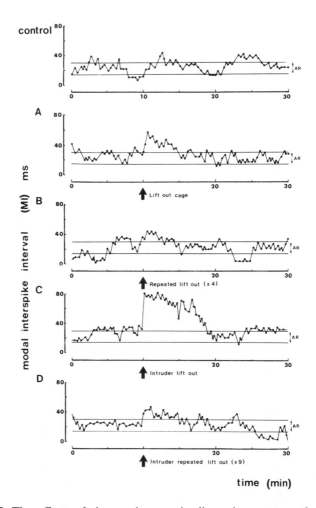

Figure 8.7. The effects of changes in an animal's environment on the temporal pattern of discharge of a neuron in the association cortex of a conscious, unrestrained rat. The same animal was subjected to a number of tests: (a) lifting out of the cage, (b) repeatedly lifting out of the cage, (c) an intruder lifting the rat out of the cage, and (d) the intruder repeatedly lifting the rat out of the cage (see text for details). For each experiment the effect on the temporal pattern of discharge is illustrated by a short record of the changing values of the modal intervals for trains of action potentials. The range of modal intervals when the rat was awake and relaxed (AR) is shown on each graph. Note that each novel experience induced a lengthening of the modal interspike interval, which is akin to an increase in level of arousal compared with the control period. The most profound effect is seen when an intruder lifts the animal from the cage but the animal seems to habituate to the experience. After nine trials of lifting the animal from the cage, the effect on the modal interval is minimal.

148

progressively less pronounced and the scatter about the mode more variable. As the animals settled down on the lap of the intruder, the value of the modal interval started to fall toward control values at rest. Again, the speed with which these shifts occurred depended on the species (rats accommodated more quickly than rabbits) and varied among animals. It even varied among different intruders working with the same animals. Eventually after at least 10 trials, the intruder could lift an animal out of the run and hold it on his or her lap without inducing a major shift in either modal interval or scatter about the mode.

These data imply that the environment of the experimental animal can have profound affects on the activity of cortical neurons and presumably therefore on the level of arousal of the animal. Consider the implications of these findings for experiments on conscious animals. Simply taking an animal from its run, despite the close relationship that often develops between these experimental animals and the experimenter, has substantial effects on the ongoing discharge of neurons recorded from sites all over the cerebral cortex. Moreover, the same procedure carried out by an unfamiliar person induces changes in the temporal pattern of discharge of these neurons that almost parallels some of the most severe changes seen in alarm. Worse still, consider some of the manipulations that are carried out on conscious animals, which are often restrained. What effects must these procedures have on cortical neuron discharge? If the effects are so profound and widespread, what distortions of normal physiology might be wrought under these conditions?

Changes in temporal discharge pattern of cortical cells and the experimenter

The human ear is particularly acute at detecting changes in patterns – more effective in detecting change than the eyes. With limited practice, it is quite possible to identify quite subtle changes in the temporal pattern of discharge of cortical cells simply by listening to the rhythm of discharges. Once the characteristic patterns that identify low and high levels of arousal were keyed into the human brain, we noticed that it became difficult for the experimenter to divorce these auditory cues from the course of the experiment. For example, when we became aware of sudden switches in the pattern of firing, we often found ourselves checking on the animal. If the modal interval increased, we would wonder why the level of arousal had increased. Commonly we associated these flutterings of higher level of arousal with our behavior in the laboratory – for example, we had shuffled papers, or turned the page of a file too loudly (rats are particularly sensitive to rustling sounds), or the animal detected an unusual sound from outside the room. It was also possible to detect the opposite effect. During a series of experiments, in which we demanded attention from rats who were trained to watch a series of rotating patterns of light, we started to observe changes in level of arousal. The rats were rewarded with food treats when they performed the task correctly. As they approached satiety, their level of attention fell and there were clearly detectable changes in the temporal pattern of firing of control neurons in the cerebral cortex. It was tempting to make a loud noise in the laboratory and wake up the rats to restore their interest. The reports of these affects on the experimenter are anecdotal but nonetheless powerful. The inevitable bond was complete, and we were under the spell of the animals in our charge.

Acknowledgments

The author is grateful to Alison Webb and Ben Delisle Burns for their support and encouragement; Andrew Paisley, Sarah Slaney, and Allison Emerton for help in the laboratory; and Dorothy Pharoah for help in preparing the manuscript. The work was supported by the MRC & AFRC (United Kingdom).

References

Burns, B. D., Stean, J. P. B., & Webb, A. C. (1974). Recording for several days from single cortical neurons in completely unrestrained cats. *Electroencephalography and Clinical Neurophysiology 36*: 314–18.
   (1979). The effects of sleep on neurons in isolated cerebral cortex. *Proceedings of the Royal Society of London, Series B 206*: 281–91.
Burns, B. D., & Webb, A. C. (1976). The spontaneous activity of neurons in the cat's cerebral cortex. *Proceedings of the Royal Society of London, Series B 194*: 211–23.
Evarts, E. V. (1962). Activity of neurons in visual cortex of the cat during sleep with low voltage fast EEG activity. *Journal of Neurophysiology 25*: 812–16.
   (1964). Temporal patterns of discharge of pyramidal tracts neurones during sleep and waking in the monkey. *Journal of Neurophysiology 27*: 152–71.
Foster, R. G. (1984). An investigation into the extraretinal photoreceptors mediating photoperiodic induction in Japanese quail (*Coturnix coturnix japonica*). Unpublished PhD. dissertation, University of Bristol.
Foster, R. G., Paisley, A. C., & Summerlee, A. J. S. (1982). The effects of arousal on the spontaneous activity of single neurones in the tectum of freely-moving quail (*Coturnix coturnix japonica*). *Journal of Physiology 327*: 5P.
Hobson J. A. (1969). Sleep: Physiologic aspects. *New England Journal of Medicine 281*: 1343–5.

Hobson, J. A., & McCarley, R. W. (1971). Cortical unit activity in sleep and waking. *Electroencephalography and Clinical Neurophysiology 30*: 97–112.
Jouvet, M. (1967). Neurophysiology of the states of sleep. *Physiological Reviews 47*: 117–77.
Klemm, W. R., & Sherry, C. J. (1981). Serial ordering in spike trains: What's it "trying to tell us"? *International Journal of Neuroscience 14*: 15–33.
   (1982). Do neurons process information by relative intervals in spike trains? *Neuroscience Biobehavioural Reviews 6*: 429–38.
Lincoln, D. W. (1974). Suckling: A time constant in the nursing behaviour of the rabbit. *Physiology & Behaviour 13*: 711–14.
Lincoln, D. W., Hentzen, K., Hin, T., van der Schoot, P., Clarke, G., & Summerlee, A. J. S. (1980). Sleep: A prerequisite for reflex milk-ejection in the rat. *Experimental Brain Research 38*: 151–62.
Lomo, T., & Mollica, A. (1962). Activity of single units in the primary optic cortex in unanaesthetized rabbit during visual, acoustic, olfactory and painful stimuli. *Archives of Italian Biology 101*: 86–102.
Neve, H. A., Pasiley, A. C., & Summerlee, A. J. S. (1982). Arousal: A prerequisite for suckling in the conscious rabbit? *Physiology and Behaviour 28*: 213–17.
Newton, M., & Newton, N. R. (1948). The let-down reflex in human lactation. *Journal of Paediatrics 33*: 698–704.
Noda, H., & Adey, W. R. (1970). Firing variability in cat association cortex during sleep and wakefulness. *Brain Research 18*: 513–26.
   (1973). Neuronal activity in association cortex of the cat during sleep, wakefulness and anaesthesia. *Brain Research 54*: 243–59.
O'Byrne, K. T., Ring, J. P. G., & Summerlee, A. J. S. (1986). Plasma oxytocin and oxytocin neurone activity during delivery in rabbits. *Journal of Physiology 370*: 501–13.
Paisley, A. C., & Summerlee, A. J. S. (1983). Suckling and arousal in the rabbit: Activity of neurons in the cerebral cortex. *Physiology and Behaviour 31*: 471–5.
   (1984a). Relationships between behavioural states and activity of the cerebral cortex. *Progress in Neurobiology 22*: 155–84.
   (1984b). Activity of putative oxytocin neurons during reflex milk ejection in conscious rabbits. *Journal of Physiology 347*: 465–78.
Poulain, D. A., Rodriguez, F., & Ellendorff, F. (1981). Sleep is not a prerequisite for the milk-ejection reflex in the pig. *Experimental Brain Research 43*: 107–110.
Slaney, S. F., & Summerlee, A. J. S. (1983). A simple description of reticular neuron activity in unanaesthetized rabbits. *Neuroscience Letters 14*: S346.
Steriade, M., Deschenes, M., & Oakson, G. (1974). Inhibitory processes and interneuronal apparatus in motor cortex during sleep and waking: I. Background firing and responsiveness of pyramidal tract neurons and interneurons. *Journal of Neurophysiology 37*: 1065–92.
Summerlee, A. J. S. (1981). An electrophysiological study of the functions of the oxytocin-producing hypothalamic cells in the conscious rat. Unpublished PhD dissertation. University of Bristol.
Summerlee, A. J. S., & Paisley, A. C. (1982). The effect of behavioural arousal upon the spontaneous activity of hypothalamic neurons in unanaesthetized, freely-moving rats and rabbits. *Proceedings of the Royal Society of London, Series B 214*: 263–74.
Trulson, M. E., & Jacobs, B. I. (1979). Raphe unit activity in freely moving cats: Correlation with level of behavioural arousal. *Brain Research 163*: 135–50.

Webb, A. C. (1976a). The effect of changing levels of arousal on the spontaneous activity of cortical neurons: I. Sleep and wakefulness. *Proceedings of the Royal Society of London, Series B 194*: 225–37.

——— (1976b). The effect of changing levels of arousal on the spontaneous activity of cortical neurons: II. Relaxation and alarm. *Proceedings of the Royal Society of London, Series B 194*: 239–51.

—9—

Practice makes predictable: the differential effect of repeated sampling on behavioral and physiological responses in monkeys

Maria L. Boccia, Christy Broussard, James Scanlan, and Mark L. Laudenslager

Editor's introduction

Boccia et al. describe a research program with monkeys in which they have taken major steps to standardize their techniques and eliminate the confounding influences of human contact. The primary focus of this research is the effects of the subjects' expectations on endocrine and immunological responses. Despite all their efforts, Boccia et al. still report that experience with the researchers' regime has unmistakable effects on their subjects. This chapter raises the question of how such a research setting might realistically minimize the effects of the scientist–animal bond on the research outcome and whether such a goal is possible.

Introduction: handling effects

A good research design controls extraneous variables so that the strongest possible inference can be made that differences in outcome are produced by the independent variables only. The difficulty has always been to identify all the extraneous variables that affect the dependent variables. Even minimal components of experimenter–subject interactions that are necessary for data collection (e.g., moving an animal to a test chamber or collecting a blood sample) can alter the results. This phenomenon is also known as "handling effects."

153

Handling effects have been studied extensively and have been found to affect a wide variety of behavioral and physiological responses in many different species, including rats and mice (e.g., Bodnoff, Suranyi-Cadotte, Quirion, & Meaney, 1987; Boix, Teruel, & Tobena, 1989; Denenberg & Whimbey, 1963; Doty, 1968; Joffe & Levine 1973), monkeys (e.g., Elvidge et al., 1976; Heath, 1989; Tarantino, 1970), dogs (Lynch & Gantt, 1968; Lynch & McCarthy, 1967; Thomas, Murphree, & Newton, 1972), chickens (Gross, 1980), pigs (Hemsworth, Barnett, & Hansen, 1981), and even moose (Franzmann, Flynn, & Arneson, 1975).

The most extensive studies of handling have been conducted with mice and rats, examining both behavioral and physiological effects of early (neonatal) and later (adulthood) handling. For example, these manipulations affect both basal and stressed corticosterone levels. The immediate response to initial exposure to novel stimuli is an elevation of corticosterone levels (e.g., Hennessy & Levine, 1978), but early handling (first 20 to 30 days of life) results in lower corticosteroid responses to exposure to novel stimuli in adulthood (e.g., Joffe & Levine, 1973; Levine, 1967). Because the neuroendocrine system (in this case particularly the hypothalamic–pituitary–adrenal system) interacts with many other physiological systems, including other aspects of the nervous system as well as the immune system (see Bateman, et al., 1989, for review), these handling effects can alter the outcome of research in a variety of areas, in some cases even to the point of affecting the survival of subjects in certain experimental procedures (e.g., Sachs & Lumia, 1981).

Overview of the immune system

Of particular concern to our own research program, evidence has been accumulating that handling can have an impact on immune function. The immune system is composed of a number of different components, which together function to (1) recognize anything foreign or not part of the organism ("non-self") and (2) destroy and eliminate it from the system (for elaboration see Laudenslager, 1988; Maier, Laudenslager, & Ryan, 1985). Relevant to the measures discussed here is the function of the lymphocytes, which comprise approximately 20% to 45% of the total white blood cells.

The two major classes of lymphocytes are T-cells and B-cells. The B-cells produce antibodies, also called immunoglobulins, which are complex protein molecules able to bind to specific foreign proteins such as those found on the surfaces of bacteria or viruses. T-cells also recognize foreign proteins, or antigens, and are involved in the immune response at several levels, such as regulation, recognition, and elimination of foreign material.

154

There are several types of T-cells, including helper cells, which enhance antibody production; suppressor cells, which suppress helper cells; and cytotoxic cells, which destroy target cells, such as viruses and malignant cells. There are two types of cytotoxic cells: natural killer cells and killer cells. Killer cells are produced after initial interaction with a novel antigen and are specific to that antigen. Natural killer cells are relatively non-specific, and will attack and destroy a wide variety of tumor cells, for example. Natural killer cells, therefore, constitute the first line of defense, while other aspects of the immune response become activated within 1 or 2 weeks.

The immune measures used in this study represent several aspects of the immune responses just described. Mitogens non-specifically stimulate lymphocytes to proliferate. Specific mitogens can differentially stimulate either T-cells (phytohemagglutinin and concanavalin A) or B-cells (poke-weed). Natural cytotoxicity tests assess the ability of natural killer cells to lyse, or destroy, foreign tumor cells. Differential cell counts reveal the proportions of the general classes of lymphocytes and other types of white blood cells, such as neutrophils, in peripheral blood. These procedures reflect the in vitro function of these cells. How in vitro function characterizes in vivo function is, however, far from clear.

Some precautions are necessary regarding the interpretation of studies examining the relationship between behavior and immune function. First, there is no one good measure of immunity. Multiple measures are necessary to sample the different aspects of immune function. Second, the decline of an immune measure in association with some experimental manipulation or event is not equivalent to a suppression of the immune system as a whole. It indicates that the immune system has been modified in some way. When only peripheral blood lymphocytes are assessed, it is not known what is happening to immunocompetent cells at other sites, such as lymph nodes, spleen, or bone marrow. This is not to say that significant suppression of the immune system does not occur in stressful situations. Rather, measures currently in use may not reflect the actual risk to the organism.

Whether one measures the proliferative responses of T- or B-cells to mitogen stimulation, antibody production, natural killer cell activity, incidence and development of mammary tumors in susceptible strains of mice, or other aspects of immunity or host defense, the immediate effect of handling is to lower the immune response (Moynihan, Koota, Brenner, Cohen, & Ader, 1989; Moynihan et al., 1990; Odio, Brodish, & Ricardo, 1987; Riley, 1975). Early handling, however, is correlated with higher antibody responses in the presence of stressors to which the animal is later exposed (Solomon, Levine, & Kraft, 1968).

155

Primate handling effects

The problems associated with the effect of experimental protocols and handling by experimenters become more acute when the subjects are nonhuman primates. Primates are highly intelligent, highly social species. Their phylogenetic proximity to human beings results not only in close physiological similarities to us but also close behavioral and social similarities. For example, human beings share many facial expressions with nonhuman primates (Chevalier-Skolnikoff, 1973; Darwin, 1872; van Hooff, 1962), and every researcher who has worked closely with primates has learned that they monitor and respond to one's facial expressions. Thus, nonhuman primates respond not only to what you do, but how you do it and what emotions you express while you're doing it! Furthermore, because they are long-lived (many macaque species can live an average of 20 to 25 years or more in captivity), they are often used repeatedly in many experiments, especially when those experiments involve noninvasive protocols. This increases the likelihood of handling effects emerging in studies of nonhuman primates.

The studies of early handling in primates have demonstrated effects on both immediate and long-term responses to novelty or stressors (e.g., Hillman, Fuselier, & Riopelle, 1973; Sackett, 1972). Later handling also affects primate physiological responses. Elvidge et al. (1976), for example, compared cortisol measures in rhesus macaques who had been habituated to being placed in a squeeze cage and having blood samples taken, with unexperienced, awake animals placed in a squeeze cage for sampling and unexperienced, anesthetized animals. The habituated animals had lower basal cortisol than the other groups. In this case, incorporating habituation trials into the experimental protocol reduced the glucocorticoid response to the procedure. In contrast, Adams et al. (1988) found that a procedure often used to reduce animal handling in certain types of experimental protocols (tethering with chronic catheterization) actually resulted in chronic sympathetic nervous system arousal.

Normal colony management, as well as the implementation of experimental protocols, can affect the outcomes of research. Hill, Greer, and Felsenfeld (1967) simultaneously manipulated a number of stimuli that occur routinely in primate laboratories, including type of housing, sensory (lights, noise) and physical stimuli (dropping the animals' cages), and feeding time. They found that this manipulation affected both cortisol and antibody production. Cortisol levels were elevated over those of control subjects in the first 2 weeks of stimulus presentation, and antibody production was measurably different by the fifth week (lower than controls). This is a particularly significant result because all of these features are

156

probably quite common in most primate laboratories, suggesting significant consequences for research results.

In our laboratory, as in others, we have made it a point to minimize the human–animal interactions, limiting those interactions to what is necessary for caretaking, veterinary care, and the necessary experimental manipulations (e.g., Reite, Short, Seiler, & Pauley, 1981). We have assumed that by minimizing interactions, we optimize experimental conditions by eliminating confounding handling effects. Even so, we have noticed that the monkeys respond to changes in necessary handling procedures and personnel. For example, we have noted that they respond differentially to male versus female caretakers and experimenters. We have also noted that when a new person comes to work in the lab, the animals respond differently (usually with greater agitation and difficulty in handling), until they adjust to the new person.

In the context of a series of studies examining the impact of dominance status on cortisol and immune responses in macaques (Boccia, Laudenslager, Broussard, & Hijazi, 1989; Laudenslager, Boccia, & Held, 1988), we had the opportunity to examine the effect of repeated handling on cortisol and immune measures, as well as the relationship of these responses to handling with dominance status, age, and the stressors we have studied.

The Colorado Psychiatric Hospital Primate Laboratory environment

In the CPH Primate Laboratory, we maintain two species of macaques, *Macaca nemestrina* (the pigtail macaque) and *M. radiata* (the bonnet macaque). Virtually all of our animals are socially housed in harem groups containing one adult male, six or seven adult females, and a variable number of juveniles and infants. These groups live in large pens (2.1 × 2.5 × 4 m) with glazed cinderblock walls, and shelves and pipes providing additional climbing and sitting space. Animals are fed fruit and standard laboratory primate chow once daily and have continuous access to water through a standard watering system. The animals in the experiments described herein were born and raised in this colony, in the above-described social groups, similar to the ones in which they currently reside.

Until recently, the primary research of this laboratory has been the study of mother–infant separation (e.g., Boccia et al., 1989; Laudenslager et al., 1990; Reite, 1977; Reite et al., 1981). As a consequence of this, the adult females in this colony have been repeatedly used in experimental protocols involving their infants. Some have been used in separation studies as many as four or five times (unpublished data). Some of these studies have included the biweekly separation of these mother–infant dyads for the purpose of obtaining a blood sample from the infant. We noted that the subjects seemed to adapt to this procedure, so that sampling became faster and more efficient over the course of the experiments. The data presented here are the results of a systematic examination of this anecdotal observation.

The procedures used in each of these experiments are standard in our laboratory, and most of the procedures are also used during routine husbandry and caretaking (see Figures 9.1a–e). Animals are first transferred to a "gang cage" (roughly 0.5 × 1 × 1 m). The experimenter enters the pen, via a door, and the animals run (or, in many cases, walk) out through a small guillotine door into the gang cage. The animals have experienced this procedure for years and typically run out with minimal encouragement from the experimenter. A small, individual transport box is then placed between the gang cage and the pen's guillotine door. The animals then run or walk, one by one, into the transport box. Those who will not be used are immediately released back into their home pen, and those who will be used are transferred to an individual restraint cage for sampling.

In most mother–infant separation studies, only two mother–infant dyads at most are sampled from any one group, and they can typically be isolated and sampled within 15 minutes of the time the experimenter first enters the pen. The male is usually the first one out, but while we had noticed that the females tended to come out of the gang cage in a fairly consistent order, we had not examined this systematically.

Experimental protocol

We examined cortisol and immune measures in the adult females in six groups of monkeys (three bonnet [total N = 17] and three pigtail [total N = 21] groups) in response to a mild stressor (see dominance test described later). We collected blood samples from each female by following the procedure outlined earlier. In the present study, all the adult females in the group were sampled. Once the female was in the restraint cage, she was restrained and a 3- to 5-ml blood sample was obtained from the saphenous vein. The group remained in the gang cage while the sampling procedure was in progress. Each female underwent venipuncture, was offered some fruit, and then returned to her home cage. The sampling was done by two experimenters per animal, two animals at a time, within 2 to 5 m of the gang cage. A stopwatch was started when the experimenter first entered the pen to chase the group into the gang cage, and the time at which each blood sample was collected was recorded. Blood samples were immediately placed on ice and transported to the laboratory for analysis.

158

Each sample of blood was centrifuged at 300 g for 15 minutes to separate the cells, and the serum was frozen at $-20°C$ for later cortisol analysis. A series of immune assays were then conducted, including B-cell and T-cell responses to mitogen stimulation (pokeweed [PWM] for B-cells, and phytohemagglutinin [PHA] and concanavalin A [ConA] for T-cells), natural cytotoxicity, differential cell counts, and total white blood cell counts (WBC) (see Laudenslager et al., 1990, for details).

Cortisol was assayed by routine radioimmunoassay using rabbit antisera (kindly supplied by Steven Calvano of Cornell Medical College). One hundred microliters unextracted plasma, diluted 1:200, and $100\,\mu l$ of ^{3}H-labeled cortisol tracer were heat-inactivated at 80°C for 30 minutes. The tracer and plasma were incubated with 100 μl antisera diluted 1:4,000 for 2 hours at room temperature and overnight at 5°C. Unbound tracer was brought down with dextran-coated charcoal. The supernatant was decanted into plastic scintillation vials and counted in Optifluor on a liquid scintillation counter.

Each group was dominance-tested. This involved turning off the water supply for 24 hours, then restoring it, and monitoring each animal's time to first drink for more than 5 seconds and its displacement from the water spout. This method of testing has been in use in our laboratory for more than 10 years and has consistently produced reliable dominance orderings that replicate across tests and are correlated with the direction of aggressive behaviors such as threat (+.35, $N = 39, p = .$ 005), bite (+.39, $N = 39$, $p = .008$), and displace (+.41, $N = 39, p = .005$) at times outside the dominance testing period.

Blood samples were collected for two baselines before the test, separated by a week; another sample was taken 48 hours after water was restored; and two more posttest samples were taken at 1 and 2 weeks after the water deprivation interval. Thus, there were five samples, at approximately 1-week intervals, with the dominance test occurring roughly in the middle of the series.

We examined the relationship between the dominance status and age of the female, and time taken to collect her blood sample, and cortisol and immune responses. The animals ranged in age from 2.5 years to 18 years, with a mean of 11.1 years. The youngest animals of both species had never been in any experiments, and the oldest were highly experienced.

Time to collect blood samples, age, and outcomes

The longest total time to collect a blood sample from an individual was 84 minutes, the shortest time was 4.0 minutes, and the mean was 25.28 minutes. These intervals include time the animal spent in the gang cage

(a)

(b)

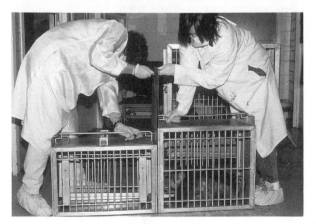

(c)

160

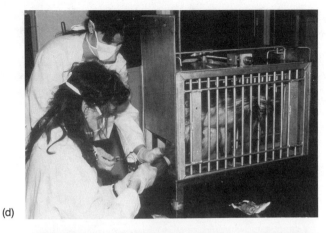

(d)

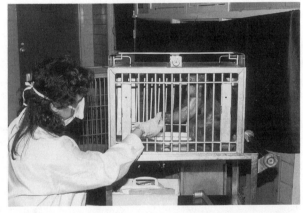

(e)

Figure 9.1. (a) Typical monkey living pen at CPH Primate Laboratory. (b) Monkeys moving from gang cage to individual transport box for separation of monkeys for sampling. (c) Monkey moved to restraint cage for sampling. (d) Monkey restrained for sampling. (e) Monkey receiving fruit reward.

waiting to be bled. There was no change in the speed of sampling or time to the last sample over the course of the experiment. Times for the first sample averaged 26.9 minutes, with a range of 5.0 to 76.0, and times for the last sample averaged 23.0 minutes, with a range of 4.0 to 54.0. Thus, the changes seen were not due to gross changes in absolute time to sample, as the experiment progressed. The absolute sequence in which the individual monkeys exited the gang cage was fairly consistent from one occasion to

Table 9.1. *Mean sequence orders in which individual female pigtail macaques in each group exited the gang cage for blood sampling, with results of Friedman ANOVA*

Group 1			Group 2			Group 3		
Age	Rank	Sequence	Age	Rank	Sequence	Age	Rank	Sequence
9.62	7	1.4	9.83	1	1.8	5.74	6	2.1
10.98	2	2.4	16.81	7	2.4	15.98	7	2.7
11.34	1	3.9	4.51	4	2.6	16.81	3	3.0
10.01	3	4.1	10.48	2	3.6	17.89	1	4.8
9.92	4	4.4	5.28	5	5.6	4.67	2	4.9
11.45	6	5.7	4.94	6	5.8	12.57	5	4.9
5.17	5	6.1	9.88	3	6.2	9.89	4	5.6
X^2		18.00			21.60			11.91
d.f.		6			6			6
p		.006			.001			.064

Table 9.2. *Mean sequence orders in which individual female bonnet macaques in each group exited the gang cage for blood sampling, with results of Friedman ANOVA*

Group 1			Group 2			Group 3		
Age	Rank	Sequence	Age	Rank	Sequence	Age	Rank	Sequence
17.71	6	1.25	17.57	2	1.4	17.02	3	1.8
5.23	4	1.75	17.60	1	2.8	6.36	6	2.3
15.54	5	3.25	16.32	3	3.8	15.67	4	2.4
13.91	1	3.75	17.36	4	3.8	17.25	1	4.4
2.35	3	5.25	6.44	5	4.2	5.25	5	4.5
4.53	2	5.75	17.79	6	5.0	3.56	2	5.6
X^2		18.71			11.17			16.80
d.f.		5			5			5
p		.002			0.48			.005

another, as indicated by a nonparametric Friedman ANOVA for each group (see Tables 9.1 and 9.2 for data on pigtail and bonnet monkeys, respectively).

The relationship between age and time to collect blood samples was related to the repetition of the sampling procedure (see Tables 9.1, 9.2,

Table 9.3. *Pearson correlations between age of the monkeys and time taken to collect blood over five successive samples*

Sample	All monkeys	Pigtails only	Bonnets only
1	− .145	− .404*	− .018
2	− .244	− .199	− .325
3	− .285[++]	− .196	− .394[+]
4	− .289[++]	− .097	− .462*
5	− .462**	− .307	− .592**

Significance levels: [+] $p < .10$, [++] $p < .07$, *$p < .05$, **$p < .01$.

Table 9.4. *Correlation between age of monkeys and cortisol levels*

Sample	All monkeys	Pigtails only	Bonnets only
1	− .225	− .018	− .241
2	− .261[+]	− .153	− .360
3	− .319*	− .024	− .580**
4	− .327*	− .117	− .655**
5	− .400**	− .222	− .662**

Significance levels as in Table 9.3.

and 9.3). Initially there was no correlation between the age of the monkey and the time the blood sample was taken. However, over repeated samplings, these correlations became larger and larger, and highly significant. When the species were treated separately, this effect was due primarily to the change in the behavior of the bonnet macaques. There was no difference in the range of ages between the species, and this result must have been a consequence of differences in their behavior.

Cortisol measures were also highly correlated with both age and time to bleed (see Tables 9.4 and 9.5 and Figure 9.2). Although these two variables are obviously related, they do not account for all these effects, as cortisol is significantly correlated with time to bleed during the first two samples, when age is not correlated with time to bleed. Although the correlations with time to bleed and cortisol were present in both bonnets and pigtails, the relationship between age and cortisol was present only in the bonnet macaques.

Interestingly, examination of the relationship between cortisol and the immune panel indicates that there were no significant correlations between

Table 9.5. *Correlations between time taken to collect blood over successive samples and cortisol levels*

Sample	All monkeys	Pigtails only	Bonnets only
1	+.362	+.210	+.742**
2	+.489**	+.709**	+.283
3	+.246	+.226	+.383
4	+.292+	+.305	+.310
5	+.422**	+.459*	+.506**

Significance levels as in Table 9.3.

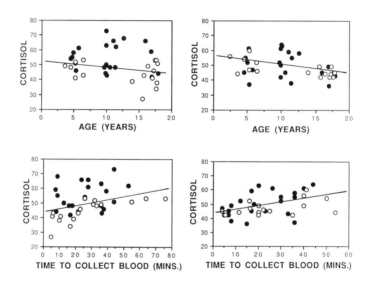

Figure 9.2. Scatterplot with regression lines for relationship between cortisol and time to bleed (lower frames) and age (upper frames) for first and last sample. Left frames: first sample; right frames, last sample. Closed circles, pigtail monkeys; open circles, bonnet monkeys.

them. Although cortisol administered exogenously has been shown to alter immune function (Cohen & Crnic, 1984), the acute, short-term elevations in cortisol seen here were unrelated to lymphocyte activation, natural cytotoxicity, differential cell counts, or total WBC. There were, however, significant correlations between the age of the monkeys and these immune measures during the baseline period (mean of the first two blood samples) (see Table 9.6 and Figure 9.3). Age was negatively correlated with both

164

Table 9.6. *Correlations between age and immune measures*

Immune measure	All monkeys	Pigtails only	Bonnets only
Mitogen responses			
PHA 1[a]	− .322*	− .535**	− .521*
2.5	− .320*	− .432*	− .515*
5	− .343*	− .345	− .498*
ConA 1[a]	− .260	− .566**	− .357
5	− .252	− .499*	− .434
7.5	− .215	− .310	− .424[++]
10	− .161	− .210	− .379
PWM 10[a]	− .391**	− .477*	− .422[+]
20	− .484**	− .460*	− .577**
30	− .492**	− .414	− .741**
Cytotoxicity [b]			
Lytic units	− .236	− .371[+]	+ .267
LYM	− .008	− .290	+ .040
SEG	− .025	− .290	− .006
WBC	− .321*	− .280	− .655**

Abbreviations: PHA, phytohemagglutinin; Con A, concanavalin A; PWM, pokeweed; LYM, lymphocytes; SEG, segmented neutrophils; WBC, white blood cells.
[a] Numbers indicate mitogen concentrations in the assay.
[b] Results from lytic unit analysis where the effector–target ratios required to produce 30% lysis were calculated.
Significance levels as in Table 9.3.

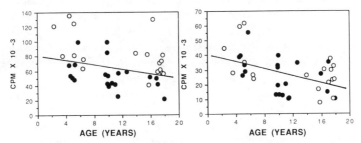

Figure 9.3. Scatterplot with regression lines for relationship between age and B- and T-cell mitogens. Left: Response to phytohemagglutinin; right, response to pokeweed; closed circles, pigtail monkeys; open circles, bonnet monkeys.

B-cell and T-cell responses to mitogen stimulation, and total WBC, indicating that all these immune functions were lower in older monkeys than younger monkeys. Furthermore, while the mitogen effects were present in both the bonnet and the pigtail monkeys, the effects on cytotoxicity were limited to pigtails only and the effects on WBC were limited to bonnets.

To summarize, monkeys that are older and more experienced with experimental protocols respond more quickly over the course of an experiment than do younger, less experienced monkeys. This affects the order in which the animals come out of the gang cage for sampling. Interestingly, the dominance status of the monkey, which we expected to influence order out of the cage, did not affect this behavior. The physiological measures (cortisol levels in particular) were effected by this self-sorting by the monkeys, in that cortisol was strongly correlated with time to bleed at most stages of the experiment.

It was reassuring, however, that the acute immune measures in which we were primarily interested in the studies relating dominance, stress, and immunity were independent of the cortisol levels at the time of blood sampling. There was the expected negative correlation with age (with older animals showing lower immune responses than younger), but no relationship between immunity and the self-sorting of the monkeys consequent to the experimental protocol.

Conclusions

Monkeys that are older and more experienced with laboratory procedures adapt to new experimental protocols more quickly than younger, less experienced monkeys. Consequently, over the course of an experiment the monkeys engage in self-sorting, whereby over a series of samples, the older animals are consistently sampled sooner in the session than the younger ones. Given the known effect of handling time on a variety of physiological measures, as well as the effects of age on these same measures, this self-sorting represents a significant confound in experiments that require repeated sampling of a group of monkeys of mixed ages.

Dominance status was unrelated to age in these monkeys, as young adult daughters of dominant females attained similar high ranks in their social groups. The observation that dominance status did not control the self-selection process suggests that it is indeed the older monkeys' experience and expectations related to handling and experimental protocols that produced the above-described effects. In a stable situation (where group membership has remained constant for at least 6 months and no other manipulations of the groups have occurred), dominance status also did not affect the physiological measures studied here (Boccia et al., 1989). Dominance status appears to influence these measures only in response to a stressor. Thus, data from the sample 48 hours after the dominance test (where the age effects are noticeably absent) suggest that this is the time at which dominance influences immunity.

Age effects on immune function are well known (Erschler et al., 1988; Short, England, Bridson, & Bowden, 1989; for review, see Makinodan & Kay, 1980). On most measures of immune function, including the ones described here, older animals often show lower responses than younger monkeys. Odio et al. (1987) found that older rats, which already have lowered immune responses, do not show an additional lowering of response to a stressor. We have found a similar effect in these monkeys (unpublished data). Age effects on cortisol are also known for other species of macaques (e.g., Scanlan, 1984; Suomi et al., 1989).

An interesting species difference was also found in this study: Bonnet monkeys showed more significant correlations between age and both time to bleed and cortisol measures. Most experimenters in our lab would classify pigtail monkeys as more stress responsive, behaviorally, than bonnet monkeys during blood-sampling procedures, and pigtail cortisol measures are elevated at the earliest sampling times in this study. Pigtails appear not to adapt readily to the experimental protocol. Bonnet monkeys, especially the older, more experienced monkeys, adapted more quickly to the protocol. The differences between the older and the younger bonnet monkeys, which have not been in enough experiments to habituate to handling procedures, would be more dramatic than between young and old pigtails, all of which are more reactive.

As already noted, in our laboratory, we have made it a point to minimize the human–animal interactions, assuming that by minimizing interactions, we optimize experimental conditions by eliminating confounding handling effects. The results presented here, however, suggest that it may be necessary to reverse this thinking (see Wolfle, 1985).

Minimization of the confounding effects of handling must take into account that the monkeys develop expectations about experimental protocols based on their prior experience. It is now well known that the predictability of noxious stimuli (which, it must be admitted, many experimental procedures are) affects the behavior, endocrine, and immune responses to those stimuli (Maier et al., 1985; Mineka & Hendersen, 1985). Since primates live a long time and participate in many experiments, the development of expectations to predictable protocols varies with the age and experience of the individual. To minimize these confounding effects on research outcomes, it would be necessary to have a sufficiently long series of habituation trials (which would vary with the species involved) before the beginning of each experiment. The changes in behavior resulting from their experience would then be complete before the onset of the experimental manipulation.

In conclusion, these results indicate that monkeys are affected by handling, in both their behavioral and physiological responses. Older, more experienced monkeys adapt more quickly and change their behavior more in response to repeated experimental manipulations than do younger, less experienced monkeys. These differences affected some but not all physiological measures: Cortisol but not immune measures varied with time taken to collect the sample. Finally, these effects vary with the species under study, so that any considerations for particular experiments must be based on results with the particular species being studied.

Acknowledgments

This research was supported by USPHS Grants MH44131 (M.L.B.) and MH37373 (M.L.L.). We thank Amal Hijazi and S. Blane Durrance for their assistance with blood sample collection.

References

Adams, M. R., Kaplan, J. R., Manuck, S. B., Uberseder, B. & Larkin, K. T. Persistent sympathetic nervous system arousal associated with tethering in cynomolgus macaques. *Laboratory Animal Science 38*: 279–81.

Bateman, A., Singh, A., Kral, T. & Solomon, S. (1989). The immune–hypothalamic–pituitary–adrenal axis. *Endocrine Reviews 10*: 92.

Boccia, M. L., Laudenslater, M. L., Broussard, C., & Hijazi, A. (1989, August). Dominance, immune function, and the impact of stressors in socially housed macaques. Paper presented at the 21st International Ethological Conference, Utrecht, The Netherlands.

Boccia, M. L., Reite, M., Kaemingk, K., Held, P., & Laudenslager, M. (1989). Behavioral and autonomic responses to peer separation in pigtail macaque monkey infants. *Developmental Psychobiology 22*: 447–61.

Bodnoff, S. R., Suranyi-Cadotte, Quirion, R., & Meaney, M. J. (1987). Postnatal handling reduces novelty-induced fear and increases [³H]flunitrazepam binding in rat brain. *European Journal of Pharmacology 144*: 105–7.

Boix, F., Teruel, A. F., & Tobena, A. (1989). The anxiolytic action of benzodiazepines is not present in handling-habituated rats. *Pharmacology, Biochemistry and Behavior 31*: 541–6,

Chevalier-Skolnikoff, S. (1973). Facial expression of emotion in nonhuman primates. In *Darwin and Facial Expression: A Century of Research in Review*, P. Ekman (Ed.). New York: Academic Press, pp. 11–89.

Cohen, J. J., & Crnic, L. S. (1984). Behavior, stress, and lymphocyte recirculation. In *Stress, Aging, and Immunity*, E. L. Cooper (Ed.). New York: Dekker, pp. 61–91.

Darwin, C. (1872). *The Expression of the Emotions in Man and Animals*. London: John Murray.

Denenberg, V. H., & Whimbey, A. E. (1963). Infantile stimulation and animal husbandry: A methodological study. *Journal of Comparative and Physiological Psychology 56*: 877–8.

Doty, B. A. (1968). Effects of handling on learning in young and aged rats. *Journal of Gerontology 23:* 142–4.

Elvidge, H., Challis, J. R. G., Robinson, J. S., Roper, C., & Thorburn, G. D. (1976). Influence of handling and sedation on plasma cortisol in rhesus monkeys (*Macaca mulatta*). *Journal of Endocrinology 70*: 325–6.

Ershler, W. B., Coe, C. L., Gravenstein, S., Schultz, K. T., Kloop, R. G., Meyer, M., & Houser, W. D. (1988). Aging and immunity in nonhuman primates: I. Effects of age and gender on cellular immune function in rhesus monkeys (*Macaca mulatta*). *American Journal of Primatology 15*: 181–8.

Franzmann, A. W., Flynn, A., & Arneson, P. D. (1975). Serum cortisol levels relative to handling stress in Alaskan moose. *Canadian Journal of Zoology 53*: 1424–6.

Gross, W. B. (1980). The benefits of tender loving care. *International Journal of the Study of Animal Problems 1*: 147–8.

Heath, M. (1989). The training of cynomolgus monkeys and how the human/animal relationship improves with environmental and mental enrichment. *Animal Technology 40*: 11–22.

Hemsworth, P. H., Barnett, J. L., & Hansen, C. (1981). The influence of handling by humans on the behavior, growth, and corticosteroids in the juvenile female pig. *Hormones and Behavior 15*: 396–403.

Hennessy, M. B., & Levine, S. (1978). Sensitive pituitary–adrenal responsiveness to varying intensities of psychological stimulation. *Physiology and Behavior 21*: 295–7.

Hill, C. W., Greer, W. E., & Felsenfeld, O. (1967). Psychological stress, early response to foreign protein, and blood cortisol in vervets. *Psychosomatic Medicine 29*: 279–83.

Hillman, N. M., Fuselier, P. H., & Riopelle, A. J. (1973). Protein deprivation in primates: II. Effects of fondling on feeding behavior of discomforted infant rhesus monkeys. *Developmental Psychobiology 7*: 369–74.

Joffe, J. M., & Levine, S. (1973). Effects of weaning age and adult handling on avoidance conditioning, open-field behavior, and plasma corticosterone of adult rats. *Behavioral Biology 9*: 235–44.

Laudenslager, M. L. (1988). The psychobiology of loss: Lessons from humans and nonhuman primates. *Journal of Social Issues 44*: 19–36.

Laudenslager, M. L., Boccia, M., & Held, P. (1988). Dominance and immunity in nonhuman primates: Some pilot observations. *Society for Neurosciences Abstracts 14*: 1281.

Laudenslager, M. L., Held, P. E., Boccia, M. L., Reite M. L. & Cohen, J. J. (1990). Behavioral and immunological consequences of brief mother-infant separation: A species comparison. *Developmental Psychobiology 23*: 247–64.

Levine, S. (1967). Maternal and environmental influences on the adrenocortical response to stress in weanling rats. *Science 156*: 258–60.

Lynch, J. J., & Gantt, W. H. (1968). The heart rate component of the social reflex in dogs: The conditional effects of petting and person. *Conditional Reflex 3*: 69–80.

Lynch, J. J., & McCarthy, J. F. (1967). The effect of petting on a classically conditioned emotional response. *Behavior Research and Therapy 5*: 55–62.

Maier, S. F., Laudenslager, M. L., & Ryan, S. M. (1985). Stressor controllability, immune function, and endogenous opiates. In *Affect, Conditioning, and Cog-*

nition: Essays on the Determinants of Behavior, F. R. Brush & J. B. Overmier (Eds.). Hillsdale, NJ: Erlbaum, pp. 183–201.

Makinodan, T., & Kay, M. M. B. (1980). Age influence on the immune system. *Advances in Immunology 29*: 287–330.

Mineka, S., & Hendersen, R. W. (1985). Controlability and predictability in acquired motivation. *Annual Review of Psychology 36*: 495–529.

Moynihan, J., Brenner, G., Koota, D., Breneman, S., Cohen, N., & Ader, R. (1990). The effect of handling on immune function, spleen cell number, and lymphocyte subpopulations.

Moynihan, J., Koota, D., Brenner, G., Cohen, N., & Ader, R. (1989). Repeated intraperitoneal injections of saline attenuate the antibody response to a subsequent intraperitoneal injection of antigen. *Brain, Behavior, and Immunity 3*: 90–6.

Odio, M., Brodish, A., & Ricardo, M. J. (1987). Effects of immune responses by chronic stress are modulated by aging. *Brain, Behavior, and Immunity 1*: 204–15.

Reite, M. (1977). Maternal separation in monkey infants: A model of depression. In *Animal Models in Psychiatry and Neurology*, I. Hanin & E. Usdin (Eds.). Elmsford, NY: Pergamon, pp. 127–40.

Reite, M., Short, R., Seiler, C., & Pauley, D. J. (1981). Attachment, loss, and separation. *Journal of Child Psychiatry 22*: 141–69.

Riley, V. (1975). Mouse mammary tumors: Alteration of incidence as apparent function of stress. *Science 189*: 465–7.

Sachs, B. D., & Lumia, A. R. (1981). Is stress due to shipment of animals a confounding variable in developmental research? *Developmental Psychobiology 14*: 169–71.

Sackett, G. P. (1972). Exploratory behavior of rhesus monkeys as a function of rearing experience and sex. *Developmental Psychology 6*: 260–70.

Scanlan, J. M. (1984). Adrenocortical and behavioral responses to acute novel and stressful conditions: The influence of gonadal status, timecourse of response, age, and motor activity. Unpublished Master's thesis, University of Wisconsin.

Short, R., England, N., Bridson, W. E., & Bowden, D. M. (1989). Ovarian cyclicity, hormones, and behavior as markers of aging in female pigtailed macaques (*Macaca nemestrina*). *Journal of Gerontology 44*: B131–8.

Solomon, G. F., Levine, S., & Kraft, J. K. (1968). Early experience and immunity. *Nature 220*: 821–2.

Suomi, S. J., Scanlan, J. M., Rasmussen, K. L. R., Davidson, M., Boinski, S., Higley, J. D., & Marriott, B. (1989). Pituitary–adrenal response to capture in Cayo-derived M-troop rhesus monkeys. *Puerto Rico Health Sciences Journal 8*: 171–6.

Tarantino, S. J. (1970). Effect of cage confinement on social behavior in squirrel monkeys. *Psychonomic Science 20*: 294–5.

Thomas, E. J., Murphree, O. D., & Newton, J. E. O. (1972). Effect of person and environment on heart rates in two strains of pointer dogs. *Conditional Reflex 7*: 74–81.

van Hooff, J. A. R. A. M. (1962). Facial expression in higher primates. *Behaviour 8*: 97–125.

Wolfle, T. (1985) Laboratory animal technicians: Their role in stress reduction and human–companion animal bonding. *Veterinary Clinics of North America: Small Animal Practice 15*: 449–54.

—10—

Improved handling of experimental rhesus monkeys

Viktor Reinhardt

Editors' introduction

Like Boccia et al., Reinhardt works with a population of captive primates from which blood samples must be obtained. However, Reinhardt's chapter stands in marked contrast to that of Boccia et al. Both Reinhardt and Boccia et al. recognize the importance of the scientist–animal bond, but deal with it in different ways. Whereas Boccia et al. try to avoid the effects of the bond, Reinhardt attempts to exploit its benefits in order to expedite data collection and minimize handling-associated excitation.

When physical strength, heavy leather gloves, and/or mechanical squeeze devices are used to control experimental rhesus monkeys, the handling is associated with considerable commotion. Flight attempts, threatening behavior, overt aggressive defense behavior, struggling, crouching, fear grinning, and psychogenic diarrhea are common reactions to such procedures and are indicative of the restrained animals' intense aversion (Figure 10.1). Handling-associated excitation, furthermore, is likely to affect the basal physiology of the research subject and hence the validity of the research data collected on it.

In an attempt to minimize handling-associated excitation and thereby to enhance the quality of research conducted on rhesus monkeys, the following techniques were developed at the Wisconsin Regional Primate Research Center.

1. *Venipuncture in the home cage.* Research subjects are transferred to squeeze-back cages. After an adequate period of familiarization, they

171

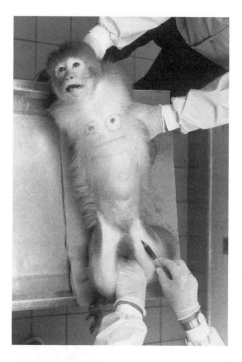

Figure 10.1. Conventional venipuncture techniques often require two or three caretakers to control the resisting experimental subject.

become used to having their cage space periodically reduced by about 75%. In this situation, they are trained to permit the caretaker to touch and groom them and finally to show no resistance to having a leg gently pulled out of the partially opened cage door for venipuncture (Figure 10.2). To avoid panic reactions, the experimental animal is not mechanically squeezed during this procedure.

2. *Intramuscular injection in the home cage.* Animals that cooperate during in-home-cage venipunture are readily conditioned to permit intramuscular injection without being mechanically squeezed (Figure 10.3a,b).

3. *Removal from home cage.* Laboratory rhesus monkeys quickly learn to enter transport cages on command when they have to be temporarily removed for treatment, examination, weighing, TB testing, or breeding, or when they are transferred from their home cage to another cage. No squeeze-backs are necessary to achieve this.

The training of experimental rhesus monkeys is based on positive reinforcement; cooperation is consistently rewarded with favored food such

172

Figure 10.2. In-home-cage venipuncture offers a well-controlled condition (the animal remains in its familiar environment and shows no signs of aversion) for the collection of scientific data. The experimental animal is not mechanically squeezed during this procedure. The animal's cooperation during blood collection guarantees that the results of the experiment will not be unnecessarily confounded by distress reactions (no significant cortisol response during this procedure).

as raisins, peanuts, fruit, or bread (Figure 10.3b). Failure to cooperate is not punished but is never allowed to end a training session, because if it were the animal would quickly learn to avoid any kind of discomfort associated with the handling procedure. Intimate knowledge of the animals, kindness, self-confidence, and patience are key attributes of the caretakers who train the animals at the Wisconsin Regional Primate Research Center. The time investment for the training is minimal (Vertein & Reinhardt, 1989) and quickly pays off in shorter handling times and a more controlled environment for collecting scientific data.

The impact of the training on the quality of data was evaluated for the in-home-cage venipuncture procedure. Significantly elevated peripheral

(a)

(b)

Figure 10.3. (a) An intramuscular injection does not have to be a disturbing event for an experimental monkey. There is no need to use physical strength, leather gloves, and/or mechanical squeeze devices to achieve this simple handling procedure. (b) Cooperation is consistently rewarded.

174

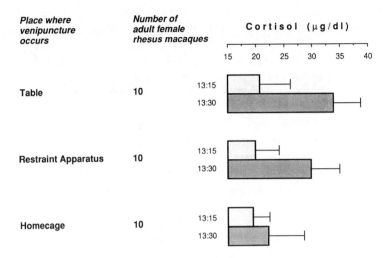

Figure 10.4. Serum cortisol response to different venipuncture procedures in adult female rhesus monkeys.

coritsol concentrations were taken as indicators of distress (Clarke, Mason, & Moberg, 1988; Line, Clarke, & Markowitz, 1987; Sassenrath, 1970; Selye, 1971; Tapp, Holaday, & Natelson, 1984; Udelsman & Chrousos, 1988). Cortisol values of 20 singly caged adult female rhesus monkeys routinely subjected to one of two conventional venipuncture techniques (animal is restrained on a table [Figure 10.1] or in a restraint apparatus) were compared with values of 10 singly caged adult female rhesus monkeys subjected to the improved in-home-cage venipunture technique. Each animal was bled by the attending caretaker at 13:15 (assessment of basal cortisol level) and again at 13:30 (assessment of the cortisol response to the preceding venipunture at 13:15). The actual venipuncture occurred for all animals 60 to 90 seconds after the caretaker had entered the animal room. Blood samples were analyzed with a Clinical Assays Gamma Coat Cortisol Kit (Dade, Baxter Travenol Diagnoistics, Cambridge, Mass.)

The three categories of animals did not differ in their mean basal cortisol levels at 13:15 (p always >.1; Figure 10.4). At 13:30, however, the mean cortisol concentration of animals bled on the table was significantly higher than that of animals bled in the restraint apparatus ($p < .1$; Figure 10.4) or in the home cage ($p < .001$; Figure 10.4). Being manually immobilized on a table is obviously particularly distressing for rhesus monkeys.

All three categories of test subjects showed a coritsol response to blood collection. The magnitude of cortisol increase, however, was significant only in animals venipunctured on the table ($+63\%$, $p < .001$) or in the

175

restraint apparatus ($+50\%$, $p < .001$), but not in animals venipunctured in their home cage ($+16\%$, $p > 0.1$; Figure 10.4). A lack of a significant cortisol response to in-home-cage venipuncture in adult female rhesus monkeys was also reported by Line and collaborators (1987). Apparently, venipunture per se need not be a distressing experience for a rhesus monkey provided that the animal is not removed from its familiar home cage for this procedure.

The present findings indicate that training rhesus monkeys to cooperate during in-home-cage venipuncture is a simple way to increase the validity of research data collected because it helps to avoid undue excitation and associated alteration in basal physiology of the research subjects.

With the refined techniques described here, an experimental monkey can easily be handled by one caretaker, whereas conventional techniques usually require two or three caretakers to control the resisting animal. Once trained, the animals cooperate with any familiar or unfamiliar person who is experienced in working with rhesus monkeys with gentleness and firmness.

The improved handling techniques are based on the experimental subject's cooperation; as such, they are safer for the animal caretaker than the conventional procedures because the monkey no longer has to be forcefully subdued and hence is not unpredictable and ready to defend itself by biting and scratching whenever it can.

Acknowledgments

I am very grateful to Mr. Doug Cowley, Mr. Steve Eisele, and Mr. Russell Vertein for training the animals in their charge to cooperate during routine handling procedures. Thanks are also due to Mrs. Joan Scheffler for analyzing blood samples for cortisol and to Dr. Bill Bridson, Dr. Dan Houser, and Mr. John Wolf for providing valuable comments on this manuscript. The training program is supported by NIH Grant RR00167 to the Wisconsin Regional Primate Research Center.

References

Clarke, A. S., Mason, W., & Moberg, G. P. 1988. Differential behavioral and adrenocortical responses to stress among three macaque species. *American Journal of Primatology 14*, 37–52.

Line, S. W., Clarke, A. S., & Markowitz, H. 1987. Plasma cortisol of female rhesus monkeys in response to acute restraint. *Laboratory Primate Newsletter 26*(4), 1–4.

Sassenrath, E. N., 1970. Increased responsiveness related to social stress in rhesus monkeys. *Hormones and Behavior 1*, 283–98.

Selye, H. 1971. *Hormones and Resistance*. Springer, New York.
Tapp, W. N., Holaday, J. W., & Natelson, B. H. 1984. Ultradian glucocorticoid rhythms in monkeys and rats continue during stress. *American Journal of Physiology 247*, 866–871.
Udelsman, R., & Chrousos, G. P. 1988. Hormonal responses to surgical stress. In G. P. Chrousos, D. L. Gold, & P. W. Gold (eds.), *Mechanisms of Physical and Emotional Stress*, pp. 265–72. Plenum, New York.
Vertein, R., & Reinhardt, V. 1989. Training female rhesus monkeys to cooperate during in-homecage venipuncture. *Laboratory Primate Newsletter 28*(2), 1–3.

—11—

Social interaction as a condition for learning in avian species: a synthesis of the disciplines of ethology and psychology

Irene M. Pepperberg

Editors' introduction

Pepperberg's research program, which has evolved over the past decade, has been unique in several ways. For one thing, she works with a rather uncommon subject, an African Grey parrot. For another, she has stretched and even redefined the boundaries of animal cognition with her research on numerical competence and conceptual behavior. Pepperberg's training method, described in the following chapter, approaches the scientist–animal bond in a creative way. It involves a method based on a natural form of intraspecific communication in which humans act as substitutes for conspecifics.

There is another, if more subtle, dimension to Pepperberg's work that is also relevant to our book. Pepperberg has developed a very intense relationship with her subject, and arguably much of her success depends on this bond. It has maximized her subject's motivation to participate in a test regime in a way that a stranger offering food rewards could never do.

Introduction

Although my research does not directly examine the nature of the social bond between experimenters and animals, I do study the effect of social interaction on the way individuals learn to communicate. Two links exist,

178

therefore, between this chapter and others in the book. First, communication is generally viewed as a social act and often an expression of a social bond. Second, the study of how social factors affect the development of behavior can link the disciplines that study learning.

Psychologists and ethologists concur on their need to discover the how, what, why, and when of animal learning, but rarely agree on procedural questions or the primary focus of such investigations. Psychologists, for example, often argue for rigorously controlled laboratory settings in which to search for the most general, basic processes underlying learning in all creatures. Ethologists counter with arguments for field studies that examine the effects of species-specific and environmental constraints on such learning. Although this dichotomy is nearly a century old (e.g., Kline, 1898, 1899; Small, 1900a,b; Thorndike, 1911; review by Galef, 1984), only during the past decade or so has the importance of reconciling these approaches become apparent (see discussions in Kamil, 1988; Macphail, 1987; Snowdon, 1988). Researchers now realize that the study of an animal's behavior in arbitrary situations imposed in the laboratory might not lead to complete understanding (or even documentation) of capacities that are observable in natural situations (see, e.g., Beer, 1973, 1976; Breland & Breland, 1961; Kroodsma, 1982; Menzel & Juno, 1982, 1985; Smith, 1977) and, alternatively, that field experiments may lack sufficient rigor to answer specific questions concerning the processes underlying these capacities (Kroodsma, 1989a,b, 1990).

The issue, then, no longer is whether a synthesis of ethology and psychology is necessary, but rather how best to proceed. Although several papers and conference proceedings have been devoted to this topic (e.g., Chiszar, 1981; Galef, 1984; Leger, 1988; Marler & Terrace, 1984), few researchers have addressed, much less recognized, the potential problems of such a synthesis: how the superficial use of a discipline other than one's own can lead to serious misunderstandings, with respect to both the context *in* which and the design *with* which experiments ought to be carried out (see discussion in Kamil, 1984). Clearly, the wholesale trade of techniques between psychology and ethology is untenable because the basic approaches of these two disciplines to the study of learning are often still so different and the differences so poorly understood by many practitioners. Possibilities nevertheless exist for an exchange of ideas between these fields that could revitalize the study of animal behavior.

In this chapter, I therefore focus on three issues concerning the synthesis of ethology and psychology. I will (1) briefly describe some of the differences in both approaches and techniques between these fields, (2) suggest how researchers in each field can *adapt*, rather than *adopt*, one another's methodologies by studying the influence of social environment on learning, and (3) describe two sets of experiments involving avian communication in which a synthesis of methodologies has indeed led to increased knowledge of what affects and effects learning. These experiments are of direct relevance to the book's theme because they underscore the importance of social interaction between subjects or between scientist and animal subject.

Psychological and ethological perspectives on techniques to study learning

Psychologists' and ethologists' differences are often expressed in the way they approach the study of learning. Their approaches differ even in the amount of initial communication between the experimenter and the subject concerning the task to be learned. There is, consequently, a basic difference in the experimental environments in which learning is to take place. Although often overlooked, this difference should be a central issue in the study of learning competency because the experimental environment can affect not only *how* but also *whether* an animal can learn a given task (note Pepperberg, 1985, 1988a,b, 1990c, 1991a,b; Pepperberg & Neapolitan, 1988).

General differences in experimental approach

Psychology. The most commonly employed techniques of the psychologist – operant and classical conditioning – isolate a subject from almost all stimuli other than those the experimenter wishes the subject to learn. The animal subject is placed in some form of apparatus and then presented with a task or series of tasks to be performed in social isolation. This isolation extends to interactions with a natural environment, including interactions with other organisms. The rationale is that such procedures elucidate learning proclivities so basic as to exist in the absence of environmental facilitators.

Because the animal can communicate directly with neither the experimenter nor with conspecifics, there exists no explicit transfer of information about the task. Instead, the animal must determine, through trial and error, not only the appropriate response but also the question the experimenter wishes to communicate to it and to which it must respond, that is, the *nature* of the task. The way in which the animal performs – whether or not it succeeds both in decoding and answering the question – communicates, in turn, certain information to the researcher. Thus, there does exist, although only indirectly, some form of communicative act between the two participants in the experiment (see Boring, 1950).

Ethology. Such a situation differs considerably from the communicative act generally engaged in and described by ethologists. In most, if not all, cases, ethologists view communication specifically as a social act – a direct transfer of information based on a number and range of vocal and nonvocal signals (Smith, 1977). The ethologists' views come from observations of animal subjects interacting with, and learning from, one another in social situations in their natural environment – that is, for the most part, outside of the laboratory. Ethological techniques were designed to distinguish innate predispositions for specific types of development, examine effects of environment on this development, and determine the types of adaptations that occur. Ethologists view trial and error not as a means of ab initio learning about the nature of a task, but rather as a way to refine an innate predisposition, a prepared association, or an observed behavior (West & King, 1985, 1987). Thus woodpeckers may use trial-and-error techniques to improve their skills at finding, wedging, and cracking nuts, but they come to the task with certain innate motor patterns, and observe where and how adults obtain, wedge, and break apart food (Chauvin & Muckensturm-Chauvin, 1980). Most ethologists, therefore, design their techniques initially to examine, through observation, the overall behavioral repertoire in which learning is but a single element. Only after such observations are complete do ethologists experiment by manipulating the environment. An ethologist would, for example, first observe how an avian species responded to songs of their neighbors before studying what – if anything – affected these birds' responses to repeated (artificial) playbacks of a specific song thought to be involved in territorial defense (Petrinovich & Patterson, 1979).

Is a synthesis possible?

These conflicting views and methodologies would thus seem to provide little basis for synthesis. In sum, ethologists question whether forms of training and testing that are far removed from the animals' natural environment, and that in particular eliminate opportunities for social interaction, are the most effective ways of inculcating information or of examining evidence for learning (Kroodsma, 1982). Ethologists are quick to refer to instances in which animals that have been subjected to psychological techniques often failed to learn (e.g., Menzel & Juno, 1982; Mowrer, 1950, 1952) and emphasize that the lack of contextual variables (e.g., environmental support) is itself a contextual variable that can affect not only what is learned but also the learning process (Lehrman, 1953; West & King, 1985). Psychologists counter, with some justification, that focusing on what appears to be particularly effective natural learning

processes may prevent investigation of other mechanisms that are potentially equally serviceable (Galef, 1984). Specifically, ethological protocols
do not necessarily provide the most efficient means of teasing apart the
multiple processes capable of supporting development of the observed
behavior (Chauvin, 1977). The resolution of the conflict, however, is possible because some researchers have realized that a synthesis does not
require that they *adopt* one another's methodologies "en masse," but
rather that they merely *adapt* specific procedures to answer specific
questions.

Social interaction: a common focus for studies of learning

Rationale

I will focus on only one aspect of this complex issue of synthesis. I suggest how social modeling theory, with its possible applications to both
psychology and ethology, might provide a synthetic approach to learning.
The specific research that I describe involves determining the extent of,
and the natural constraints on, vocal and conceptual learning in some avian
species. This area of study might appear to favor an ethological approach,
but is one in which comparative psychologists have taken considerable
interest (e.g., Gossette, Gossette, & Riddell, 1966; Gossette & Gossette,
1967), often with controversial consequences (see review by Lea, 1984). I
discuss two cases that demonstrate the types of problems that can arise and
how the synthesis of ethological and psychological techniques have led
to appropriate solutions. In these sets of studies, one on white-crowned
sparrows (*Zonotrichia leucophrys*) and the other on an African Grey parrot
(*Psittacus erithacus*), the inclusion of a naturalistic environment of social
interaction into the experimental design has led to data that have significantly revised our notions of the extent of avian learning capacities. In both
cases, the procedures are consistent with social modeling theory, which
predicts that a live interacting tutor can effect otherwise inhibited learning
(Bandura, 1971, 1977).

Some principles of social modeling

Social modeling theory developed from an analysis of the mechanisms
that enabled human subjects to overcome strong inhibitions or phobias
(Bandura, 1971), that is, to engage in *exceptional learning*. Subsequent
studies examined the relevance of these principles to tasks involving human
communicative competence (e.g., Brown, 1976; Snow & Hoefnagle-Höhle,
1978). The principles relevant to exceptional *avian*[1] vocal learning are as

follows (summarized from Bandura, 1971, quoted from Pepperberg, 1986): (1) Changes in behavior are most likely to occur if the targeted behavior is carefully demonstrated by a live tutor who adjusts the demonstration to the responses of the observer. (2) Learning is most effective when observers see and practice the targeted behavior under conditions similar to those they face in their regular environment; thus, a demonstration must be *contextually relevant*. A corollary is that the modeled behavior is more likely to be acquired if it has functional value for the observer, and this functionality (i.e., *referentiality*) is also demonstrated. (3) The more intense the contact between the observer and the model, the more likely is the observer to learn the targeted behavior. (4) The more resistance the observer has toward acquiring the targeted behavior, whatever the reason, the more intense must be the interaction between the observer and the model, and the more referential and contextually relevant must be the demonstration. Thus, in cases where minor inhibitions or no inhibitions exist, an audiotaped demonstration of the behavior or verbal instruction on how to perform without explicit demonstration may be sufficient (Bandura & Walters, 1965). Teaching a strongly inhibited behavior or one for which the subject is not developmentally ready, however, requires intense interaction with a live tutor who is performing in a contextually applicable, referential manner (Bandura, 1977). Interestingly, although modeling theory does suggest that the identity of the tutor can affect the learning process, the extent of interaction appears to be more important than whether the interaction involves conspecifics or allospecifics (see Pepperberg, 1991b). By implication this allows a human to be an acceptable tutor for an avian subject.

How social modeling theory provides a synthesis of techniques in two studies

Several studies have shown that for certain tasks, particularly those involving vocal abilities, the extent of avian learning can differ significantly with small changes in the experimental design (Kroodsma et al., 1985; Kroodsma et al., 1984). In many cases, the more the experimental situation resembles the optimal natural environment, including the presence of partners with whom to interact, the faster learning occurs and the more extensive it appears to be (Baptista & Petrinovich, 1984, 1986; Baptista & Schuchmann, 1990; Pepperberg, 1981, 1988b; Petrinovich, 1985; Petrinovich & Baptista, 1987). Yet few studies in avian learning have, until recently, taken such findings into account. Problems occur not just when the effects of social interaction are ignored by psychologists who try to examine animal learning without a thorough grounding in animal behavior, but also when ethologists fail to recognize how the move into the lab can affect the behavior of a species they know well in the field (Beer, 1973; Kroodsma, 1982).

Studies on song acquisition in the white-crowned sparrow

The current reexamination of song acquisition in various birds, particularly research concerning the white-crowned sparrow (Baker & Cunningham, 1985; Baptista & Petrinovich, 1984, 1986; Cunningham & Baker, 1983; Kroodsma et al., 1985; Petrinovich & Baptista, 1987), provides a good illustration of how research design can affect results and conclusions. The study of exceptional song learning in the white-crowned sparrow involves a large number of experiments, and I can only summarize the complexities involved. (Readers interested in specifics should refer to the original cited articles; for a review, see Pepperberg, 1991b). In brief, by readjusting the laboratory environment to reflect more closely the natural social environment (and, inadvertently, the principles of modeling theory; see Pepperberg, 1985), researchers demonstrated an unexpected flexibility in the vocal abilities of these birds (Baptista & Petrinovich, 1984, 1986; DeWolfe, Baptista, & Petrinovich, 1989; Petrinovich & Baptista, 1987).

All of the numerous studies on white-crowned sparrow song development sought to determine those factors critical for vocal learning. Birds in the original experiments were raised in social and acoustic isolation (Marler, 1970). These subjects were played *tapes* of conspecific and allospecific vocalizations during various time frames, but were rarely, if ever, allowed to interact with one another, and never allowed to interact with a live tutor. In this way, innate *predispositions* toward conspecific song and optimal periods for learning were indeed uncovered; but as the recent work of Baptista and Petrinovich (1984, 1986; Petrinovich, 1985, 1988; see also Kroodsma et al., 1985) has shown, the extent to which these predispositions regulated song acquisition was a function of the experimental social isolation. When the laboratory situation reflected instead the interactive nature of song acquisition, the results were strikingly different: White-crowned sparrows with exposure to live tutors neither acquired conspecific song exclusively nor learned only during a limited sensitive phase, but were able, if social interaction with the live tutor was strongly enough focused, to learn allospecific song, significantly extend the period for song acquisition, and even alter their dialect in subsequent years (Baptista & Petrinovich, 1984, 1986; DeWolfe, Baptista, & Petrinovich, 1989). The newer findings did not call into question the *basic* validity of the earlier ones; rather, such data pointed out how a predisposition in a natural setting can become a rigid selective filter in the laboratory.

The goal of researchers must, therefore, be to determine the interplay of the social and genetic factors that are involved in the developmental process. Thus, although the emphasis in this chapter is on situations in which social context affect learning, there clearly exist many factors that influence learning abilities. Specifically, even though the effect of social interaction on song acquisition has been demonstrated for species as varied as song sparrows (*Melospiza melodia*; Kroodsma, 1977), marsh wrens (*Cistothorus palustris*, Kroodsma & Pickert; 1984a,b), and even parasitic cowbirds (*Molothrus ater*; West & King, 1985, 1987), there exist species for which social factors appear less important. Some suboscine birds (e.g., flycatchers: *Empidonax alnorum, E. traillii, Sayornis phoebe*), for example, develop normal vocalizations even if they hear no adult song after their eighth day posthatching (Kroodsma, 1984, 1985; Kroodsma & Konishi, 1991). Observational learning and the shaping of communicative exchanges by social interaction *are* likely widespread phenomena (see Pepperberg, 1991b), but the success of a research project is related to the extent to which all factors are taken into account.

Learning to communicate about concepts: research with an African Grey parrot

Given that many factors influence the ability to learn, the effects of social context have nevertheless been particularly apparent in my research, where a knowledge of the subject's natural environment and the means to adapt this environment to a laboratory situation have also meant the difference between previous failures and my own (if somewhat limited) success. My research, like that on the white-crowned sparrow, similarly involves the teaching of an allospecific form of communication (referential use of English speech) to an avian subject. The subject is an African Grey parrot named Alex, and I use this interspecies communication as a tool for assessing his competence on various cognitive tasks. This parrot has demonstrated considerable competence on these tasks, many of which were previously thought to be beyond the capacity of any species other than humans and, possibly, nonhuman primates (see Premack, 1978).

Background. My goal of using an interactive communication system as a tool for examining avian intellingence meant that I needed a technique to teach such a code. Devising a procedure was not as simple as one might suppose. Although parrots were known for their vocal productivity in the wild and for their acquisition of human speech patterns in the informal setting of a home, various researchers had found mimetic birds incapable of significant vocal learning in the laboratory (Gossette, 1967; Grosslight, Zaynor, & Lively, 1964; Mowrer, 1950, 1952, 1954). The question was, why? As early as the 1940s, several scientists had reasoned that the vocal ability of mimetic birds, coupled with their considerable intelligence, should enable them to engage in referential two-way communication with humans (see review in Mowrer, 1950). Armed with the standard psychological techniques of the day, but with little knowledge of how their procedures might correspond to the way these birds learned in the wild, several researchers attempted such experiments. Their techniques included vocal repetition in the absence of any referent whatsoever and extrinsic (i.e., nonreferential) rewards: a single reinforcer, generally food, that had no direct relation to the utterance being taught. Thus, a bird that did manage to say "hello" upon the appearance of a trainer, a targeted task, would be rewarded with a peanut. It somehow never occurred to the researchers that the bird would consider the vocalization "hello" to be a request for the nut rather than a comment on the entrance of the trainer. When the bird's subsequent (and, with respect to the behavior of the trainer, often inappropriate) vocalizations of "hello" failed to produce a nut, the frequency of the vocalization decreased. Thus, the procedure actually *delayed* the acquisition of referential communication (Bruner, 1977; Greenfield, 1978; Pepperberg, 1978). Too, although Mowrer did attempt to socialize his birds, the learning tasks were unrelated to this socialization. For example, he never tried to teach the bird referential use of "Come here." Subsequent research on another mimid, the mynah, proved equally frustrating: Grosslight and his colleagues (Grosslight et al., 1964; Grosslight & Zaynor, 1967), using audiotapes in a nonsocial, nonreferential setting (much the same as was used with those birds in studies of song acquisition), also found that their subjects acquired very little in the way of trained allospecific vocalizations (note Pepperberg, 1988c).

Although some of the shortcomings of their techniques are clear, these researchers could not really be faulted for failing to take into account the behavior of these birds in the wild. Little was known of the details of the vocal behavior of mimids, including ontogeny, until work in the late 1960s and early 1970s by researchers such as Bertram (1970), Nottebohm (1970; Nottebohm & Nottebohm, 1969), and Todt (1975). Nottebohm, for example, learned that orange-winged Amazons, living in two different habitats, had two distinct dialects. He also found a "babbling" stage in these birds, comparable to that of songbirds that learn their songs. Given such evidence for flexible productivity and early practice, he proposed that these parrots learned their vocalizations through social interactions with their parents and flock members. At about the same time, Bertram (1970) showed that mynahs predominantly mimic other *mynahs* in the wild, presumably learning these vocalizations during social interactions. These studies thus suggested a role for social interaction in the development of communicative behavior.

One of the first studies on the role of social interaction in vocal learning in mimids tested the effects of a particular *type* of interaction. The premise of Todt's (1975) study was that auditory exposure would facilitate imitation if presented in *conjunction* with social interaction. He therefore attempted to replicate, in a laboratory situation, the duets between adults that formed part of the natural vocal environment of the Grey parrot. Todt used humans to demonstrate to the bird the types of interactive vocalizations that were to be acquired. In his procedure, called the model/rival, or M/R technique, one human acted as a principal trainer, asking questions and providing increased visual attention for appropriate responses; another human acted as a model for the parrot's behavior and as a rival of the parrot for the attention of the principal trainer. So, for example, the trainer would say, "What's your name?" and the M/R would respond, "My name is Lora." Todt's parrots would often learn their parts of the duet (that of the M/R) in less than a day, in considerable contrast to the earlier studies.

The behaviors of birds trained with his techniques were not, however, entirely consistent with my goal of teaching a parrot a referential, contextually applicable, interactive code. Todt was more interested in behavioral than intellectual development (e.g., he wanted to determine optimal conditions for acquisition of mimetic vocalizations) and his birds (1) might not have learned anything more than a human-imposed form of duetting (e.g., Thorpe, 1974) or a simple conditioned response (e.g., Lenneberg, 1971, 1973), (2) would engage in vocal interactions only with their particular trainers, and (3) gave no evidence of having acquired any contextual or referential understanding of their learned vocalizations. Despite these limitations, Todt, unlike the psychologists, *had* succeeded in teaching birds to reproduce the targeted vocalizations of an allospecific code.

Moreover, Todt's results were, with respect to both the success and the limits of his techniques, consistent with the predictions of human social modeling theories (e.g., Bandura, 1971). Specifically, the extent of his birds' learning reflected the extent to which his protocols took into account the principles of social modeling theory described earlier. It was therefore my belief that Todt's ethologically based technique, if modified to reflect fully the theories of the social psychologists, could be used to teach a parrot a referential, contextually relevant interactive communication code.

Developing a technique to train a communication code. Whereas Todt had demonstrated how to train the *form* of an allospecific code, researchers in other areas, most notably social psychology (Bandura, 1971, 1977), child development (Piaget, 1952), and linguistics (Vygotsky, 1962), had described, respectively, the types of social, intellectual, and environmental situations

187

that facilitate referential, contextually applicable learning. The social modeling theories of Bandura (1971, 1977) discussed earlier thus suggested the appropriate protocols for these adaptations, and the work of Piaget and Vygotsky provided the theoretical basis for these adaptations. For Piaget (1952), intellectual development, which includes the acquisition of referential communication, arises from continuous interaction between the subject and the specifics of its environment. Development proceeds as novel experiences are assimilated into and modify preexisting behaviors or concepts, which themselves were constructed during the subject's reactions to prior experiences. Crucial to this process are contexts that encourage such "assimilation" and "accommodation" (e.g., Doré & Dumas, 1987). Although not directly stated by Piaget, his writings thus suggest (see Bandura, 1971; Wadsworth, 1978) that a concept or behavior is more likely to be acquired if it is expressed in a context that has functional value for the student, and if this functionality (i.e., referentiality) is explicitly demonstrated.

In a related sense, some interpreters of Vygotsky's writings (1962, 1978) propose that a subject requires appropriate contextual support, or "scaffolding" (e.g., Bruner, 1977), in order to use its general intellectual skills to solve problems that are not obviously related to one another (e.g., Rogoff, 1984; Wertsch, 1985). "Scaffolding" is anything that *shows* the subject how to transfer its skills between situations by making explicit the features that may be shared (Rogoff & Gardner, 1984). This idea is also expressed by Bandura (1971), who suggests that learning is most effective when subjects observe demonstrations of, and themselves practice, a novel, targeted behavior under conditions that show how the behavior relates to their regular environment.

The work of all these researchers clearly suggested that the optimal conditions for teaching a parrot an interactive, allospecific communication code would involve re-creating, with some modification, its natural learning environment – a situation in which the referential use of the communication code was demonstrated by others in its social community. The knowledge necessary for creating this environment could not have been gained from either ethological or psychological studies alone; rather, an appropriate experimental paradigm required the synthesis of information from both disciplines.

Details of the model/rival technique and the system of intrinsic rewards

The training system, because of its relationship to that of Todt, is also called the model/rival, or M/R, technique. My M/R procedure involves three-way interactions between two competent human speakers and the avian student, Alex. M/R training is used primarily to introduce new labels and concepts, but also aids in shaping correct pronunciation. Because the details of these procedures have previously been published (see Pepperberg, 1981, 1987a,b, 1988b, 1990a,b), only a summary will be provided here (quoted in part from Pepperberg, 1990c).

During M/R training, humans demonstrate to Alex different types of targeted interactions. A typical interaction begins with the parrot observing two humans handling objects in which he has already demonstrated some interest. One human, who acts as a trainer, shows the object(s) to the second human, who acts both as a model for the bird's responses and as the bird's rival for the trainer's attention. The trainer asks the M/R questions about the object(s) (e.g., "What's here?" "What color?" "How many?") and gives praise and the object(s) as a reward for a correct answer. The technique thus demonstrates referential and contextual use of labels for observable objects, qualifiers, quantifiers, and, on occasion, actions. Sometimes the M/R's response is (deliberately) incorrect or garbled (e.g., unclear vocalizations, partial identifications, similar to errors being made by Alex at the time), and the trainer demonstrates disapproval of such incorrect responses by scolding and temporarily removing the object(s) from sight. Thus, the bird also observes the consequences of an error. Because the M/R is, however, encouraged to try again or talk more clearly (e.g., "You're close; say better"), Alex observes the type of "corrective feedback" that may assist acquisition (see Goldstein, 1984; Pepperberg, Brese, & Harris, 1991; Vanayan, Robertson, & Biederman, 1985).

Three actions on the part of the trainers help ensure that Alex does indeed attend to such sessions; all these precautions are consistent with the principles of human social modeling theory described earlier in this chapter. First, trainers adjust the level of their interactions to reflect Alex's current capacities. If, for example, the target utterance resembles an existent vocalization, trainers are careful to praise (but not reward) Alex's likely initial use of this term and clearly demonstrate how the existent and targeted utterances differ. Trainers then consistently adjust their rewards as Alex practices his utterances to challenge him to achieve the correct pronunciation. Second, the exemplars that are involved must be objects that Alex has some interest in obtaining. Trainers working with Alex on a numerical task, for example, who choose to use corks rather than keys are more likely to engage his attention. Third, trainers must act as though they themselves find the task of interest. Alex is less likely to ignore the session and begin to preen if the emotional content of the trainers' interactions suggests that there is real relevance to the task, and trainers who actively engage him in the task are more likely to succeed in teaching the targeted label or concept.

As part of the attempt to demonstrate functionality and relevance, the M/R protocol also involves repeating the interaction while *reversing* the roles of the human trainer and M/R, and occasionally includes Alex in the interactions. Thus, unlike subjects in the studies of Todt (and several other researchers; see Goldstein, 1984), Alex does not simply hear stepwise vocal duets, but rather observes a communicative process that involves reciprocity and can be used by either party to request information or to effect changes in the environment. It was my belief, based on Bandura's (1971, 1977) findings on the efficacy of interactive modeling, that Todt's lack of demonstrated role reversibility between the trainer and M/R was the source of two of the limits on functional learning in his birds, that is, their inability to transfer their responses to anyone other than the particular human who posed the questions and their failure to learn both parts of the interaction.

My students and I use a technique in addition to the M/R procedure to encourage Alex's correct pronunciation of a new utterance. We present the new exemplar along with a set of contextually applicable phrases ("sentence frames"; de Villiers & de Villiers, 1978; Pepperberg, 1978, 1990b) – for example, "Such a big piece of *paper!*" "Here's your *paper!*" that include the targeted vocalization. This combination of frequent vocal repetition and the physical action of presenting the object, as well as the lack of single, repetitive utterances, not only is consistent with modeling theory but also resembles the behavior parents sometimes use when introducing labels for new items to very young children (Berko-Gleason, 1977; de Villiers & de Villiers, 1978). By hearing the label employed in normal, productive speech, Alex, like children, experiences – and appears to learn (see later) – not only the label but the context in which it is to be used. The varied presentation of the targeted vocalization also appears to forestall any direct word-for-word imitation of the trainers.

On occasion, Alex's experimentation with the sounds in his repertoire have led to novel vocalizations; to encourage such productivity and enlarge his referential repertoire, we reward him (when possible) with appropriate objects and use a variant of the M/R technique to associate the novel utterances and objects (Pepperberg, 1990b). The novel vocalizations are often recombinations of, or variants on, parts of existent labels. Thus, after Alex acquired the label "gray," he produced (in the absence of any object) grate, grape, grain, chain, and cane. After each utterance, we presented him with, and used the M/R technique to discuss, respectively, a nutmeg grater (which could be used to trim his beak), the appropriate fruit, some parakeet treat, and a circle of paper clips. We were initially unable to devise an appropriate referent for "cane" and ignored that utterance. All but "cane" remained in his repertoire and subsequently became referential vocalizations (Pepperberg, 1983, 1990b). "Cane," however, reappeared in Alex's repertoire less than a day after we obtained sections of sugarcane in a local market and reintroduced the label (Pepperberg, 1990c). Our data suggest that the frequency with which these spontaneous vocalizations arise and stabilize in Alex's repertoire reflects the rapidity with which the trainers respond and the desirability of the objects or action chosen as a referent (Pepperberg, 1990b). "Banacker" (presumably an elision of banana + cracker), for example, disappeared soon after we introduced, but Alex rejected, dried banana chips.

This technique, which I have called "referential mapping" (Pepperberg, 1990b), demonstrates another way in which my study uses modeling techniques to train a parrot to communicate with humans. The procedure also provides for the integration of the principles of psychology and ethology. Like a psychologist, I use a specific, detailed protocol for training. But like an ethologist, I attempt to reproduce an appropriate social context. Thus, in contrast to many laboratory researchers, I emphasize *intrinsic* rather than *extrinsic* reinforcers.[2] In programs using extrinsic rewards, all correct identifications for food or nonfood items or appropriate responses to various specific commands are rewarded with a single desirable item (generally food) that neither directly relates to the skill being taught nor varies with the specific task being targeted. In a natural environment, however, most such interactions are referential: A child that says "oogie" is likely to be given a cookie, and one that identifies the family dog as "wau-wau" is likely to be allowed to pet it; neither behavior is likely to be rewarded with a piece of chocolate. Several researchers have shown that a child's competence often advances when adults interpret and respond to its utterances as "intentional" (i.e., present a cookie for "oogie") even before there is evidence to support such intentionality (see review in Pepperberg, 1990b). In contrast, extrinsic rewards may, as already noted, act to delay label or concept acquisition by confounding the label or concept to be learned with some aspect of the reward item (Greenfield, 1978; Miles, 1983; Pepperberg, 1978). Interestingly, human subjects (e.g., severely retarded children) in programs that stress learning through extrinsic rewards and nonreferential, noninteractive drills may have difficulty extrapolating what has been learned to situations that differ in any manner from the one in which they were trained (e.g., Brown & Campione, 1986; Lovaas, 1977; reviews in Pepperberg, 1988a,b, 1990b): The contextual support that helps the subjects assimilate the new information into existing structures (see Piaget, 1952) and provides an appropriate "scaffolding" to transfer information between domains (sensu Vygotsky, 1962) is missing.

In our procedure, however, each interaction has always been rewarded by the object to which the label or concept refers, rather than any single, extrinsic item. My students and I began training the initial set of object labels, for example, by determining which exemplars Alex preferred to manipulate and thus might be most eager to accept for reward. He consequently received a piece of an unlined index card, a favored item, for his first attempt at "paper" ("ā-uh"), a metallic key for "kuh," and a wooden plant stake for "uut." Continued rewards then depended on his progress toward the targeted pronunciation. For all subsequent tasks, Alex's initial reward continues to be whatever was the object of the question – for example, the *blue* key or *six* corks. If Alex erroneously produces a vocalization in his repertoire that resembles the targeted utterance (e.g., "grate" for "grape"), he is shown, but does not receive, the object to which his utterance referred ("grate"). Every interaction thus provides the closest possible association of the label or concept that is being taught and the object(s) or task to which it refers (Pepperberg, 1978, 1981).

Because it is sometimes difficult to maintain Alex's interest in the particular set of objects that, for example, are being used for training a new task, he may also be rewarded with the right to request vocally a more desirable item than that which he has identified ("I want *X*"; see Pepperberg, 1988b). Alex is, for example, more likely to engage in various displacement behaviors (e.g., preening, requests for numerous unseen objects, and changes of location) when tasks involve previously used rather than novel exemplars, even if the task itself is novel (Pepperberg, 1987b). Allowing Alex to request his own reward not only provides some flexibility, but also maintains the use of intrinsic rewards: He will never, for example, automatically receive a key when he identifies a cork. The key must specifically be requested ("I want key"), and trainers will not respond to such a request until the appropriate prior task is completed.

Summary of the rationale for the M/R technique. Although little had been published on psittacine learning when this project began, my procedures were designed to provide contexts consistent with those known to promote learning in both field and laboratory. In sum (quoted from Pepperberg, 1991a): (1) Communication among parrots in both wild and aviary settings appeared to be primarily in the vocal mode (Busnel & Mebes, 1975; Dilger, 1960; Mebes, 1978; Power, 1966a,b; Serpell, 1981) and to be learned through social interaction (Nottebohm, 1970; Nottebohm & Nottebohm, 1969). Our training demands were thus closely related to naturally occurring tasks. (2) The conversational turn taking we used in both our training and testing procedures resembled natural duetting behavior (Mebes, 1978; Wickler, 1976, 1980). Alex was therefore exposed to, and required to participate in, behaviors similar to those in which parrots engage in the wild; that is, his responses would be within his normal range of behaviors. (3) All training procedures involved the kinds of referential, contextually relevant situations that facilitate learning, in both birds and mammals, of behaviors that can otherwise be difficult to establish (Pepperberg, 1985, 1987a,b).

Summary of results. Using these techniques, my students and I have, over the course of several years, taught a parrot tasks that were once thought beyond the capability of all but humans or, possibly, certain nonhuman primates (Premack, 1978; the following is cited from Pepperberg, in press). Alex has learned labels for about 40 different objects: paper, key, wood, hide (rawhide chips), kiwi, grain, peg wood (clothes pins), cork, corn, nut, walnut, showah (shower), wheat, pasta, box, banana, gym, cracker, scraper (a nail file), chain, shoulder, block, rock (lava stone beak conditioner), carrot, gravel, back, chair, chalk, water, nail, grape, cup, grate, treat, cherry, wool, popcorn, citrus, green bean, and banerry (apple). We have tentative evidence for labels such as bread and jacks. He has functional use of "no," phrases such as "Come here," "I want X," and "Wanna go Y" where X and Y are appropriate labels for objects or locations (see Pepperberg, 1990c, in press). Incorrect responses to his requests by a trainer (e.g., substitution of something other than what he requested) generally results (~75% of the time) in his saying no and repeating the initial request (Pepperberg, 1987a, 1988b). He has acquired labels for seven colors: rose (red), blue, green, yellow, orange, gray, and purple. He identifies five different shapes by labeling them as two, three, four, five, or six-cornered objects (Pepperberg, 1983). He uses the labels "two," "three," "four," "five," and "sih" (six) to distinguish quantities of objects up to six, including collections made up of novel objects, heterogeneous sets of objects, and sets in which the objects are placed in random arrays (Pepperberg, 1987b). He combines all the vocal labels to identify proficiently, request, refuse, categorize, and quantify more than 100 different objects, including those that vary somewhat from training exemplars. His accuracy has averaged ~80% when tested on these abilities (Pepperberg, 1981, 1983, 1987a,b, 1990b).

We have also examined Alex's capabilities for comprehending the concept of "category." We have taught him not only to label any one of a number of different hues or shapes, but also to understand that "green," for example, is a particular instance of the category "color," and that for any object that is both colored *and* shaped, the specific instances of these attributes (e.g., "green" and "three-corner") represent *different* categories. Thus, he has learned to categorize objects having both color and shape with respect to either category based on a vocal query of "What color?" or

"What shape?" (85.5%, all trials; Pepperberg, 1983). Because the protocol often requires Alex to categorize the same exemplar with respect to shape at one time and color at another, the task involves flexibility in changing the basis for classification. Such flexibility, or capacity for reclassification, is thought to indicate the presence of "abstract aptitude" (Hayes & Nissen, 1956/1971).

Alex has also learned abstract concepts of "same" and "different" and to respond to the absence of information about these concepts if nothing is same or different. Such faculties were once thought to be beyond the capacity of an avian subject (note Premack, 1978, 1983; but see Zentall, Hogan, & Edwards, 1984). Thus, when presented with two objects that are identical or that vary with respect to some or all of the attributes of color, shape, and material, Alex can respond with the appropriate *category* label as to which attribute is "same" or "different" for any combination (80.8%, all trials; 76.0%, first trials; Pepperberg, 1987c). If, however, nothing is same or different, he has learned to reply "none" (83.9%, all trials; 80.9%, first trials; Pepperberg, 1988a). He can respond equally accurately to instances involving objects, colors, shapes, and materials not used in training, including those for which he has no labels. Furthermore, we have shown that Alex is indeed responding to the specific questions, and not merely responding on the basis of his training and the physical attributes of the objects: His responses were still above chance levels when, for example, the question "What's same?" was posed with respect to a green wooden triangle and a blue wooden triangle. If he were ignoring the question and responding on the basis of his prior training, he would have determined, and responded with the label for, the one anomalous attribute (in this case, "color"). Instead, he responded with one of two appropriate answers (in this case, "shape" or "mah-mah" [matter]; Pepperberg, 1987c, 1988a).

Although our research on numerical concepts does not demonstrate that Alex has an understanding of "number" comparable to that of a human child (e.g., Fuson, 1988), the data suggest that he does comprehend some concept of quantity (Pepperberg, 1987b). Thus, although we have yet to show conclusively that Alex can, for example, actually count (see Davis & Pérusse, 1988), he can recognize and label different quantities of physical objects up to and including six (78.9%, all trials; Pepperberg, 1987b). The sets of objects need not be familiar, nor need they be placed in any particular pattern, such as a square or triangle. Furthermore, if presented with a heterogeneous collection (of *X*s and *Y*s), he can respond appropriately to questions of either "How many *X*?" *or* "How many *Y*?" (62.5%, all trials; 70.0%, first trials; Pepperberg, 1987b). Comparable abilities, when demonstrated by primates, have been offered as evidence for advanced levels of intelligence (note Pepperberg, 1990c, 1991a).

Our most recent study was to determine formally how similar Alex's abilities are to those of marine mammals that have also been trained to use a system of interspecies communication (Pepperberg, 1990a). Most of the work with cetaceans and pinnipeds uses the comprehension mode; that is, researchers assess competence in cognitive and communicative skills by demonstrating how well their animal subjects understand the communication code (e.g., Herman, 1987). In contrast, much of the work with nonhuman primates and all of the prior work with Alex, although clearly involving comprehension, emphasized instead the productive mode; that is, how accurately and appropriately the subjects can *produce* the code. To maintain our vocal paradigm but provide the necessary comparisons, my students and I chose to train and test Alex on a recursive task similar to those used with other animals. In a recursive task, a subject is presented with several different objects and one of several different possible questions or commands concerning the attributes of these objects. Each question or command contains several parts, the combination of which uniquely specifies which object is to be targeted and what action is to be performed. The complexity of the question is determined by its context (the number of different objects from which to choose) and the number of its parts (e.g., the number of attributes used to specify the target and the number of actions from which to choose). The subject must divide the question into these parts and (recursively) use its understanding of each part to answer correctly. The subject thus demonstrates its competence by reporting on only a single aspect (e.g., color, shape, or material) of, or performing one of several possible actions (fetching, touching) on, an object that is one of several differently colored and shaped exemplars of various materials (see Granier-Deferre & Kodratoff, 1986). Alex was therefore shown trays of seven unique combinations of exemplars and asked questions such as "What color is object X?" "What shape is object Y?" "What object is color A?" or "What object is shape B?" His accuracy on all questions, which was better than 80% (84.2%, all trials; 81.3%, first trials; Pepperberg, 1990a), was comparable to that of marine mammals (and also nonhuman primates) that had been tested on similar tasks.

Concluding remarks

Demonstrating that a parrot has all these abilities, however, is *not* the point of this chapter; the findings of Koehler (1950, 1953), Braun (1952), Lögler (1959), and Krushinskii (1960) suggested these capabilities decades ago. The point I *am* trying to make is that it was knowledge of the bird's environment and natural social system (i.e., the role of social interaction with, and observation of, conspecifics) that enabled my students and me to

195

(1) choose the most appropriate techniques from psychologists and ethologists and (2) most effectively adapt them to our circumstances. Thus, we could *design* situations for observational learning and *apply* techniques of social modeling to demonstrate most effectively the parrot's abilities. I thus wish to propose that the experimenter–subject interaction is indeed a special case of learning through general social interaction: Almost all of the animal subjects described in this volume learn significant material from parents and peers in the wild. In our laboratories, this role of social tutor has, by default, been taken over by the human experimenter. I suggest that it is precisely because such interactions become so analogous to the natural situation that they are such effective means of inculcating information or training behaviors.

Even though it would be ideal to study learning processes and cognitive abilities in nature, in many cases the full extent of an animal's capacities can be uncovered only through detailed study in the controlled laboratory situation. Discovery of the *multiple* processes underlying homing in pigeons (Keeton, 1974; Kreithen & Keeton, 1974) or the acquisition of diet preferences in rats (Galef, 1984), and the separating out of the various environmental constraints on song acquisition in birds (e.g., Krebs & Kroodsma, 1980; Slater, Eales, & Clayton, 1988), are good examples of how detailed laboratory work can tease apart the often overlapping mechanisms responsible for the observed behaviors. Other chapters in this volume also present evidence for the elucidation of behavioral capacities not usually observed in the wild. Additional examples exist throughout the literature (e.g., Holmes & Sherman, 1982, 1983; Labiale, 1977).

The point that I, and others before me (e.g., Menzel & Juno, 1982; Smith, 1977), have been trying to make is that studies that fail to take into account the ecology and ethology of the animal, either by design or default, are unlikely to provide knowledge of the complete extent of the animal's learning capabilities. Experiments performed in the absence of such relevant information can be interpreted only in their narrowly defined context. An animal placed in an impoverished environment is unlikely to exhibit the full range of its typical behavior patterns: Those portions of its natural repertoire that are observed are likely to provide information that is irrelevant or misleading as to the full extent of its abilities (see Baptista & Petrinovich, 1984; West & King, 1985). The widespread application of such findings to other situations, particularly to the "real world," must be called into question (e.g., Baptista & Petrinovich, 1986; Galef, 1984; Menzel & Juno, 1982). If the aim of the researcher is indeed to determine the maximal capability of the animal subject, rather than its ability to react under a specified limited range of conditions, then the ecological and ethological relevance of the laboratory paradigms (particularly the role of the experimenter) must not only be taken into account, but carefully integrated into the protocol of the investigation (Kamil, 1988).

Notes

1. Exceptional *avian* communication is usually characterized (see Pepperberg, 1985) by vocal learning that, in the normal course of development, is thought unlikely to occur: (a) use of non-species-specific ("allospecific") vocalizations by subjects generally expected to acquire functional use of only conspecific vocalizations (e.g., contextual use of song of unrelated species; see Baptista & Petrinovich, 1984) and (b) age-independent acquisition of vocalizations in species generally recognized as having a limited "sensitive phase" for vocal learning (Baptista & Morton, 1982; Petrinovich, 1988; see also Marler, 1970).
2. Because of the stress I have placed on the importance of social interaction in training, our reasons for using any reinforcements other than social may not be clear. With children, for example, parental approbation ("yes," "good," even "hmm") can sometimes act as an effective nonreferential ("extrinsic") reward (see Bowerman, 1978, for a review). My students and I do, in fact, find such approbation useful in maintaining an established behavior and in directing Alex's attention during training. Approbation alone, however, does not provide the contextually relevant rewards that appear necessary to effect acquisition of a targeted behavior in a non-species-specific communication code.

References

Baker, M. C., & Cunningham, J. A. (1985). The biology of birdsong dialects. *Behavioral and Brain Sciences, 8*: 85–133.

Bandura, A. (1971). Analysis of social modeling processes. In A. Bandura (Ed.), *Psychological modeling* (pp. 1–62). Chicago: Aldine-Atherton.

(1977). *Social modeling theory*. Chicago: Aldine-Atherton.

Bandura, A., & Walters, R. H. (1965). *Social learning and personality development*. New York: Holt, Reinhart, & Winston.

Baptista, L. F., & Morton, M. L. (1982). Song dialects and mate selection in montane white-crowned sparrows. *Auk, 92*: 537–47.

Baptista, L.F., & Petrinovich, L. (1984). Social interaction, sensitive phases, and the song template hypothesis in the white-crowned sparrow. *Animal Behaviour, 32*: 172–81.

(1986). Song development in the white-crowned sparrow: Social factors and sex differences. *Animal Behaviour, 34*: 1359–71.

Baptista, L. F., & Schuchmann, K. L. (1990). Song learning in Anna's hummingbird. *Ethology, 84*: 15–26.

Beer, C. G. (1973). A view of birds. In A. Pick (Ed.), *Minnesota Symposium on Child Psychology* (Vol. 7, pp. 47–86). Minneapolis: Minnesota University Press.

(1976). Some complexities in the communication behavior of gulls. *Annals of the New York Academy of Sciences, 280*: 413–32.

Berko-Gleason, J. (1977). Talking to children: Some notes on feedback. In C. E. Snow & C. A. Ferguson (Eds.), *Talking to children* (pp. 199–205). Cambridge University Press.

Bertram, B. C. R. (1970). The vocal behavior of the Indian Hill mynah, *Gracula religiosa*. *Animal Behaviour Monographs, 3*: 79–192.

Boring, E. G. (1950). *A history of experimental psychology*. New York: Appleton-Century-Crofts.

Bowerman, M. (1978). The acquisition of word meaning: An investigation of some current conflicts. In N. Waterson & C. E. Snow (Eds.), *Proceedings of the Third International Child Language Symposium: The development of communication* (pp. 263–87). New York: Wiley.

Braun, H. (1952). Uber das Unterscheidungsvermögen unbenannter Anzahlen bei Papageien. *Zeitschrift für Tierpsychologie, 9*: 40–91.

Breland, K., & Breland, M. (1961). The misbehavior of organisms. *American Psychologist, 16*: 681–4.

Brown, A. L., & Campione, J. C. (1986). Psychological theory and the study of learning disabilities. *American Psychologist, 14*: 1059–68.

Brown, I. (1976). Role of referent concreteness in the acquisition of passive sentence comprehension through abstract modeling. *Journal of Experimental Child Psychology, 22*, 185–99.

Bruner, J. S. (1977). Early social interaction and language acquisition. In H. R. Schaffer (Ed.), *Studies in mother-infant interaction* (pp. 271–89). New York: Academic Press.

Busnel, R. G., & Mebes, H. D. (1975). Hearing and communication in birds: The cocktail party effect in intra-specific communication of *Agapornis roseicollis*. *Life Science, 17*: 1567–70.

Chauvin, R. (1977). *Ethology: The biological study of animal behavior*. New York: International Universities Press.

Chauvin, R., & Muckensturm-Chauvin, B. (1980). *Behavioral complexities*. New York: International Universities Press.

Chiszar, D. (1981). Learning theory, ethological theory, and developmental plasticity. In E. S. Gollin (Ed.), *Developmental plasticity: Behavioral and biological aspects of variation in development* (pp. 71–99). New York: Academic Press.

Cunningham, M. A., & Baker, M. C. (1983). Vocal learning in white-crowned sparrows: Sensitive phase and song dialects. *Behavioral Ecology and Sociobiology, 13*: 259–69.

Davis, H., & Pérusse, R. (1988). Numerical competence in animals: Definitional issues, current evidence, and a new research agenda. *Behavioral and Brain Science, 11*, 561–615.

de Villiers, J. G., & de Villiers, P. A. (1978). *Language acquisition*. Cambridge, MA: Harvard University Press.

DeWolfe, B. B., Baptista, L. F., & Petrinovich, L. (1989). Song learning and territory establishment in the Nuttall's white-crowned sparrow. *Condor, 91*: 397–407.

Dilger, W. C. (1960). The comparative ethology of the African parrot genus *Agapornis*. *Zeitschrift für Tierpsychologie, 17*: 649–85.

Doré, F.Y., & Dumas, C. (1987). Psychology of animal cognition: Piagetian studies. *Psychological Bulletin, 102*: 219–33.

Fuson, K. C. (1988). *Children's counting and concepts of number*. New York: Springer.

Galef, B. G., Jr. (1984). Reciprocal heuristics: A discussion of the relationship of the study of learned behavior in laboratory and field. *Learning and Motivation, 15*: 479–93.

Goldstein, H. (1984). The effects of modeling and corrected practice on generative language and learning of preschool children. *Journal of Speech and Hearing Disorders*, 49: 389–98.

Gossette, R. L. (1967). Successive discrimination reversal (SDR) performances of four avian species on a brightness discrimination task. *Psychonomic Science*, 8: 17–18.

Gossette, R. L., & Gossette, M. F. (1967). Examination of the reversal index (RI) across fifteen different mammalian and avian species. *Perceptual Motor Skills*, 27: 987–90.

Gossette, R. L., Gossette, M. F., & Riddell, W. (1966). Comparisons of successive discrimination reversal performances among closely and remotely related avian species. *Animal Behaviour*, 14: 560–4.

Granier-Deferre, C., & Kordratoff, Y. (1986). Iterative and recursive behaviours in chimpanzees during problem solving: A new descriptive model inspired from the artificial intelligence approach. *Cahiers de Psychologie Cognitive*, 6: 483–500.

Greenfield, P. M. (1978). Developmental processes in the language learning of child and chimp. *Behavioral and Brain Sciences*, 4: 573–4.

Grosslight, J. H., & Zaynor, W. C. (1967). Verbal behavior in the mynah bird. In K. Salzinger & S. Salzinger (Eds.), *Research in verbal behavior and some neurophysiological implications* (pp. 5–19). New York: Academic Press.

Grosslight, J. H., Zaynor, W. C., & Lively, B. L. (1964). Speech as a stimulus for differential vocal behavior in the mynah bird (*Gracula religiosa*). *Psychonomic Science*, 1: 7–8.

Hayes, K. J., & Nissen, C. H. (1956/1971). Higher mental functions of a home-raised chimpanzee. In A. Schrier & F. Stollnitz (Eds.), *Behavior of nonhuman primates* (Vol. 4, pp. 59–115). New York: Academic Press.

Herman, L. M. (1987). Receptive competencies of language-trained animals. In J. S. Rosenblatt, C. Beer, M.-C. Busnel, & P. J. B. Slater (Eds.), *Advances in the study of behavior* (Vol. 17, pp. 1–60). New York: Academic Press.

Holmes, W. G., & Sherman, P. W. (1982). The ontogeny of kin recognition in two species of ground squirrels. *American Zoologist*, 22: 491–517.

(1983). Kin recognition in animals. *American Scientist*, 71: 46–55.

Kamil, A. C. (1984). Adaptation and cognition: Knowing what comes naturally. In H. L. Roitblat, T. G. Bever, & H. S. Terrace (Eds.), *Animal cognition* (pp. 533–44). Hillsdale, NJ: Erlbaum.

(1988). A synthetic approach to the study of animal intelligence. In D. W. Leger (Ed.), *Nebraska Symposium on Motivation: Comparative perspectives in modern psychology* (Vol. 35, pp. 257–308). Lincoln: University of Nebraska Press.

Keeton, W. T. (1974). The mystery of pigeon homing. *Scientific American*, 231: 96–107.

Kline, L. W. (1898). Methods in animal psychology. *American Journal of Psychology*, 10: 256–79.

(1899). Suggestions toward a laboratory course in comparative psychology. *American Journal of Psychology*, 10: 399–430.

Koehler, O. (1950). The ability of birds to "count." *Bulletin of Animal Behaviour*, 9: 41–5.

(1953). Thinking without words. *Proceedings of the XIV International Congress of Zoology*. Copenhagen, 75.

Krebs, J. R., & Kroodsma, D. E. (1980). Repertoires and geographical variation in bird song. In J. R. Rosenblatt, R. A. Hinde, C. Beer, & M.-C. Busnel (Eds.), *Advances in the study of behaviour* (Vol. 11, pp. 143–77). New York: Academic Press.

Kreithen, M. L., & Keeton, W. T. (1974). Detection of changes in atmospheric pressure by the homing pigeon, *Columba livia. Journal of Comparative Physiology, 89*: 73–84.

Kroodsma, D. E. (1977). A re-evaluation of song development in the song sparrow. *Animal Behaviour, 25*: 390–9.

(1982). Learning and the ontogeny of sound signals in birds. In D. E. Kroodsma & E. H. Miller (Eds.), *Acoustic communication in birds* (Vol. 2, pp. 1–23). New York: Academic Press.

(1984). Songs of the alder flycatcher (*Empidonax alnorum*) and willow flycatcher (*Empidonax traillii*) are innate. *Auk, 101*: 13–24.

(1985). Development of two song forms by the eastern phoebe. *Wilson's Bulletin, 97*: 21–9.

(1989a). Suggested experimental designs in song playbacks. *Animal Behaviour, 37*: 600–9.

(1989b). Inappropriate experimental designs impede progress in bioacoustic research. *Animal Behaviour, 38*: 717–19.

(1990). How the mismatch between the experimental design and the intended hypothesis limits confidence in knowledge, as illustrated by an example from birdsong dialects. In M. Bekoff & D. Jamieson (Eds.), *Interpretation and explanation in the study of animal behavior: Comparative perspectives* (pp. 226–45). Boulder, CO: Westview Press.

Kroodsma, D. E., Baker, M. C., Baptista, L. F., & Petrinovich, L. (1985). Vocal "dialects" in Nuttall's white-crowned sparrows. *Current Ornithology, 2*: 103–33.

Kroodsma, D. E. (rapporteur), Bateson, P. P. G., Bischoff, H.-J., Delius, J. D., Hearst, E., Hollis, K. L., Immelmann, K., Jenkins, H. M., Konishi, M., Lea, S. E. A., Marler, P., & Staddon, J. E. R. (1984). Biology of learning in nonmammalian vertebrates: Group report. In P. Marler & H. S. Terrace (Eds.), *The biology of learning* (pp. 399–418). Berlin: Springer.

Kroodsma, D. E., & Konishi, M. (1991). A suboscine bird (eastern phoebe, *Sayornis phoebe*) develops normal song without auditory feedback. *Animal Behaviour, 42*: 477–87.

Kroodsma, D. E., & Pickert, R. (1984a). Sensitive phases for song learning: Effects of social interaction and individual variation. *Animal Behaviour, 32*: 389–94.

(1984b). Repertoire size, auditory templates, and selective vocal learning in songbirds. *Animal Behaviour, 32*: 395–9.

Krushinskii, L. V. (1960). *Animal behavior: Its normal and abnormal development.* New York: Consultants Bureau.

Labiale, G. (1977). La reconnaissance visuelle des formes chez le merle *Turdus merula*. Unpublished doctoral thesis, Université René Descartes.

Lea, S. E. G. (1984). Complex general process learning in nonmammalian vertebrates. In P. Marler & H. Terrace (Eds.), *The biology of learning* (pp. 373–97). New York: Springer.

Leger, D. W. (Ed.) (1988). *Nebraska Symposium on Motivation: Comparative perspectives in modern psychology* (Vol. 35). Lincoln: University of Nebraska Press.

Lehrman, D.S. (1953). Can psychiatrists use ethology? In N. F. White (Ed.), *Ethology and psychiatry* (pp. 187–96). Toronto: University of Toronto Press.

Lenneberg, E. (1971). Of language, knowledge, apes, and brains. *Journal of Psycholinguistic Research, 1*: 1–29.

(1973). Biological aspects of language. In G. A. Miller (Ed.), *Communication, language, and meaning* (pp. 49–60). New York: Basic Books.

Lögler, P. (1959). Versuche zur Frage des "Zähl"-Vermögens an einen Grau-papagein und Vergleichsversuche an Menschen. *Zeitschrift für Tierpsychologie, 16*: 179–217.

Lovaas, O. I. (1977). *The autistic child: Language development through behavior modification.* New York: Irvington.

Macphail, E. M. (1987). The comparative psychology of intelligence. *Behavioral and Brain Sciences, 10*: 645–95.

Marler, P. (1970). A comparative approach to vocal learning: Song development in white-crowned sparrows. *Journal of Comparative and Physiological Psychology, 71*: 1–25.

Marler, P., & Terrace, H. S. (Eds.). (1984). *The biology of learning.* New York: Springer.

Mebes, H. D. (1978). Pair-specific duetting in the peach-faced lovebird, *Agapornis roseicollis. Naturwissenschaften, 65*: 66–7.

Menzel, E. W., Jr., & Juno, C. (1982). Marmosets (*Saguinus fuscicollis*): Are learning sets learned? *Science, 217*: 750–2.

(1985). Social foraging in marmoset monkeys and the question of intelligence. In L. Weisenkrantz (Ed.), *Animal intelligence* (pp. 145–57). Oxford: Clarendon Press.

Miles, H. L. (1983). Apes and language: The search for communicative competence. In J. de Luce & H. T. Wilder (Eds.), *Language in primates* (pp. 43–61). New York: Springer.

Mowrer, O. H. (1950). *Learning theory and personality dynamics.* New York: Ronald Press.

(1952). The autism theory of speech development and some clinical applications. *Journal of Speech and Hearing Disorders, 17*: 263–8.

(1954). A psychologist looks at language. *American Psychologist, 9*: 660–94.

Nottebohm, F. (1970). Ontogeny of bird song. *Science, 167*: 950–6.

Nottebohm, F., & Nottebohm, M. (1969). The parrots of Bush Bush. *Animal Kingdom, 72*: 19–23.

Pepperberg, I. M. (1978, March). *Object identification by an African Grey parrot* (*Psittacus erithacus*). Paper presented at the midwestern meeting of the Animal Behavior Society, West Lafayette, IN.

(1981). Functional vocalizations by an African Grey parrot (*Psittacus erithacus*). *Zeitschrift für Tierpsychologie, 55*: 139–60.

(1983). Cognition in the African Grey parrot: Preliminary evidence for auditory/vocal comprehension of the class concept. *Animal Learning & Behavior, 11*: 179–85.

(1985). Social modeling theory: A possible framework for understanding avian vocal learning. *Auk, 102*: 854–64.

(1986). Acquisition of anomalous communicatory systems: Implications for studies on interspecies communication. In R. J. Schusterman, J. A. Thomas, & F. G. Wood (Eds.), *Dolphin cognition and behavior: A comparative approach* (pp. 289–302). Hillsdale, NJ: Erlbaum.

(1987a). Interspecies communication: A tool for assessing conceptual abilities in the African Grey parrot (*Psittacus erithacus*). In G. Greenberg &

E. Tobach (Eds.), *Language, cognition and consciousness: Integrative levels* (pp. 31–56). Hillsdale, NJ: Erlbaum.

(1987b). Evidence for conceptual quantitative abilities in the African Grey parrot: Labeling of cardinal sets. *Ethology, 75*: 37–61.

(1987c). Acquisition of the same/different concept by an African Grey parrot (*Psittacus erithacus*): Learning with respect to color, shape, and material. *Animal Learning & Behavior, 15*: 423–32.

(1988a). Comprehension of "absence" by an African Grey parrot: Learning with respect to questions of same/different. *Journal of the Experimental Analysis of Behavior, 50*: 553–64.

(1988b). An interactive modeling technique for acquisition of communication skills: Separation of "labeling" and "requesting" in a psittacine subject. *Applied Psycholinguistics, 9*: 31–56.

(1988c). The importance of social interaction and observation in the acquisition of communicative competence: Possible parallels between avian and human learning. In T. R. Zentall & B. G. Galef, Jr. (Eds.), *Social learning: Psychological and biological perspectives* (pp. 279–99). Hillsdale, NJ: Erlbaum.

(1990a). Cognition in an African Grey parrot (*Psittacus erithacus*): Further evidence for comprehension of categories and labels. *Journal of Comparative Psychology, 104*: 41–52.

(1990b). Referential mapping: A technique for attaching functional significance to the innovative utterances of an African Grey parrot. *Applied Psycholinguistics, 11*: 23–44.

(1990c). Some cognitive capacities of an African Grey parrot (*Psittacus erithacus*). In P. J. B. Slater, J. S. Rosenblatt, & C. Beer (Eds.), *Advances in the study of behavior* (Vol. 19, pp. 357–409). New York: Academic Press.

(1991a). A communicative approach to animal cognition: A study of the conceptual abilities of an African Grey parrot. In C. Ristau (Ed.), *Cognitive ethology: The minds of other animals* (pp. 153–86). Hillsdale, NJ: Erlbaum.

(1991b). Learning to communicate: The effects of social interaction. In P. P. G. Bateson & P. H. Klopfer (Eds.), *Perspectives in ethology* (Vol. 9, pp. 119–64). New York: Plenum.

(In press). Studies to determine the intelligence of an African Grey parrot. In W. J. Rosskopf & R. W. Woerpel (Eds.), *Petrak's diseases of cage and aviary birds*. Philadelphia: Lea & Febiger.

Pepperberg, I. M., Brese, K. J., & Harris, B. J. (1991). Solitary sound play during acquisition of English vocalizations by an African Grey parrot (*Psittacus erithacus*): Possible parallels with children's monologue speech. *Applied Psycholinguistics, 12*: 151–78.

Pepperberg, I. M., & Neapolitan, D. M. (1988). Second language acquisition: A framework for studying the importance of input and interaction in exceptional song acquisition. *Ethology, 77*: 150–68.

Petrinovich, L. (1985). Factors influencing song development in the white-crowned sparrow (*Zonotrichia leucophrys*). *Journal of Comparative Psychology, 99*: 15–29.

(1988). The role of social factors in white-crowned sparrow song development. In T. R. Zentall & B. G. Galef, Jr. (Eds.), *Social learning: Psychological and biological perspectives* (pp. 255–78). Hillsdale, NJ: Erlbaum.

Petrinovich, L., & Baptista, L. F. (1987). Song development in the white-crowned sparrow: Modification of learned song. *Animal Behaviour, 35*: 961–74.

Petrinovich, L., & Patterson, T. L. (1979). Field studies of habituation: I. The effects of reproductive condition, number of trials, and different delay intervals on the responses of the white-crowned sparrow. *Journal of Comparative and Physiological Psychology, 93*: 337–50.

Piaget, J. (1952). *The origins of intelligence in children* (M. Cook, trans.). New York: International Universities Press.

Power, D.M. (1966a). Agnostic behavior and vocalizations of orange-chinned parakeets in captivity. *Condor, 68*: 562–81.

(1966b). Antiphonal duetting and evidence for auditory reaction time in the orange-chinned parakeet. *Auk, 83*: 314–19.

Premack, D. (1978). On the abstractness of human concepts: Why it would be difficult to talk to a pigeon. In S. H. Hulse, H. Fowler, & W. K. Honig (Eds.), *Cognitive processes in animal behavior* (pp. 421–51). Hillsdale, NJ: Erlbaum.

(1983). The codes of man and beast. *Behavior and Brain Sciences, 6*: 125–67.

Rogoff, B. (1984). Introduction: Thinking and learning in a social context. In B. Rogoff & J. Lave (Eds.), *Everyday cognition: Its development in social contexts* (pp. 1–8). Cambridge, MA: Harvard University Press.

Rogoff, B., & Gardner, W. (1984). Adult guidance of cognitive development. In B. Rogoff & J. Lave (Eds.), *Everyday cognition: Its development in social contexts* (pp. 95–116). Cambridge, MA: Harvard University Press.

Serpell, J. (1981). Duets, greetings, and triumph ceremonies: Analogous displays in the parrot genus *Trichoglossus*. *Zeitschrift für Tierpsychologie, 55*: 268–83.

Slater, P. J. B., Eales, L. A., & Clayton, N. S. (1988). Song learning in zebra finches (*Taeniopygia guttata*): Progress and prospects. In J. S. Rosenblatt, C. Beer, M.-C. Busnel, & P. J. B. Slater (Eds.), *Advances in the study of behavior* (Vol. 18, pp. 1–34). New York: Academic Press.

Small, W. S. (1900a). An experimental study of the mental process of the rat. *American Journal of Psychology, 11*: 133–65.

(1900b). An experimental study of the mental processes of the rat. II. *American Journal of Psychology, 12*: 206–39.

Smith, W. J. (1977). *The behavior of communicating*. Cambridge, MA: Harvard University Press.

Snow, C. E., & Hoefnagel-Höhle, M. (1978). The critical period for language acquisition: Evidence from second language learning. *Child Development, 49*: 1114–28.

Snowdon, C. T. (1988). A comparative approach to vocal communication. In D. W. Leger (Ed.), *Nebraska Symposium on Motivation: Comparative perspectives in modern psychology* (Vol. 35). Lincoln: University of Nebraska Press.

Thorndike, E.L. (1911). *Animal intelligence*. New York: Macmillan.

Thorpe, W.N. (1974). *Animal and human nature*. New York: Doubleday.

Todt, D. (1975). Social learning of vocal patterns and modes of their application in Grey parrots. *Zeitschrift für Tierpsychologie, 39*: 178–88.

Vanayan, M., Robertson, H. A., & Biederman, G. B. (1985). Observational learning in pigeons: The effects of model proficiency on observer performance. *Journal of General Psychology, 112*: 349–57.

Vygotsky, L. (1962). *Thought and language*. Cambridge, MA: MIT Press.

(1978). *Mind in society: The development of higher mental processes*. Cambridge, MA: Harvard University Press.

Wadsworth, B. J. (1978). *Piaget for the classroom teacher*. New York: Longman.

Wertsch, J. V. (1985). Introduction. In J. V. Wertsch (Ed.), *Culture, communication, and cognition: Vygotskian perspectives* (pp. 1–18). Cambridge University Press.

West, M. J., & King, A. P. (1985). Learning by performing: An ecological theme for the study of vocal learning. In T. D. Johnston & A. T. Pietrewicz (Eds.), *Issues in the ethological study of learning* (pp. 245–72). Hillsdale, NJ: Erlbaum.

(1987). Settling nature and nurture into an ontogenetic niche. *Developmental Psychobiology, 20*: 549–62.

Wickler, W. (1976). The ethological analysis of attachment. *Zeitschrift für Tierpsychologie, 42*: 12–28.

(1980). Vocal duetting and the pairbond: I. Coyness and the partner commitment. *Zeitschrift für Tierpsychologie, 52*: 201–9.

Zentall, T. R., Hogan, D. E., & Edwards, C. A. (1984). Cognitive factors in conditional learning by pigeons. In H. L. Roitblat, T. G. Bever, & H. S. Terrace (Eds.), *Animal cognition* (pp. 389–405). Hillsdale, NJ: Erlbaum.

—12—

Pongid pedagogy: the contribution of human–chimpanzee interactions to the study of ape cognition

Sarah T. Boysen

Editors' introduction

There are several parallels between Boysen's chapter and the preceding contribution by Pepperberg. The research in both cases focuses on the animal's cognitive abilities, and like Pepperberg, Boysen has made impressive strides in that direction with chimpanzees.

Boysen provides a vivid account of the day-to-day concerns in socializing and teaching her animals. The chapter is rich in detail and offers unique insights into the development of bonding and the establishment of relationship ground rules. Boysen's is another case in which there would be no data collection without close bonding between scientist and animal.

Introduction

There is little question that, by a variety of indices, the chimpanzee (*Pan troglodytes*) shares a considerable portion of its evolutionary past with humans (*Homo sapiens*) (Goodman, Braunitzer, Stangl, & Shrank, 1983; Goldman, Giri, & O'Brien, 1987). These include immunological, genetic, anatomical, and morphological characteristics, as well as numerous features of its complex and dynamic social structure (de Waal, 1982; Goodall, 1986). This puts the chimpanzee in an interesting position when placed in an experimental setting, such as a cognition laboratory, where the two species must interact. And the evidence to date still supports the proposition that humans do differ significantly from other species, including the chimpanzee. This is no simple allusion to the obvious morphological differences, but rather more fundamental features of their developmental, social, and evolutionary history.

Despite one suggestion to the contrary (Fouts, Fouts, & Van Cantfort, 1989), a significant difference is that most chimpanzees are not natural teachers (Premack, 1984). For example, in their native environment, there is no substantive documentation that chimpanzees intentionally teach one another, even mothers their infants. This difference sets apart from humans the chimpanzee. It has been the purposeful transmission of information from one generation to another on which humans have both challenged and capitalized individuals' limitations and capabilities. In the chimpanzee, what has been documented is the strong contribution of observational learning to emerging social behavior, a significant behavioral adaptation toward flexible group living, as well as more dialect-like cultural traditions, including specific types of tool use or utilization of food types (McGrew, 1983). Clearly, "teaching," in the tutorial sense, and learning, even in a complex social context, are quite different processes, and humans appear, as a species, to be among the best true teachers.

Limited teaching skills are also seen among the other great apes, including gorillas and orangutans. However, when provided with the opportunity to acquire skills and concepts that would never be required in its natural environment, the chimpanzee typically excels. Moreover, this potential for *learning* has hardly been tapped in the several hundred years that chimps have been maintained in captivity, and only in the past two decades have we acquired a significant understanding of their cognitive potential (e.g., Premack, 1976, 1986). Important contributions have come from numerous projects focusing on the study of language-like processes, numerical competence, and other cognitive abilities in chimpanzees and other great apes (e.g., Boysen & Berntson, 1989; Matsuzawa, 1985; and Premack, 1986, with chimpanzees; Miles, 1983, with orangutans; Patterson, 1978, with gorillas). Significant information has also been contributed by efforts at resocializing chimpanzees (Fritz, 1986), as well as by zoos and research facilities where chimpanzee social groups have been carefully studied (Bloomstrand, 1989; de Waal, 1982; van Hooff, 1972). From these various sources, the chimpanzee emerges as capable of complex social manipulation and intricate political maneuvering, combined, particularly in the case of males, with enormous physical strength. Such a combination of attributes presents a challenge to the behavioral scientist interested in their behavior and cognitive abilities in a captive setting. Yet it is precisely this combination of characteristics that has provided the raw material for some of the most ingenious experiments in animal cognition (Gardner & Gardner, 1984; Gardner, Gardner; &

Van Cantfort, 1989; Gillan, Premack, & Woodruff, 1981; Menzel, 1973; Premack, 1986).

The nature and establishment of relationships with captive chimpanzees

After only a few minutes of observing chimpanzees, the striking overlap with respect to facial expressions, gestures, and social interaction among and between the animals, and similar behaviors observed in humans, is quite apparent (Ploiij, 1978; van Hooff, 1972). While similarities can appear striking, the differences between the two species are equally critical. For example, chimpanzee groups often exhibit sudden and violent changes in behavior, with most members of the group reacting, often in a dramatic manner, to the precipitating event and/or individual. This emotional lability not only is observed in captive groups, but has been frequently noted in wild chimpanzees (Goodall, 1986). Behaviors range from explosive rage to, typically, reconciliation behaviors among the principal parties (S. T. Boysen, personal observations; de Waal, 1982; Goodall, 1986). Such potential for reactivity makes working with captive chimps extremely difficult and demanding, and requires a form of daily "negotiation" between subjects and experimenters. In our laboratory, for example, much time is devoted to social interaction and play between individual chimps and teacher/experimenters, before, during, and after daily testing. For the most part, training and experimental evaluation of the animals' skills are but one facet of the daily routine, and are not typically set apart in any respect from other laboratory activities that occur on a regular basis – cleaning of the cages, provision of foliage for nest building, distribution of other enrichment items, such as drawing materials or toys, and similar routine activities.

The emergence of the inevitable bond

Despite their emotional lability, chimpanzees are, in general, very gregarious, and intense relationships with their human care givers not only unfold, but are pivotal to the very nature of the research questions we are pursuing regarding chimpanzee cognitive abilities. As noted in this volume, other captive experimental animals, such as laboratory rats, are quite sensitive to a variety of experiences, including handling and other types of interactions with the experimenter (Davis & Pérusse, 1988), and such factors could no doubt, in some cases, contribute to experimental outcomes. With the chimpanzee, a close, stable relationship with the human teacher is crucial. In a very real sense, no viable data will likely be collected without such a relationship. In some animal research, it is possible to structure the task such that minimal human interaction is necessary, and the animal is still able to perform and/or learn effectively. For example, it is possible to place a rat in a maze and for the subject eventually to find its way to the goal box. Experimenters typically do not monitor the rat's behavior, pointing this way or that, or verbally encourage the rat to turn left here, right there. The task demands are such that the rat is fully capable of performing adequately on its own. For chimpanzees immersed in the types of studies conducted in our laboratory, however, the relationship with the experimenter/teacher is important for several reasons. First, striking individual differences in the animals' personalities sometimes require differential teaching strategies, or particular tasks may prove more difficult for one animal than another and similarly require slightly different approaches to training (more repetition or different correction procedures during initial acquisition, etc.). In addition, one of the exciting features of working with chimpanzees is the significant influence that the social context has on the animals' ability to learn. With a chimpanzee subject, the experimenter is challenged with providing a sufficiently complex task that will reveal the range and limitations of the animal's cognitive abilities, yet continue to encourage active participation by the animal. This is not, in many cases, an easy task, as noted for other chimpanzees who were studied over a long period of time, such as Viki (Hayes & Hayes, 1951; Hayes & Nissen, 1971) and Sarah (Premack, 1986), even though it has been documented for some time that the chimpanzee is nevertheless quite responsive in a tutorial situation (Kohts, 1923; Premack & Premack, 1972). Nurturing the level of competence required for complex tasks, particularly those involving symbol manipulation, requires a strong attachment between pupil and teacher, and feedback and guidance provided to the chimp in a social-interactional approach makes a significant difference.

It is clear that relationships among and between the chimps, and between the chimps and their teachers, grow and change. The development of such relationships has also been noted in other great ape laboratories with goals similar to ours (Gardner, Gardner, & Van Cantfort, 1989). A limited group of people work with the chimps, since it takes a considerable period of adjustment for the animals to accept new teachers, and the disruption that results from new staff is expressed in all aspects of the animals' behavior, including task performance. Thus, a stable group of familiar individuals whose behaviors become predictable to the animals are necessary for the animals to direct their attention to the demands made on them during training.

Chimpanzee pragmatics

The easiest way to elaborate on how chimp–human relationships in the laboratory affect the research effort is to describe our experimental subjects. They are Sarah, Darrell, Kermit, Sheba, and Bobby – individual animals with names, unique personalities, idiosyncrasies, and what might be described as "opinions." This description clearly characterizes the chimpanzee as a species and thus would hold in any laboratory setting where chimps and humans work in close proximity. The animals look very different from one another, act differently from one another, and require attention as individuals. The list of personal likes and dislikes, ranging from play activities to foods to reactions to new people joining the laboratory, is extensive among our five chimps. Each animal is truly an individual, and we must often tailor our teaching approaches to such individual variability while holding experimental manipulations constant.

Maintaining chimpanzees in captivity is quite labor intensive, and as already alluded to, we have found it important for the same individuals who provide daily care (cleaning and feeding) to be the same people who serve as the animals' teachers. This permits a more thorough appreciation of a particular animal's needs and establishes opportunities for social interaction around a wide range of activities. Experimenters do not simply arrive at the laboratory at a given time to run a certain number of test trials per day, or run a study for a few weeks, but rather are in the lab with the chimps throughout the day, seven days a week. Human teachers therefore provide a large portion of the significant relationships in the animals' lives. One of the easiest ways to establish such a relationship with a chimpanzee is to grow up with the animal – that is, to begin to interact with the chimp when he or she is perhaps 1 or 2 years old and to maintain a long-term relationship. This is particularly important through the difficult years of emerging adolescence. Nonetheless, given experience with chimpanzee captive behavior (because of the obvious potential for aggressive behavior), it is also possible to establish a stable relationship with adult animals.

Pongid pedagogy

In addition to animal management difficulties associated with large apes, there are a host of practical problems relative to engaging the animals as experimental subjects. With the exception of Sarah, who arrived at age 28, the other four animals were introduced to the project between the ages of $2\frac{1}{2}$ and $3\frac{1}{2}$. Because of the complex nature of the questions we hoped to address with the animals, as well as successful experience with similar approaches in other laboratories, we have elected to maintain an interactive teaching situation that focuses on gamelike tasks and extensive social interaction between the chimps and their teachers. This initially required clear and deliberate instructions from humans as to the range and limits of the chimps' behavior within the lab. In many instances, the rules were very specific and continually enforced. For example, no chimpanzees were permitted to jump down from elevated sites onto an unsuspecting human, *unless* the human had indicated his or her willingness to engage in a specific game that somehow required this activity. While not so critical when the animals were small, it was an important rule to establish early in the animals' training.

It was also necessary to encourage the curtailment of other species-typical behaviors, such as using teeth during play (which chimps readily do with one another). Because we planned to work with food as incentives throughout the day, it was also important to establish rules about such resources. The chimps were taught (via verbal commands) never to help themselves freely to any type of edible (or perceived edible, such as disinfectant) that might be available in the laboratory. After a number of years, with the rule established with the chimps as youngsters, they will now sit seemingly unconcerned, even though food may be easily within reach, and never attempt to pilfer. This holds even if the teacher gets up and leaves the testing area for a period of time. Coupled with the great variety of foods the animals receive throughout the day and no food deprivation, the rules surrounding access to food may have deemphasized the significance of food as a reinforcer in some unique ways. In one case, when the delivery and ingestion of incentives interfered with our ability to collect reaction-time data, the use of food reinforcement was abandoned and proved unnecessary for maintaining attention and for motivating the animals to perform (Berntson & Boysen, 1990). Social praise alone was sufficient, and thus the acquisition of food-related rules may have resulted in greater flexibility in adjusting to novel experimental procedures that were less constrained than a strict reinforcement approach might dictate.

Such rules of social conduct within the laboratory must always be conveyed to new project personnel to ensure a consistent set of working guidelines for interacting with the animals. These include reasonably defined limits for human and chimp conduct and some standards for dealing with infractions. In turn, the animals have established expectations about the experimenters that likely provide them with some control over their surroundings. Their environment is fairly predictable and, as such, perhaps less stressful than it might otherwise be. For example, larger chimps that no longer work unrestrained outside their home cages are theoretically free to leave a teaching/testing session at any time simply by moving away from the front of the cage. All the animals have access to outdoor play areas 24 hours a day, 365 days a year. But with rare exceptions, the animals readily engage in working on tasks and express great excitement when materials are being readied in front of their home cages.

Some rules of interaction have been initiated with specific animals or particular situations in mind. For example, our two subadult males, Kermit (11 years) and Darrell (11½ years), have been housed together since infancy. Shortly after their arrival (at ages 3 and 3½), we instituted a policy of turn taking with them. By encouraging cooperative sharing of the opportunity to work on a task, the teacher did not have to separate the two animals physically during teaching sessions. Separation would have been very stressful for both animals and would likely have resulted in greater difficulty or the complete inability to attend to training. Once the animals had been taught to take turns, one chimp was free to play, while the other could devote his attention to the task. After two or three trials, the animals switched places. During the initial training on turn taking, Darrell, who was the dominant animal and also physically larger, was reluctant to permit Kermit to have an opportunity to work with the teacher. However, this tendency was actively discouraged, again through vocal commands and encouragement by the teacher, and eventually subsided. Currently, while both animals are quite large (approximately 160 pounds) and, as mentioned, free to leave the testing situation at any time, Kermit and Darrell actively take turns by physically exchanging places. This may mean working at a testing apparatus that fits against the front of the cage or moving aside to permit the other chimp to interact with the experimenter through the front of the cage on tasks that require manipulation of objects. They respond to verbal requests to allow the other to have a "turn" or will exchange places in response to the teacher's gestures to "move aside."

Another practice with Kermit and Darrell, which is idiosyncratic to them and their relationship, concerns most tasks related to numerical competence. We are currently exploring a range of issues related to counting with the chimps (Boysen & Berntson, 1989) and began training several years ago with Kermit, Darrell, and Sheba. While individual differences emerged early in numbers training, all animals were progressing along comparable levels until the introduction of the Arabic numerals beyond 2. At this point, Kermit's performance deteriorated dramatically. We have worked around Kermit's difficulties with numbers so that effective interaction with Darrell might proceed (Darrell's current counting repertoire includes 0 through 7). Thus, on all number tasks, even though Darrell takes all the turns, Kermit sits and watches, and receives his own portion of any food rewards that Darrell earns throughout the session. Kermit is not required to attend, however, and he sometimes elects to move outside to play or interact with other animals in adjacent cages while Darrell and the teacher work together. Either way, Darrell is free to engage in the task, and Kermit is not left to distract or otherwise attempt interaction with him, which would interfere with the teaching situation.

Teaching versus training animals

While lacking in purposeful teaching capacities, the chimpanzee is quite adept at acquiring complex concepts under the guidance of a skilled human teacher. Given its extraordinary capabilities for observational learning, coupled with a generally social nature and spontaneous inquisitiveness, the chimpanzee is, above all else, "teachable." It is one thing to create a structured situation for an animal to engage in some circumscribed behavior, such as pressing a bar in an operant chamber, and quite another for an animal to be both receptive to, and capable of, incorporating feedback from a human teacher within a highly dynamic, interactional context. In this regard, with the chimpanzee one is faced with a nonhuman species that is capable of a remarkable form of dialogue – not verbal, but clearly marked by the direct exchange of information between the two individuals (chimp and human teacher) and the incorporation of such feedback into the stream of behavior exhibited by the animal. These receptive capacities are powerful tools, for they permit the sharing of knowledge at a rapid pace and within a specific task framework. This type of exchange can thus circumvent inappropriate response strategies, classically described for nonhuman primates by Harlow (1950) and Levine (1965), and help redirect attention to critical features of the task that might otherwise be overlooked.

Teaching humans is so second nature that one may forget its tremendous impact in directing attention away from less efficient approaches to solving a problem. Persistence by the chimp with ineffective strategies such as perseveration, resulting in errors, increased frustration, or perhaps failure at the task, can be minimized with appropriate guidance from the teacher. With the chimpanzee as an experimental subject, the inevitable relationship that develops between two highly similar species encourages a kind of synergistic collaboration that permits the simultaneous flow of information through several channels. Not only does the animal extract information from context via multiple sensory channels, but the human teacher specifically directs and redirects attention to focal aspects of the exchange, often employing concurrent modalities (vocal commands or verbal highlights, gestures, and other nonverbal feedback). By verbally directing attention, in concert with gestures such as pointing or indicating, the teacher accentuates critical features of the task, and other noise in the learning situation can be minimized. A teaching approach such as this, with multiple opportunities for clarification, directed attention, and different modalities utilized to focus continually on crucial task features, provides significant redundancy (Campbell, 1982). An important adjunct to redundancy is the convergence of efforts to direct the chimp's attention to the salient feature(s) of the task. More important, it is this convergent redundancy that can be organized by the human teacher specifically for an individual learner. It can also be continuously updated and redirected as the dialogue of learning between the chimp and the human continues.

The nature of such interaction between chimpanzees and their teachers is, in our experience, one of acquired synchrony. Initially, early in their training, our animals' attention spans were exceptionally short. In retrospect, it seems that the process of socialization between chimpanzees and humans in this type of teaching situation may be a combination of entrainment of attention and developmental changes associated with a maturing brain. Over time, the animals were capable of attending for longer periods of time and integrating more information simultaneously. Thus, while initial tasks revolved around simple match-to-sample tasks that could be readily accomplished (i.e., matching colors and shapes), more recent studies have introduced complex tasks that include novel test demands designed to challenge the animal's attentional and conceptual capabilities (e.g., functional counting; see Boysen & Berntson, 1989).

One successful teaching strategy, which might be described as "cognitive scaffolding," is similar to an approach described by Wood, Bruner, and Ross (1976) for tutoring children. As applied to the chimps, previously taught skills are used as building blocks for new tasks that retain some of the original training features, but whose structures are parsed into the smallest components possible to ensure successful performance early in the task. For example, all the animals initially learned to match colors, and this match-to-sample strategy was then used to present shapes in a novel testing situation. The chimps readily transferred their skills at matching colors to the new shape stimuli.

From shapes, a similar task was devised that required the animals to match photographs of foods to actual foods presented. Once again, the basic match-to-sample features were preserved, but in this case the sample stimuli were presented as two-dimensional photographs, rather than color placards or wooden shapes. Gradually, new and more divergent tasks were introduced that shared and/or overlapped other skills and task parameters. This approach encouraged successful acquisition by the animals, while minimizing frustration and maintaining the gamelike structure of the teacher–student interaction. The animals were therefore not experimentally naïve when challenging new tasks were presented, and their training history and successful integration of familiar concepts and skills were of critical importance. In this regard, they are reminiscent of children who are not stripped of all previous learning experiences when they encounter new task demands or novel opportunities to take in new information. Children's ability to tackle such situations takes advantage of their existing repertoire of experiences and skills, and our chimps are similarly endowed. In this respect, they are quite different from many chimpanzees, even most laboratory chimps, since they bring a very real cultural heritage to each learning situation. Yet we believe they also represent a measure of the potential for learning that any chimpanzee who has undergone similar kinds of purposeful teaching might demonstrate.

The inevitable bond that develops between chimpanzees and their human teachers may, at first glance, seem somewhat unidirectional. However, it is just as important for the chimpanzees to feel confident in their roles as students as it is for their human teachers to be effective in their capacity. The development of teachers' roles has a decided time course with chimps (see Pepperberg, Chapter 11; Oden & Thompson, Chapter 13, this volume), and the changing status of these roles directly affects the animals' performance. Humans must first become the chimpanzees' friends, and much time has to be devoted to social interaction, including play, grooming, and other interactions, before a comfortable working relationship can emerge. Even the role of friend is not readily accomplished by just any human, however. Some qualities of individual humans' personalities that may be important for working with chimpanzees (i.e., the ability to readily disinhibit and the related ability to be highly animated in some situations) are somewhat definable, while still other qualities are almost elusive. Whatever the traits, skills, and inclinations may be, they are rare in humans and are reflected in the small number of laboratories where successful chimpanzee–human teaching situations occur.

Acquiring the facility to play the role of a chimpanzee's friend, however, is easy compared with making the transition to the role of teacher. Until the animal's relationship with a human friend is entirely resolved – that is, the chimp is comfortable in a play situation, readily inhibits aggressive tendencies, and exhibits numerous prosocial behaviors such as grooming directed at the individual – that person will not able to *teach* the chimp anything, much less anything new or difficult. Rather, the chimp will spend enormous amounts of energy challenging the new person. Once the transition to teacher is accomplished, however (and this phase may take months), a remarkable exchange between student and teacher can proceed. Hence, regardless of how committed and enthusiastic the teacher might be, he or she may not accomplish anything unless the chimp is comfortable in the relationship. A chimpanzee has to want to be engaged with a human as a facilitator, and there is almost nothing that can be done to otherwise engender cooperation, short of investing the hours and months necessary to nurture the bond. Food deprivation is seldom used with any primates; it is simply not effective (see also Davis, 1984). Without question, the relationship between the chimpanzee and his or her teacher is pivotal, and bidirectional. Over time, both chimp and human develop strong bonds of friendship, trust, and affection that support the critical relationship necessary for pongid pedagogy, and it is time well invested.

Acknowledgments

This research was supported by grants from the National Science Foundation (BNS-8820027) and the National Institute of Mental Health (RO3 MH44022). Two of the chimpanzees are on permanent loan from Yerkes Regional Primate Research Center, Emory University, which is supported by Base Grant RR-00165. The Yerkes Center and the Ohio State University are fully accredited by the American Association for Laboratory Animal Science. The continued cooperation of Jack Hanna, director of the Columbus Zoo, and Dr. Lynn Kramer, D.V.M., director of Research, is gratefully acknowledged. The contributions of many supportive colleagues, including Gary G. Berntson, Barbara Thomson, Juanita Mays, and Karen Trego, are deeply appreciated. Numerous students and volunteers, including Traci Shreyer, Kurt Nelson, Daniel J. Povinelli, Lisa Raskin, Karen Henderson, Diane Donahue, Kirstin Bryan, Diane Jones, Shana Weber, Nicole Adimey, Lisa Cannon, and Yvonne Jorrey, have contributed generously to the day-to-day care and study of the chimpanzees.

References

Berntson, G. G., & Boysen, S. T. (1990). Cardiac correlates of cognition in chimpanzees and children. In L. Lipsett & C. Rovee-Collier (Eds.), *Advances in infancy* (pp. 187–220). New York: Ablex.

Bloomstrand, M. A. (1989). Interactions between adult male and immature captive chimpanzees: Implications for housing chimpanzees. *American Journal of Primatology*, Supplement, *1*, pp. 93–99.

Boysen, S. T., & Berntson, G. G. (1989). Numerical competence in a chimpanzee (*Pan troglodytes*). *Journal of Comparative Psychology, 103*, 23–31.

Campbell, J. (1982). *Grammatical man*. New York: Simon & Schuster.

Davis, H. (1984). Discrimination of the number three by a raccoon (*Procyon lotor*). *Animal Learning & Behavior, 12*, 409–13.

Davis, H., & Pérusse, R. (1988). Human-based social behavior can reward a rat's behavior. *Animal Learning and Behavior, 16*, 89–92.

de Waal, F. (1982). *Chimpanzee politics*. New York: Harper & Row.

Fouts, R. S., Fouts, D. H. & van Cantfort, T. E. (1989). The infant Loulis learns signs from cross-fostered chimpanzees. In R. A. Gardner, B. T. Gardner, & T. E. Van Cantfort (Eds.), *Teaching sign language to chimpanzees* (pp. 280–92). Albany: State University of New York Press.

Fritz, J. (1986). Resocialization of asocial chimpanzees. In K. Benirschke (Ed.), *Primates: The road to self-sustaining populations* (pp. 352–9). New York: Springer.

Gardner, R. A., & Gardner, B. T. (1984). A vocabulary test for chimpanzees (*Pan troglodytes*). *Journal of Comparative Psychology, 98,* 381–404.

Gardner, R. A., Gardner, B. T., & van Cantfort, T. E. (Eds.). (1989). *Teaching sign language to chimpanzees.* Albany: State University of New York Press.

Gillan, D., Premack, D., & Woodruff, G. (1981). Reasoning in the chimpanzee: I. Analogical reasoning. *Journal of Experimental Psychology: Animal Behavior Processes, 7,* 1–17.

Goldman, D., Giri, P. R., & O'Brien, S. J. (1987). A molecular phylogeny of the hominoid primates as indicated by two-dimensional protein electrophoresis. *Proceedings of the National Academy of Sciences of the United States of America, 84,* 3307–11.

Goodall, J. (1986). *The chimpanzees of Gombe: Patterns of behavior.* Cambridge, MA: Harvard University Press.

Goodman, M., Braunitzer, G., Stangl, A., & Shrank, B. (1983). Evidence on human origins from haemoglobins of African apes. *Nature, 303,* 546–8.

Harlow, H. F. (1950). Analysis of discrimination learning by monkeys. *Journal of Experimental Psychology, 40,* 26–39.

Hayes, K. V., & Hayes, C. H. (1951). *The ape in our house.* New York: Harper Bros.

Hayes, K. J., & Nissen, C. H. (1971). Higher mental functions in a home-reared chimpanzee. In A. M. Schrier & F. Stollnitz (Eds.), *Behavior of non-human primates* (Vol. 4, pp. 59–115). New York: Academic Press.

Kellogg, W. N., & Kellogg, L. A. (1933). *The ape and the child.* New York: McGraw-Hill.

Kohts, N. (1923). Untersuchungen uber die Erkenntnisfahigkeiten des Schimpansen Aus dem Zoopsychologischen Laboratorium des Museum Darwinianum in Moskau [in Russian]. From the translation reported in R. M. Yerkes & A. W. Yerkes (1929), *The great apes: A study of anthropoid life.* New Haven, Conn: Yale University Press.

Levine, S. (1965). Hypothesis behavior. In A. Schrier, H. Harlow, & F. Stollnitz (Eds.), *Behavior of non-human primates* (Vol. 1, pp. 97–127). New York: Academic Press.

Matsuzawa, T. (1985). Use of numbers by a chimpanzee. *Nature, 315,* 57–9.

McGrew, W. C. (1983). Animal foods in the diet of wild chimpanzees: Why cross cultural variation? *Journal of Ethology, 1,* 46–61.

Menzel, E. W. (1973). Chimpanzee spatial memory organization. *Science, 182,* 943–5.

Miles, L. (1983). Apes and language: The search for communicative competence. In J. De Luce & H. T. Wilder (Eds.), *Language in primates: Perspectives and implications* (pp. 43–61). New York: Springer.

Patterson, F. (1978). The gestures of a gorilla: Sign language acquisition in another pongid species. *Brain and Language, 5,* 72–97.

Ploiij, F. X. (1978). Some basic traits of language in wild chimpanzees. In A. Lock (Ed.), *Action, gesture, and symbol* (pp. 111–32). New York: Academic Press.

Premack, A. J., & Premack, D. (1972). Teaching language to an ape. *Scientific American, 227,* 92–9.

Premack, D. (1976). *Intelligence in ape and man.* Hillsdale, NJ: Erlbaum.

(1984). Pedagogy and aesthetics as sources of culture. In M. Gazzaniga (Ed.), *Handbook of cognitive neuroscience* (pp. 15–35). New York: Plenum.

(1986). *Gavagai.* Cambridge University Press.

van Hooff, J. A. R. A. M. (1972). A comparative approach to the phylogeny of laughter and smiling. In R. Hinde (Ed.), *Non-verbal communication* (pp. 129–79). London: Royal Society and Cambridge University Press.

Wood, D., Bruner, J. S., & Ross, G. (1976). The role of tutoring in problem-solving. *Journal of Child Psychology and Psychiatry, 18*, 89–100.

—13—
The role of social bonds in motivating chimpanzee cognition

David L. Oden and Roger K. R. Thompson

Editors' introduction

Like Boysen, Oden and Thompson study cognition in chimpanzees. They report that their animals come to appreciate the
task-oriented character of experimental situations and distinguish
between "work" (the experiment) and "play" (informal social
interactions).

Oden and Thompson argue that a "contract" emerges between
scientist and animal that determines the subject's performance.
Like Boysen, they emphasize the uniqueness of the teacher role
and argue that animals can differentiate between scientists acting
as "mentors" and "playmates." This differentiation is reflected in
an enhanced willingness to participate in and improved performance on cognitive tasks.

Introduction

Repeated interactions between human and nonhuman primates, especially
apes, often result in the formation of a social bond. Such an affiliative
relationship is most likely to occur when the interactions are not limited to
routine caretaking and data collection, but include more "social" activities
such as vocalizations, mutual grooming, and play. There is increasing
appreciation that these human–animal social interactions are potentially as
important to the psychological well-being of the animal as are physical
factors in its environment (Segal, 1989). However, even those individuals
who are concerned with the issue of psychological well-being have not
challenged the traditional view that social interactions during data collection are anything but problematic.

Traditionally, research with animal subjects has minimized interactions between animal and human to avoid experimenter expectancy effects and incidental cuing (e.g., Clever Hans effect). One solution to these design "problems" is to automate data collection as much as possible. Placing the subject in the socially impoverished environment of an automated experimental chamber precludes any interaction between the animal subject and human experimenter during data collection.

In contrast, the fundamentally social character of cognition (Vygotsky, 1962, 1978) is recognized implicitly in many studies of human cognitive development (e.g., Lamb & Sherrod, 1981; Piaget, 1954; Walker-Andrews, 1988). Studies of intelligence, memory, and related cognitive skills in infants and children often involve social interactions between subject and experimenter that are designed to facilitate the identification and expression of cognitive skills.

In this chapter we argue that the broadest range of cognitive processes of nonhuman primates, like those of children, will be identified using procedures that involve social interactions between subject and experimenter. Investigators of human cognitive development have long recognized that the social context of a testing situation is an important determinant of cognitive performance; we believe that the same is true of apes.

Our belief is consistent with recent suggestions that the evolutionary origins of primate cognition lie in the social contingencies necessarily encountered by individuals in the daily life of their group (Byrne & Whiten, 1988; Humphrey, 1976; Jolly, 1966). The general argument is that problem-solving skills conveyed a selective advantage to those who could keep track of kinship relations and tally tit-for-tat interactions among other group members. These evolutionary approaches to animal cognition emphasize the role of the social milieu as a distal causal factor and have provided important insights into the possible origins of primate cognitive skills (see Cheney, Seyfarth, & Smuts, 1986; Kummer, 1982; Mason, 1982).

Presumably, cognitive skills that originated as adaptations within the social domain were precursors of the more general learning abilities demonstrated by social creatures in the traditional animal learning laboratory. However, practitioners of the evolutionary approach appear to have overlooked an additional, more proximate role played by social factors. These factors might make an important motivational contribution to the expression of cognitive skills that transcend social knowledge and are applied as general problem-solving strategies.

If the motivational basis for complex cognition was originally social in nature, then it is likely that such cognition continues to be influenced by social dynamics. We contend that the social milieu has a potentiating effect on cognition regardless of whether the specific "problem" facing the animal is social. Our hypothesis is that in order for the spontaneous expression of complex, broadly construed cognitive abilities to occur, a particular hedonic set must be established. Furthermore, this motivational state will be activated when the cognitive task is performed in the context of social interactions.

We have found that the chimpanzee's performance on a variety of cognitive tasks varies as a function of the social milieu, especially the animal's social relationship to the human experimenter. It does so in a way that cannot be attributed simply to experimenter bias, Clever Hans effects, or the animal's preference for the human as a companion. We believe that the type of social motivation evoked by the human directly influences the application of cognitive processes by an ape. Our ideas are based on experimental data and informal observations of infant chimpanzees who were subjects in a larger study of cognitive development.

The social milieu

The animals in question were four infant chimpanzees (*Pan troglodytes*) that were observed over a 2-year period at the University of Pennsylvania's Primate Research Center at Honeybrook. The three females (Freida, Liza, and Opal) and one male (Whiskey) were born captive and brought to the facility at approximately 4 to 5 months of age. The observations described here were begun when they were about 8 to 9 months of age.

The chimpanzees were first familiarized with any person who was to serve as either caretaker or experimenter. This familiarization took the form of simulating the normal dynamics of a captive chimpanzee troop. Humans engaged in rough and tumble play, mutual grooming, hugging, and soliciting social interactions. Humans also encouraged exploration of the indoor and outdoor environments. In the former, which consisted of toy-filled nursery-like rooms, humans encouraged manipulation of new objects, some of which were used subsequently in testing. In the latter, which was a verdant 0.2-hectare (0.5-acre) compound, humans pointed out new objects and places and encouraged active exploration and exploitation as in the case of foraging for berries. Much of the exploration of both indoor and outdoor environments, of course, was initiated by the animals. During mealtimes, especially at lunch, a human often controlled access to food and "shared" it with the chimpanzees.

A human was judged to be "familiarized" when each chimpanzee would leave her or his peers and approach the human when called. The total time required for familiarization ranged from 10 to 20 hours of interaction. The work schedules of the humans varied such that this was accomplished sometimes during a week of concentrated interaction and, at other times, over a period of 3 to 4 weeks with less frequent weekly contact.

Humans who interacted with the infant chimpanzees included senior personnel (e.g., research associates and postdoctoral students), full-time technicians, and undergraduate and high-school student volunteers. After the initial familiarization period, the degree to which humans participated in "troop" life varied. First, schedules and assigned project responsibilities were such that the total time each person spent with the chimpanzees, either conducting experiments or engaging in social interaction, varied. Some full-time personnel spent more than half of every day in the animals' company, whereas some volunteers saw the animals for only 1 to 2 hours, two to three times per week. Additionally, and largely independent of the total time spent with the animals, the proportion of time each person spent in free play, as opposed to data collection, varied. Most experimental data were collected in the late morning, after breakfast and a recreation period, as well as in the early afternoon, after lunch and another recreation period. Late afternoons and early evenings constituted another period of free play. Staff schedules were such that a person might be available primarily during experimental periods, primarily during recreation and natural observation periods, or at any time during the day.

Because of these variations, it is possible to describe two major patterns of differential responsibilities among the humans who were responsible for collecting experimental data. These patterns were independent of the total time an individual spent interacting with the chimpanzees across all situations. We labeled the two patterns "playmate" and "mentor." By "playmate" we mean a person who spent the major proportion of his or her time caretaking or in free play and who only occasionally collected data. By "mentor" we mean a person whose time was divided about equally between play and experimentation.

These differential responsibilities were partly accidents of each person's available time and partly intentional. The playmates constituted a pool of experimenters who were thoroughly familiar with the chimpanzees and with the basic procedures for data collection. However, these individuals generally were naïve with respect to any particular experimental hypothesis and an animal's performance on a particular task. They collected data during probe sessions to test for possible experimenter cuing by the mentors. Largely out of consideration for staff morale, we attempted to avoid as much as possible a third possible pattern – the "drill sergeant" – a person whose assignment was primarily experimentation.

The varied experimental tasks (e.g., see Oden, Thompson, & Premack, 1988, 1990) were all conducted in a social context. Sometimes data were collected in the course of free play (Oden et al., 1990, Expt. 1). More often,

Table 13.1. *Matching-to-sample accuracy as a function of experimenter*
(percent correct of 12 trials)

| | Experimenter | | | | | |
| | Mentor | | | Playmate | | |
Subject	BD	DO	PS	BP	JA	CD
Whiskey	100	92	100	67	75	58
Liza	92	92	100	67	67	75
Opal	83	67	92	58	50	67
Frieda	92	67	83	50	58	50
All	92***	79***	94***	60*	62**	62**

* $p < .10$, binomial test; ** $p < .05$, binomial test; *** $p < .01$, binomial test.

formal procedures were used and were conducted in rooms designated for training and testing only (Oden et al., 1988, 1990, Expt. 2). Each animal was always tested in the company of another chimpanzee in addition to any human testers and was never deprived of food (Oden et al., 1988). There were certain constraints on the animals' permissible social behaviors in the test rooms that were not enforced in the play situations. For example, during testing, theft of another animal's security blanket or food reward was prevented by verbal admonitions or gentle physical restraint. During training and testing, social rewards like hugging and verbal praise were applied as frequently as were fruit and yogurt rewards. Thus, both free play and testing situations were defined by social interactions with familiar humans and chimpanzee companions in environments associated with only one type of activity (i.e., "work" or "play").

Effect of social milieu on performance

We were first alerted to the possibility that intrinsic social motivation guided the animals' performances when a number of volunteers, with whom the chimpanzees were familiar, were given their first assignment to collect match-to-sample data. In this task the animal had learned to match a metal cup or lock with like objects (Oden et al., 1988). The data from each novice experimenter's (i.e., playmates BP, JA, and CD) first two 6-trial sessions with each subject are presented in Table 13.1 together with data collected by experienced, but equally familiar data collectors (i.e., mentors BD, DO, and PS). Matching performances by the chimpanzees varied from extraordinarily good to borderline significance, but still above pure chance. The relatively poor performances were associated with playmates, whereas good performances were associated with mentors. The playmates by definition had spent proportionately more time playing with the animals than testing them.

One might hypothesize that a particularly strong affective bond between subject and experimenter might lead to a decrement in performance. An animal's attention might be more easily distracted from the test materials and focused on the experimenter. According to this hypothesis, playmates, by virtue of their extensive informal interactions with the chimpanzees, might have established emotional bonds stronger than those established by mentors. Thus, the playmate, rather than the test materials, is more likely to be the primary focus of the chimpanzee's attention in the experimental situation. We tend to dismiss this as a full explanation. Although no formal data on emotional bonding were collected, there was general consensus among laboratory personnel regarding the "popularity" of each human. More popular persons were defined as those who were the first to be approached by the chimpanzees when two or more humans were available in a play area and those from whom the chimpanzees would most likely solicit play. In addition, a chimpanzee would tolerate brief periods of separation from her or his peers when accompanied by a highly popular human, but not if accompanied by a person who ranked low in popularity. We believe this informal consensus ranking represents a reasonable indicator of the magnitude of the chimpanzee–human emotional bond.

In general, playmates and mentors appeared at all levels of the popularity hierarchy. Referring to Table 13.1, mentor BD and playmate BP were both highly popular; mentor DO and playmate JA were both moderately popular. Furthermore, in Table 13.1, there is no systematic relation between popularity and performance. Playmates BP, JA, and CD were of high, moderate, and low popularity, respectively, but the mean performances of the chimpanzees are highly similar. Thus, popularity of a human appears to be orthogonal to the chimpanzees' test performances.

Another line of evidence further persuaded us to reject the above-stated hypothesis that differences in performance across human testers reflected differences in affective bonds and their consequent effect on the chimpanzees' distractibility. If there had been a relationship between degree of affection and attention, we should have found similar experimenter-associated differences in performances across a variety of other tasks. This was not the case. For example, the same animals and additional experimenters, also falling into the categories of playmates and mentors, were involved in a pilot familiarization/novelty procedure in which the dependent measure was "looking time." Note that in this situation there is no "correct" response, but merely an opportunity to gaze at a stimulus. We

Table 13.2. *Mean looking times in familiarization/novelty procedure
as a function of experimenter (no. seconds/15-second interval;
mean for eight intervals)*

| | Experimenter | | | | | |
| | Mentor | | | Playmate | | |
Subject	BD	PD	PS	BP	JA	DN
Whiskey						
M	13.0	11.0	—[a]	11.0	—	—
SD	3.8	3.0	—	2.8	—	—
Liza						
M	9.7[b]	9.4	—	7.5[b]	—	—
SD	1.8	2.9	—	2.1	—	—
Opal						
M	—	—	8.2[c]	—	5.1[d]	11.5[c,d]
SD	—	—	1.9[e]	—	4.3[e]	2.7
Frieda						
M	—	—	6.2	—	5.4	7.7
SD	—	—	3.3	—	2.6	2.1

Note: Statistical tests (t, homogeneity of variance) on these data were conducted separately for each row of the table (between experimenters, within subjects).
[a] Dashes indicate that no data were collected for this subject by this experimenter in the pilot study.
[b,c,d] Means with common subscript were significantly different ($t(14), p < .05$).
[e] Test for homogeneity of variance was significant ($F = 4.99, p = .02$).

should certainly expect to find differences in cumulative looking times, or in variability of looking times, across experimenters if the animals were differentially distracted by them from the stimulus materials. The data presented in Table 13.2 do not support this prediction. Neither cumulative looking time nor variability of looking time across sessions varied systematically with humans' mentor or playmate status.

There was a similar lack of relationship between playmate/mentor status and data collected in paired-comparison preference tests conducted with the same animals. These data were collected in a pretest designed to equate each animal's preferences for all items used in match-to-sample experiments. In the preference tests, there is again no correct response, merely a requirement that the chimpanzee attend to the stimulus materials and select one of a pair of items. If the playmate, and not the stimulus materials, were the chimpanzees' primary locus of attention, we would expect them to be less discriminating in their selection of items. This was not the case. The mean Pearson r of preference ratings for eight objects when tested by playmates and rankings of the same objects when tested by mentors was .88.

In sum, neither the familiarization/novelty data nor the preference data support the hypothesis that chimpanzees are less attentive to stimulus materials when tested by a playmate. Consequently, a shift in the locus of attention does not seem to be an adequate explanation of the dramatic performance decrement that occurs when a playmate conducts a match-to-sample test.

The relationship between an experimenter's playmate or mentor status and an animal's performance appeared only on those tasks where correct and erroneous responses could be defined as in the case of match-to-sample. We then addressed the more obvious explanation that we and our subjects were unwitting perpetuators of a Clever Hans effect. This possibility, however, was ruled out by explicit tests involving conceptual matching trials interspersed with matching sessions with a lock and a cup (Oden et al., 1988). In the conceptual matching task a correct response entailed placing two objects that were either identical or different with another pair of objects that instantiated the same relation (i.e., place AA with BB and place CD with EF). If the relationship between experimenter and level of matching performance with objects was attributable to a Clever Hans effect, then we expected a similar pattern to emerge during the conceptual matching trials. This was not the case; performance on the latter task was essentially at chance regardless of experimenter (Oden et al., 1988).

If excellent object-matching performances could not be attributed to Clever Hans effects, then perhaps the poorer performances, with playmate as experimenter, resulted from the animals becoming confused on any given trial by some behavior of the experimenter. Hence, the poorer performances may have reflected a lack of finesse on their part as experimenters relative to those mentors who interacted with the chimpanzees equally often in both play and test situations. We are struck by the fact that in contrast to Clever Hans controls, the latter possibility is seldom if ever investigated as an explanation for consistently poor performances by either animal or child.

To test for this possibility, those individuals associated with poorer performances (i.e., playmates JA, CD, and BP) were observed either through a one-way mirror or by an observer in the room with them and the subjects. We were looking specifically for any departure by the playmates from standard experimental protocol (postures, gestures, verbalizations, reward administrations) that might have confused or excessively aroused

Table 13.3. *Matching-to-sample accuracy obtained by playmate with and without mentor experimenter in test room (percent correct of 12 trials)*

	Experimenter					
	Window observation			In-room observation		
Subject	BP	JA	CD	BP	JA	CD
Whiskey	67	58	58	75	92	92
Liza	67	58	67	92	92	92
Opal	58	50	58	75	83	83
Frieda	58	58	58	75	75	92
All	63**	56	60*	79***	85***	90***

$* p < .10$, binomial test; $** p < .05$, binomial test; $*** p < .01$, binomial test.

the animals. Two points are important here. First, in both cases, the experimenter knew that he or she was being observed, but the chimpanzee was only aware of the observer when he or she was in the room. Second, the observer was always an experimenter consistently associated with good object matching by the chimpanzees (i.e., a mentor). When in the room with experimenter and subject, the observer sat unobtrusively at some distance from them and did not interact in any way with either subject or experimenter.

The matching data from the subjects in those sessions that were observed are presented in Table 13.3. When observations were made through a one-way mirror, performances were similar to those shown in Table 13.1. However, when observations were made in the room, performances improved substantially. There were no detectable changes in the behavior of the experimenter across the two observation conditions. And no discernible blunders or mistakes in experimental procedures were noted in either observation condition. We seriously doubt that some subtle alteration of cues from the playmate, discernible only to a chimpanzee, could have accounted for the dramatic improvement in performance when a mentor was in the room.

The requirements of the main research program precluded any further systematic study of the mentor/playmate performance differential. The playmates' schedules were adjusted so that roughly equal amounts of time were spent in free play and experimental situations. After 10 to 15 hours of work with the animals on well-practiced matching tasks, data obtained by playmates were equivalent to those obtained by mentors. We did notice, however, a similar transient disruption of performance whenever new staff members first began collecting data in any experiment requiring a "correct" response from the subjects. We noted no such disruptions when mentors were transferred between problem-solving procedures (e.g., from match-to-sample procedures to a discriminative approach–avoid procedure), even though the chimpanzees may never before have experienced a particular mentor as experimenter in the context of a particular procedure.

Distinctions between work and play

Why should the mere presence of an observer who is independently associated with good performances enhance performance levels when data are collected by another experimenter? Also, as noted earlier, why should performance differentials appear only on those tasks demanding a "correct" answer from the animals? Finally, why should the playmate/mentor distinction be salient to the chimpanzees in a testing situation? The demands of the research program mapping the cognitive development of these animals precluded systematic investigation of these questions.

In retrospect, we have concluded that the remaining invariance that can best approximate answers to these questions is the social milieu engendered by an experimenter in the test situation. We suspect that the chimpanzee makes a distinction between work and play contexts that is analogous to that made by a human experimenter. The work context requires that the chimpanzee not only inhibit its normal exuberance with companions, but also focus its attention directly on the cognitive task at hand. That is to say, the chimpanzee adopts a "task-oriented" stance. A number of informal observations provide anecdotal evidence that chimpanzees indeed do this.

We have found, for example, that a particular constellation of emotional behaviors is evoked in the experimental situation. Tantrums, whining, and rocking, which are normally evoked in the play context only by startling or threatening stimuli, or by particularly strong discipline, occur in the experimental situation after mild reproof for an incorrect response. These emotional responses cannot be attributed easily to a high level of general arousal in the work context because other autonomic indicators (e.g., piloerection, urination, and defecation) were infrequent. Furthermore, these emotional responses were not ameliorated by those solicitous gestures and sounds by the human companion that were typically effective in the play context. If anything, solicitation in the work context intensified or prolonged the emotional outburst. We believe that this aberrant emotional response pattern reflects the animal's appreciation of the task-oriented character of the work context.

227

Further evidence that the chimpanzees are intrinsically motivated to solve problems in the work situation includes the finding that differential reinforcement is not necessary for the chimpanzee to acquire certain cognitive skills. For example, the same infant chimpanzee subjects spontaneously generalized their object-matching ability to novel items in the absence of differential rewards (Oden et al., 1988). Similarly, after a period of chance performance on the matching of pictures, the same animals, again under conditions of nondifferential reward, eventually mastered the task (Oden & Thompson, 1991). There is then good evidence that the chimpanzees spontaneously seek solutions to problems, independent of external reward contingencies in the work situation.

There is additional anecdotal evidence indicating that the chimpanzees can be aware of their own proficiency on problem-solving tasks. Given the opportunity, a skilled chimpanzee who had mastered a task would attempt to displace a less proficient one from the test area and solve its problem. Conversely, the skilled chimpanzee would pout when prevented from doing so by the experimenter. At least one subject, Whiskey, was observed by several experimenters to whine and retreat at the sight of materials for a test on which he performed poorly. He responded this way consistently, despite the fact that neither he nor any of the other chimpanzees had ever been punished beyond a mild verbal reproof for incorrect responses. Conversely, Whiskey would grunt and approach materials for a test on which he was currently highly proficient. No such differences in emotional reactions were ever observed when the same materials were encountered in a free-play environment. The interested reader will find similar instances in which other chimpanzees seem sensitive to the work demands and level of proficiency documented by Premack and Premack (1983, pp. 83–97).

Social motivation of task orientation

The conventional wisdom in comparative psychology would have us believe that problem solving is merely a means to an end (the reward) and that the experimenter is merely a reward dispenser who could be better replaced by an automated device. The overall pattern of our findings run counter to this mechanistic viewpoint. We have found that a more compelling theoretical approach is provided by recent developments in experimental human social psychology (Pittman & Heller, 1987). This approach emphasizes the importance of social motivation for an understanding of human cognition. We believe that social motivation must also be considered in discussions of the cognitive processes of apes.

The evidence in the preceding section forces us to conclude that the chimpanzee knows and cares about competent performance in a work context. However, merely exposing an animal to the nonsocial cues of the test context (i.e., test materials, testing room) is not sufficient to evoke this attitudinal stance. In addition, a human (i.e., a mentor) with a particular history of interactions with the animal is essential. We believe that the differential experimenter effects documented here are attributable to differences in the humans' capacities to socially motivate this task-oriented stance in the chimpanzees.

It is important to stress here that it is not sufficient to regard the human as merely an additional conditional cue for problem solving. If this were the case, then good performances, equivalent to those obtained by a mentor, should be recorded by an experimenter whose interactions with the chimpanzees were limited to test situations. Given the general laboratory protocols described earlier, there were relatively very few such individuals and no systematic data were collected from them during testing. However, the anecdotal evidence and post hoc inspection of data sheets strongly suggest that performance levels obtained from the chimpanzees by these individuals were not optimal, but were equivalent to those obtained by playmates.

The latter evidence, although it is anecdotal, strongly suggests that social interaction outside the work context is necessary if a person is to evoke a socially motivated task-oriented stance in the animal. However, this type of social interaction is not sufficient, as is evidenced by the differences in performance obtained by mentors and playmates. We believe that at least a balance of play- and work-related interactions, as experienced by those we have described as mentors, enables the human to evoke an intrinsic level of social motivation that sustains problem solving in general, independently of any particular task.

The socially motivated task orientation described here could be construed as evidence for a first-order intentional stance (Dennett, 1983; Ristau, 1986) such that the chimpanzee *wants* to solve problems in the presence of a particular experimenter. Does the chimpanzee assume a second-order intentional stance? That is, does the chimpanzee *believe* that the experimenter *wants* him or her to work? Perhaps, but our data do not allow us to say anything about this possibility. It should be noted that our suggested hypothesis emphasizing the necessary role of social motivation does not demand this level of intentional attribution. Nevertheless, the issue is worthy of further investigation given the current receptivvvityto such notions in animal cognition (e.g., Premack & Woodruff, 1978).

David L. Oden and Roger K. R. Thompson

Conclusion

The inevitable bond that develops between human and chimpanzee should not be viewed as an unequivocal "problem" to be surmounted or avoided in empirical studies of comparative cognition. Rather, we hope this chapter persuades the reader that it may be viewed as a necessary condition for developing a comprehensive perspective on ape cognition. At the very least we should consider the possibility that placing an animal like the chimpanzee in a socially impoverished testing environment like an experimental chamber is likely to produce a restricted, if not impoverished, theory of animal mind.

Acknowledgments

The research and observations reported here were supported by a grant from the National Science Foundation (NSF: BNS 8418942) to D. Premack and grants to D. Oden and R. Thompson from the NSF program for support of small-college faculty through association with faculty at large institutions. We thank all those experimenters, mentors, and playmates whose interactions with the chimpanzees alerted us to the possibilities described here.

References

Byrne, R., & Whiten, A. (1988). *Machiavellian intelligence: Social expertise and the evolution of intellect in monkeys, apes and humans.* New York: Oxford University Press.
Cheney, D., Seyfarth, R., & Smuts, B. (1986). Social relationships and social cognition in nonhuman primates. *Science, 234*, 1361–6.
Dennett, D. C. (1983). Intentional systems in cognitive ethology: The "Panglossian paradigm" defended. *Behavioral and Brain Sciences, 6*, 343–90.
Humphrey, N. K. (1976). In P. P. G. Bateson & R. A. Hinde (Eds.), *Growing points in ethology* (pp. 303–71). Cambridge University Press.
Jolly, A. (1966). Lemur social behavior and primate intelligence. *Science, 153*, 501–6.
Kummer, H. (1982). Social knowledge in free-ranging primates. In D. R. Griffin (Ed.), *Animal mind–human mind* (pp. 113–30). New York: Springer.
Lamb, M. E., & Sherrod, L. R. (Eds.). (1981). *Infant social cognition: Empirical and theoretical considerations.* Hillsdale, NJ: Erlbaum.
Mason, W. A. (1982). Primate social intelligence: Contributions from the laboratory. In D. R. Griffin (Ed.), *Animal mind–human mind* (pp. 131–40). New York: Springer.
Oden, D. L., & Thompson, R. K. R. (1991). Analyses of pictorial-object identity judgements by infant chimpanzees (*Pan troglodytes*). Unpublished manuscript.
Oden, D. L., Thompson, R. K. R., & Premack, D. (1988). Spontaneous transfer

of matching by infant chimpanzees (*Pan troglodytes*). *Journal of Experimental Psychology: Animal Behavior Processes, 14,* 140–5.

(1990). Infant chimpanzees (*Pan troglodytes*) spontaneously perceive both concrete and abstract same/different relations. *Child Development, 61,* 621–31.

Piaget, J. (1954). *The construction of reality in the child.* New York: Basic Books.

Pittman, T. S., & Heller, J. F. (1987). Social motivation. In M. R. Rosenweig & L. W. Porter (Eds.), *Annual Review of Psychology, 38,* 461–89.

Premack, D., & Premack, A. (1983). *The mind of an ape.* New York: Norton.

Premack, D., & Woodruff, G. (1978). Does the chimpanzee have a theory of mind? *Behavioral and Brain Sciences, 1,* 515–26.

Ristau, C. A. (1986). Do animals think? In R. J. Hoage & L. Goldman (Eds.), *Animal intelligence: Insights into the animal mind* (pp. 165–85). Washington, DC: Smithsonian Institution Press.

Segal, E. (Ed.). (1989). *Housing, care and psychological well-being of captive and laboratory primates.* Park Ridge, NJ: Noyes.

Vygotsky, L. S. (1962). *Thought and language.* Cambridge MA: MIT Press.

(1978). *Mind in society: The development of higher psychological processes* (M. Cole, V. John-Steiner, S. Scribner, & E. Souberman, Eds. & Trans.). Cambridge, MA: Harvard University Press.

Walker-Andrews, A. S. (1988). Infants' perception of the affordances of expressive behaviors. In C. Rovee-Collier & L. P. Lipsitt (Eds.), *Advances in infancy research* (Vol. 5, pp. 173–221). Norwood, NJ: Ablex.

—14—

Minimizing an inevitable bond: the study of automated avoidance in rats

Morrie Baum and Laurie Hiestand

Editors' introduction

There is a continuum of strategies involving the scientist–animal bond. At one extreme are chapters such as those by Boysen and Pepperberg that document attempts to maximize the efficacy of research by exploiting the positive side of the relationship. The chapter by Baum and Hiestand lies at the other extreme of the continuum. The scientist–animal relationship is seen in negative terms, and considerable effort is directed at minimizing its occurrence. This contribution surveys the first author's attempts to develop an automated shock avoidance apparatus that maximized performance while minimizing the time-consuming and often hostile interactions between scientist and rat.

Early behavioral psychologists, such as Mowrer (1940), settled on footshock as a convenient, repeatable, and controllable source of aversion in avoidance studies using rats. Footshock could be easily delivered and quantified, and was a reliable elicitor of responses. Apparatuses for avoidance conditioning included runways and two-compartment boxes with electrifiable grid floors, in which the rat had to learn to escape the onset of shock to the feet by responding to a reliably presented stimulus. Such methods required the experimenter to pick up the rat manually after the completion of each trial and place it back in the start chamber. While these manual, one-way apparatuses were relatively humane in that the rat could quickly learn to avoid the shock, the interaction between scientist and rat was often hostile. Rats that had been recently shocked might display aggression and hypersensitivity to being handled, and quickly came to associate the scientist with the aversive procedures that he or she employed.

Initially, Baum (1965a,b) used these traditional apparatuses in attempting to verify predictions derived from the two-process theory of avoidance (Mowrer, 1951; Solomon & Wynne, 1953). This theory posited that avoidance learning involves two stages: (1) inducing a state of fear in the rat by classical conditioning methods and (2) reinforcing a response that enabled the animal to avoid the feared stimulus. While conducting these studies, Baum found that a state of fear was evident in even the tamest strains of rats, which would attempt to bite the experimenter's hand when handled between trials in the manual, one-way avoidance apparatus. It was necessary to wear heavy work gloves so that the rats' teeth could not penetrate. This was typical of the type of negative interactions between rat and experimenter that were frequently observed in such studies. Differential handling had potential influences on data obtained with these procedures. Particularly reactive rats would inevitably be treated more hesitantly or roughly than quiescent animals. Behavioral and physiological measures of this species are sensitive to even subtle variability in handling by humans (see Dewsbury, Chapter 2, this volume). It is therefore important to reduce as much as possible the necessary contact between trials. The purpose of this chapter is to review attempts to do that and to evaluate their success.

The shuttlebox

An early attempt to circumvent handling involved dividing an apparatus into two equal-sized compartments (Miller, 1948; Mowrer, 1951). To avoid shock, the rat had to cross a hurdle from one side (compartment) to the other, whenever a reliable warning signal such as a tone or light was presented. Thus, rats could be conditioned to shuttle back and forth from side to side, without having to be handled by the experimenter. Though still currently used, the shuttlebox has several disadvantages. A high proportion of rats often fail to acquire the avoidance response within a reasonable number of trials (Gordon & Baum, 1987). Even those rats that do acquire the response perform it less reliably than they do in manual one-way avoidance procedures, and strict acquisition criteria (such as 10 consecutive successful avoidance responses, or 100% avoidance) must often be relaxed (e.g., to 80 or 90% avoidance, or less). Responding in extinction is also less consistent and susceptible to disruption (Gordon & Baum, 1987). Also, since many more shocks are received by rats learning to avoid in the shuttlebox (vs. one-way manual avoidance), and it is far more difficult for the rat to meet criterion, shuttlebox procedures are more aversive for the rat, causing it unnecessary stress and pain.

Lever-press avoidance

Development of the lever-press operant chamber is, of course, associated with B. F. Skinner (1932), and the bar-press response has been successfully used to study rewarded behavior in rats for many years. However, when attempts were made to use this response for avoidance conditioning in rats, the animals found this type of task very difficult to accomplish. They did not readily learn to press the lever to avoid being shocked when the warning stimulus was presented (D'Amato & Schiff, 1964; Meyer, Cho, & Wesemann, 1960). Keehn (1967) directly compared the efficiency of running and lever pressing as avoidance responses and found running to be more easily acquired. Conversely, in the lever-press situation the rat typically remains frozen above the lever and does nothing when the warning stimulus is presented. Over the years, despite the many modifications that have been added to the lever-press avoidance paradigm to improve rats' ability to learn this particular task, none have succeeded in making lever-press avoidance as easily learned as avoidance learning with the one-way manual procedure.

The difficulties of conditioning rats to avoid being shocked in the shuttlebox and lever-press situations have been interpreted in terms of Bolles's (1970) theory of species-specific defensive reactions (SSDRs) in the rat. Basically, according to this view it is unnatural for a rat to learn to avoid danger and pain by manipulating a lever with its paws, and thus it takes many trials to teach it to do so. A rat's initial reaction to shock is to freeze, that is, to remain completely motionless (Brener & Goesling, 1970). Only if this strategy is ineffective will it flee or fight, the other two SSDRs in this species' repertoire (Bolles, 1970). Freezing, flight, and aggression are all incompatible with pressing a lever, and so it is a difficult task for a rat to learn to avoid shock. Other species have different defensive behaviors than the rat and also respond differently to avoidance conditioning (Pearl, 1963).

It is unnatural for a rat to return immediately to a place where it has received pain, such as is required in a shuttlebox. However, rats are "prepared" by their evolutionary history (Seligman, 1970) to run from a consistently dangerous place to a consistently safe place, as in a manual one-way avoidance apparatus. Unfortunately, such procedures necessitated handling before each trial, which was aversive for both subject and experimenter.

All of these considerations resulted in the development of an automated avoidance apparatus that allowed the rat to flee to a predictable place of safety and did not require the experimenter to handle the rat between

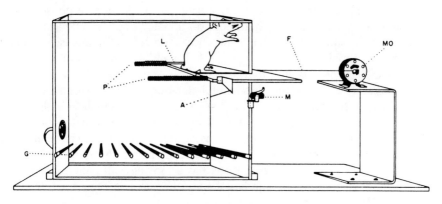

Figure 14.1. The automated ledge box (reproduced from Baum, 1965c). *Key*: G, electrifiable grid floor; L, safety ledge onto which the rat climbs to escape or avoid shock; MO, electric motor; F, filament joining the motor shaft to the safety ledge; A, aluminum armature affixed to the safety ledge; M, microscitch, which is thrown by the armature (and thus turned off the motor) when the safety ledge is retracted to the desired distance; P, springs (mounted on the outside of the box), which return the safety ledge after retraction.

trials (Baum, 1965c). This apparatus, which results in rapid learning and consistent performance, is described in the next section.

The automated one-way ledge box

The automated ledge box is depicted in Figure 14.1 (reproduced from Baum, 1965c). The rat initially escapes and later avoids footshock by climbing onto the ledge, where it sits during the intertrial interval (typically 30 or 45 seconds). A momentary retraction of the ledge presents the next trial. Note that there is a consistently dangerous place (the grid floor) and a consistently safe place (the ledge). Consequently, acquisition of the avoidance response is very rapid. After five or six shock escape trials, the typical rat achieves 100% avoidance and the response persists for a considerable number of trials in extinction (i.e., even after the grid floor is no longer electrified).

The automated ledge box has been used to study acquisition processes in avoidance conditioning (Baum, 1969a; Holloway & Baum, 1989), as well as in drug studies (e.g., Baum, 1969b; Morin & Baum, 1986). Perhaps its greatest use has been in the study of extinction of avoidance via response prevention or "flooding." This term refers to the forcible exposure of a subject to a stimulus that it fears (Baum 1970, 1976; Thyer, Baum, & Reid, 1988). For example, in the automated ledge box the ledge is retracted, preventing the rat from making an escape, or avoidance response, and forcing it to remain on the feared, but now nonelectrified grid floor. Flooding has been shown to hasten avoidance extinction in animals and aid in reducing phobias in humans (Baum, 1976) and has considerable clinical relevance (see Thyer et al., 1988).

A modified version of the automated ledge box is manufactured and sold by Lafayette Instrument Co. The Lafayette version, which is called an automated one-way shuttlebox, differs in two respects from the original ledge box. First, the interior dimensions of the box are somewhat smaller. Second, instead of the ledge being retracted momentarily from the box to present a new trial, the Lafayette box has a moving wall panel above the ledge that pushes the rat gently off the ledge and is then retracted to redisplay the ledge for the next trial.

Some infrequently used automated apparatuses

Among infrequently used automated avoidance apparatuses, the one with the longest history is undoubtedly the running wheel, first employed in the 1930s or earlier. In this procedure the rat sits inside a large wheel with a perimeter consisting of grid bars through which footshock can be delivered. When the warning signal is presented, the rat must avoid by running in the wheel, causing it to rotate. One problem associated with this apparatus is that a rat will tend to run in the wheel as an unconditioned response to shock, and this can be confounded with the conditioned avoidance response. Learning in a wheel is more easily obtained than in a shuttlebox, although not as easily as with the ledge box or one-way manual apparatuses. This may be because although a wheel permits flight, as do the more successful procedures, the rat returns to the place that it has been shocked (Bolles, 1970). While still in occasional use (e.g., Iso, Brush, Fujii, & Shimazaki, 1988), the running wheel is far less often employed than the apparatuses reviewed previously.

Baum and Bobrow (1966) developed an automated analog of the one-way manual apparatus. The two-compartment box was mounted on a turntable that could rotate horizontally through 180° to present a new trial. While there was no consistently safe and dangerous place in this apparatus, there was a consistently safe direction in which to run. This seemed to produce faster conditioning of avoidance than, for example, the shuttlebox.

Modaresi, Coe, and Glendenning (1975) reported the development of an automated apparatus with the potential of producing identical one- and two-way avoidance performance. This is a rather complex device that involves movable walls and slightly elevated platforms. It has not received much use.

These automated apparatuses result in rapid conditioning comparable to that found with manual, one-way avoidance apparatuses and produce evidence of much faster learning than the shuttlebox or lever-press procedures. They permit a natural, one-way flight response using an entirely automated procedure. As such, these apparatuses are less stressful in that the rat receives fewer footshocks before conditioning is achieved. The rat also need never be handled between conditioning trials with this type of procedure. Differences in the ways in which experimenters interact with rats and the resulting opportunity for bias are thereby minimized, although there is still the chance of experimenter influence occurring when the rat is introduced into, and removed from, the apparatus.

Recent developments

Minimizing the necessity of handling rats during avoidance conditioning has been of abiding interest to experimental psychologists for many years, and continues to be so. Recently, two novel solutions to the problem have arisen. Tikal (1988) described a pivoting runway, which can be tilted from the horizontal to the vertical position. The grid bars of this apparatus run parallel to the length of the runway (unlike most apparatuses). Upon completion of a trial and the intertrial interval, the runway is tilted nearly 90°, and the rat slides the length of the box, back to the start chamber. The runway is then returned to the horizontal position to initiate the next trial. This runway leads to very rapid avoidance conditioning and can also be used for brightness discrimination studies. Although Tikal (1988) did not actually automate the pivoting runway, this could readily be done.

Cándido, Maldonado, and Vila (1988) employed a conditioning box to study signalized avoidance using a vertical jumping response. The rat was required to jump vertically into the air, interrupting photobeams, in order to avoid shock when a warning buzzer was sounded. This procedure, requiring the rat to jump out of danger, leads to rapid acquisition of the avoidance response because the rat is able to flee from the source of the shock. Also, because the other components of the rat's defensive repertoire, freezing and aggression, do not enable it to avoid shock, the rat quickly comes to rely on its only other option, fleeing (in this case, by jumping; Bolles, 1970).

Concluding remarks

For many years, investigators of avoidance conditioning have sought to decrease the amount of handling necessary during training in order to reduce the rat's aggression toward the scientist. Only more recently has it been recognized that differential effects of handling may influence the data. Early uses of the shuttlebox and lever-press apparatuses were not efficient methods to assess learning in the rat, nor were they humane, because the required response was not readily conditioned and caused the rat to receive many aversive footshocks over the course of learning. Automated ledge boxes, Tikal's runway, and other apparatuses described in this chapter provide ready alternatives that minimize the need for handling and produce rapid conditioning, thus being less stressful to the rat. While avoidance conditioning with aversive footshock is inherently inhumane, procedures have been designed to maximize the animals' success, and the results of such studies appear to have extensive applications to the treatment of human clinical problems (e.g., Thyer et al., 1988).

References

Baum, M. (1965a). The recovery-from-extinction of an avoidance response following an inescapable shock in the avoidance apparatus. *Psychonomic Science, 2*, 7–8.

(1965b). "Reversal learning" of an avoidance response as a function of prior fear conditioning and fear extinction. *Canadian Journal of Psychology, 19*, 85–93.

(1965c). An automated apparatus for the avoidance training of rats. *Psychological Reports, 16*, 1205–11.

(1969a). Dissociation of respondent and operant processes in avoidance learning. *Journal of Comparative and Physiological Psychology, 67*, 83–8.

(1969b). Paradoxical effect of alcohol on the resistance to extinction of an avoidance response in rats. *Journal of Comparative and Physiological Psychology, 69*, 238–40.

(1970). Extinction of avoidance responding through response prevention (flooding). *Psychological Bulletin, 74*, 276–84.

(1976). Instrumental learning: Comparative studies. In M. P. Feldman & A. Broadhurst (Eds.), *Theoretical and experimental bases of the behaviour therapies* (pp. 113–31). New York: Wiley.

Baum, M., & Bobrow, S. A. (1966). An automated analogue of the one-way Miller-type avoidance box. *Psychonomic Science, 5*, 361–2.

Bolles, R. C. (1970). Species-specific defense reactions and avoidance learning. *Psychological Review, 77*, 32–48.

Brener, J., & Goesling, W. J. (1970). Avoidance conditioning of activity and immobility in rats. *Journal of Comparative and Physiological Psychology, 70*, 276–80.

Cándido, A., Maldonado, A., & Vila, J. (1988). Vertical jumping and signalled avoidance. *Journal of the Experimental Analysis of Behavior, 50*, 273–6.

D'Amato, M. R., & Schiff, D. (1964). Long-term discriminated avoidance performance in the rat. *Journal of Comparative and Physiological Psychology, 57*, 123–6.

Gordon, A., & Baum, M. (1987). Shuttlebox avoidance in rats and response prevention (flooding): Persistence of fear following reduced instrumental responding. *Journal of General Psychology, 114*, 263–72.

Holloway, F. A., & Baum, M. (1989). One-way avoidance acquisition as a function of intertrial interval and relative time in "safe" area. *Bulletin of the Psychonomic Society, 27,* 452–4.

Iso, H., Brush, F. R., Fujii, M., & Shimazaki, M. (1988). Running-wheel avoidance learning in rats (*Rattus norvegicus*): Effects of contingencies and comparison of different strains. *Journal of Comparative Psychology, 102,* 350–71.

Keehn, J. D. (1967). Running and bar pressing as avoidance responses. *Psychological Reports, 2,* 591–602.

Meyer, D. R., Cho, C., & Wesemann, A. F. (1960). On problems of conditioning discriminated lever-press avoidance responses. *Psychological Review, 67,* 224–8.

Miller, N. E. (1948). Studies of fear as an acquirable drive: Fear as motivation and fear reduction as reinforcement in the learning of new responses. *Journal of Experimental Psychology, 38,* 89–101.

Modaresi, H. A., Coe, W. V., & Glendenning, B. J. (1975). An efficient one- and two-way avoidance apparatus capable of producing identical one- and two-way avoidance performance. *Behavior Research Methods and Instrumentation, 7,* 348–50.

Morin, M. M., & Baum, M. (1986). Effects of delta^8-tetrahydrocannabinol (THC) on avoidance extinction in rats. *Bulletin of the Psychonomic Society, 24,* 385–7.

Mowrer, O. H. (1940). Preparatory set (expectancy): Some methods of measurement. *Psychological Monographs, 52,* 1–43.

  (1951). Two-factor learning theory: Summary and comment. *Psychological Review, 58,* 350–4.

Pearl, J. (1963). Avoidance learning in rodents: A comparative study. *Psychological Reports, 12,* 139–45.

Seligman, M. E. P. (1970). On the generality of the laws of learning. *Psychological Review, 77,* 406–18.

Skinner, B. F. (1932). Drive and reflex strength. *Journal of General Psychology, 6,* 32–48.

Solomon, R. L., & Wynne, L. C. (1953). Traumatic avoidance learning: Acquisition in normal dogs. *Psychological Monographs, 67* (No. 4, Whole No. 354).

Thyer, B. A., Baum, M., & Reid, L. D. (1988). Exposure techniques in the reduction of fear: A comparative review of the procedure in animals and humans. *Advances in Behaviour Research and Therapy, 10,* 105–27.

Tikal, K. (1988). New apparatus for training the avoidance reaction. *Physiology and Behavior, 43,* 677–9.

—15—

Underestimating the octopus

Jennifer Mather

Editors' introduction

Mather's chapter about octopuses is of interest not only because these animals are among the least familiar to most readers, but also because it is the only chapter that describes a scientist–invertebrate interaction. As the title implies, we have indeed underestimated the intelligence and complexity of this animal and its potential for social interaction with the humans who attempt to study it.

Most people who encounter octopuses find it difficult to see them as complex, intelligent animals, and while there are differences, their similarities to "higher" vertebrates make them fascinating to study. Octopuses are cephalopods, a specialized branch of mollusks, and, like vertebrates, have enlarged and centralized brains, a closed circulatory system, and an excellent sensory apparatus (Lane, 1960; Wells, 1978). An octopus depends to a great extent on learning about its environment, and two areas of its brain are devoted to encoding new information (Wells, 1965). This dependence on learning is particularly noteworthy because the octopus has a very short maximum life span, from the 6 months of *Octopus joubini* to the 4 years of *O. dofleini*. Octopuses hatch from eggs that are tended by their mother, but she dies soon after their hatching and the young receive neither parental care nor instruction from kin. Individuals of many species of *Octopus* are not much larger than a rice grain at birth. They float in the open water for months as plankton before settling down on the ocean bottom, still very small and very much at risk. They must learn about several different habitats and they must learn quickly (Figure 15.1). All this suggests that octopuses are also able to learn about the researcher, the

Figure 15.1. Free-ranging *Octopus dofleini*. (Photo courtesy of Jim Cosgrove.)

experimental situation, and laboratory conditions in ways that scientists may not anticipate.

Animals cannot be studied in the lab unless their habits and physiology are understood so that they will not only survive but thrive. Octopuses, being asocial and relatively inactive, are "preadapted" to the solitary confinement of the lab. They are searching predators, but sleep or rest for 75% of daylight hours (Mather, 1988). They stay in a central "home" from which they move out to hunt across the ocean floor, and similar homes can be provided in the lab. Octopuses choose a wide variety of animals as prey (a maximum of 55 species was mentioned for *Octopus bimaculatus*; Ambrose, 1984), but learn to select a different range of species depending on local availability (Nixon, 1987). They normally use chemotactile search to find prey in their natural environment (Mather, 1991a); their acute vision is probably used to detect predators or to navigate (see Mather & O'Dor, in press) and is easily tested in the lab.

Octopuses belong to the family Octopodinidae, composed of three genera – *Octopus, Eledone,* and *Haplochlaena* – all of which live on the ocean bottom as adults and whose preferred depth ranges from 100 m to intertidal. Several members of the genus *Octopus*, including *O. dofleini* (the giant Pacific octopus), *O. vulgaris* (the common octopus of the Mediterranean and Caribbean), and *O. bimaculatus* (the California octopus), have been observed temporarily coming out of the water or living in the intertidal zone. Octopuses do nevertheless require well-oxygenated water – in part because they have hemocyanin in their blood cells, which does not bind as well to oxygen as does hemoglobin. Wells (1962) suggested that octopuses have pushed the inefficient molluscan physiology as far as it goes

241

and cannot invade marginal habitats. They will die if kept out of water for long or if their water is not changed or treated to get rid of the ammonia that they excrete. In size, octopuses range from *O. joubini*, which weighs 15 g at maturity, to the giant *O. dofleini*, which weighs 20 kg at adulthood, but the species most used in the laboratory, *O. vulgaris*, weighs 500 g to 1 kg at adulthood. Despite the efforts of Voss (1973), the systematics of octopuses are not well known. Little or nothing is known of the evolutionary background of even the genus *Octopus*.

Its life history as a solitary forager, subtidal marine habitat, visual acuity, adaptability, and learning ability make the octopus a fascinating subject for laboratory research on behavior. Still, Wells (1978) extended a warning. He noted that even though our physiology and dependence on learning may be similar in some respects to those of the octopus, this does not mean that we can presume to understand the octopus; "we should not fall into the trap of supposing that we can interpret its behavior in terms of concepts derived from birds or mammals." The octopus presents the experimenter with three types of challenges, which will be discussed in the following sections: risk, difficulties of maintenance, and procedures for testing.

Risk

Octopuses are well equipped for predation, with suction cups along the arms for grasping, a cartilaginous beak for tearing flesh, a toothed papilla for drilling holes in shells, and a toxic secretion from the posterior salivary gland for paralyzing and killing prey. Among the substances in the saliva is cephalotoxin (Ghiretti, 1960), a neurotoxin that attacks the neuromuscular junction and can kill a crab in 30 seconds. The poison is not particularly specific and is known to affect humans as well as prey. This has led to the inclusion of octopuses among dangerous and venomous invertebrates (McMichael, 1971), sometimes with good reason. Wells (1978) noted that *O. vulgaris* often bites people, but apparently seldom injects saliva, and that individuals who were bitten usually had no reaction. This does not hold true for other species. Wittich (1966) reported that the bite of a small *O. joubini* left him nauseated, feverish, and in pain for 8 hours, and produced a swollen and itchy area that lasted for 3 weeks. Anderson (1987) commented that *O. rubescens* gave a bite similar to a bee sting. These effects, however, are minor compared with the results of a bite from *Haplochlaena*, which can be fatal (McMichael, 1971). The descriptions of weakness, nausea, vomiting, and partial or total paralysis are daunting – particularly since no antidote has been found. Apparently, however, *Haplochlaena* species do not bite unless "annoyed," before which they flash a warning coloration of blue rings (Tranter & Augustine, 1973). Workers familiar with this warning coloration can avoid the consequences and have cared for the animals for years without any adverse effects. Nevertheless, careless handling can have unpleasant results, even with the nonfatal species, and octopuses may therefore be regarded more as potentially dangerous "things" than as interesting animals to which the researcher can relate.

Maintaining octopuses

Because of their need for well-oxygenated water and their excretion of ammonia into the water, octopuses have been difficult to keep. The efforts of Hanlon and co-workers have changed this, not only for octopuses but for all cephalopods; Forsythe and Hanlon's (1980) description of the apparatus for keeping *O. joubini* is a good example of this progress. Even when their physical requirements have been dealt with, however, the behavior of octopuses in captivity has continued to present a challenge. Researchers tend to react to behaviors such as those described in this section either with irritation or by viewing them as "amusing idiosyncrasies," failing to see how they could acquaint us with the capacity behind the misbehavior.

Octopuses invariably "investigate" anything that is placed in their tank. Any necessary apparatus must either be physically separated from the octopus or removed immediately after testing. According to Wells (1962), "the expectation of life of a floating thermometer...is about twenty minutes." Octopuses' manipulative capacity is excellent; Wodinsky's (1977) experimental animals recovering from surgery untied knots in surgical silk and removed the stitches from their incisions. A more difficult challenge for an *O. dofleini* was described by Anderson (1987). For filtration, he had an undergravel filter constructed from a plastic egg crate glued together and covered with nylon mesh sewn over it; a thick nylon mesh was held over this with nylon cables, and all was held in place by a heavy weight of gravel. In one night the octopus had "removed the gravel from the corners, broken the egg crate, torn up the nylon screen and cut through the nylon cables"; pieces were floating in the tank the next morning. Anderson suggested that Murphy's law must have been advanced by an octopus keeper! Even experimental apparatus may be at risk; Dews's (1959) experiment in which octopuses were taught to press a lever was terminated by one octopus's habit of pulling violently on the lever: The animal eventually broke it.

Octopuses confined in aquaria also tend to escape and subsequently die from lack of oxygen. They apply a good deal of ingenuity and force to this usually self-defeating behavior. Occasionally escape may be adaptive; an octopus in the Brighton Aquarium climbed into a neighboring tank almost nightly to capture and eat lumpfish, then climbed back into its own tank (Lee, 1875). Being soft-bodied, octopuses can escape from almost any confinement, through an opening only large enough to allow passage of their eyeball. Their strength also allows them to push or lift heavy lids or weights. Dilly, Nixon, and Packard (1964) reported that octopuses of 480 and 350 g escaped from an aquarium whose lid was supplemented with bricks to a total weight of 19.8 kg. Some species may be more prone to escape attempts than others. Walker, Longo, and Bitterman (1970) suggested that an *O. vulgaris* in Naples escaped because the circulating water was of poor quality, but noted that their *O. maya* did not attempt to escape. The author has not found *O. joubini* prone to escape, but other species do so readily. In fact, there is a widespread "lab lore" about ways to keep octopuses from leaving. Boyle (1981) lined the upper edge of large tanks with plastic foam to keep *Eledone cirrhosa* within; Anderson (1987) used Astroturf to similarly confine *O. rubescens*, and Van Heukelem (1977) simply kept a lower water level in tanks enclosing *O. maya*. Researchers also weighted or locked lids on octopus tanks, or both. When the author conducted research at the Bermuda Biological Station (Mather, 1991) technicians were already expecting to construct special locked-lid tanks, having heard the legends.

Octopuses are usually solitary (Mather, 1988). Individuals of most species can seldom be kept together without larger individuals killing smaller ones or forcing them to escape, although *Eledone* are somewhat "social" (Mather, 1984) and can be kept in groups. Octopuses may escape from confinement to avoid conspecific aggression or poor water quality, but in the natural environment they also move irregularly and leave environments that appear rich in resources (Mather & O'Dor, in press). This suggests an innate tendency to wander, perhaps an adaptation allowing octopuses both to avoid predation and learn about the environment.

Another potential problem, particularly pertinent to this book, is caused by octopuses' level of "awareness" (see Griffin, 1982). Wells (1978) noted that human observers are aware that they themselves are being observed by the octopuses. The observation that octopuses frequently manage to outwait scuba divers was interpreted as evidence that octopuses did not readily habituate to observers. More recent field studies (Mather, 1988), however, have demonstrated that at least some octopus behaviors do habituate in the presence of an observer. To date, however, there have been no comparisons of the behavior of octopuses watched by visible versus hidden observers. The well-known tendency of octopuses to shoot jets of water at researchers, while clearly stemming from natural uses, also raises questions about octopuses' awareness of the human observer. Octopuses inhale by expanding the muscular mantle cavity and exhale by locking its sides and shooting water through a small but flexible funnel. This exhalation is used to some extent for locomotion, but also to move sediment when "digging out" a home, to expel pieces of uneaten prey remains from the home, and even to repel scavenging fish (Mather, in press).

An *O. rubescens* in my lab refrained from jetting at either myself or my research assistant, but aimed effectively at a visiting scientist and a photographer. While observing *E. moschata* (Mather, 1984), I was sitting on a stool in front of locked boxes containing *O. vulgaris*. One tried to push up and escape, so I banged on the lid. The action was effective, but for 2 weeks thereafter the octopus squirted water frequently either in the general direction of the stool or at me. A report from Lettvin (1961) described how octopuses prepared for single-unit recording from the brain first used their arms to pull out the probes, then, when the author cut off the arms, jetted at the recording apparatus, shorting out the equipment. Dews's (1959) uncooperative octopus "spent much time with eyes above the surface of the water, directing a jet of water at any individual who approached the tank." In a masterful understatement, Dews (1959) reported that this behavior "interfered materially with the smooth conduct of the experiments, and is . . . clearly incompatible with lever-pulling." The interaction between octopus and experimenter obviously involves more than a simple presentation of stimuli by the experimenter and the release of responses by the octopus.

Many of these "difficult" aspects of octopus keeping and testing are unavoidable. By its nature, the laboratory is restrictive and the experimental situation simplistic. The octopus will generally try to expand its range of activity and get out and on with its life, and the experimenter will try to restrain it. As yet the needs of invertebrates as research subjects have barely been considered (see Mather, 1989); while we now direct concern to the bored primate in captivity, we have not yet been asked to worry about the bored octopus.

Testing octopuses

Descriptions of "the" octopus seldom acknowledge the existence of individual variability in these unique invertebrates. Anderson (1987) wrote about individual differences from his perspective as an aquarium manager. He noted that the *O. rubescens*, which are a common display species in the Seattle Aquarium, vary so much that he divided them into "personality types." He described the "docile" animal that easily learned and accepted tank life, the "aggressive" octopus that attacked the keeper, and the "disturbed" one that swam rapidly around the tank and banged into the tank side. Such variability can be seen in the natural environment as well; differential tolerance of observers, for instance, has been described for *O*.

vulgaris in Bermuda (Mather, 1988). Over 4 years of observation, two individuals were tolerant of observers and remained faithful to a particular home site. One individual was similarly faithful to a home site, yet was *never* observed outside its home by any human.

Anderson and the author are presently cooperating on a study of the individual differences of *O. rubescens*. We have administrated three standardized tests: opening the lid of the octopuses' aquarium and looking in, feeding it a crab, and touching the octopus with a test tube brush. Preliminary results suggest that the variation in response can be partially described by three dimensions. The first could be called Withdrawal and accounted for 30% of the variance; octopuses loading high on this factor were in the home during observation, returned home with prey, and inked at the touch of the brush. The second factor, Timidity, accounted for 25% of the variance. Octopuses loading on this factor shrank away from the extended brush, did not grasp it, and were more likely to crawl slowly toward prey when capturing it. A third factor, Activity, accounted for another 15% of the variance. Octopuses loading high on this factor were seldom at rest when viewed and usually moved out to investigate any novel stimulus.

These variations make a tremendous difference in trainability; one of my *O. rubescens* spent a 4-month training period for visual landmark learning simply swimming in circles around an open circular tank, never tolerating the apparatus enough to learn the task. Dews's (1959) octopus, which could be described as "aggressive," never acquired the required behavior either. Such apparent "failures" to learn particular laboratory tasks may, of course, still reflect considerable learning – albeit not of the type desired by the experimenter. Particular "personality types" may perhaps be predisposed to work under different conditions.

Boycott's (1954) article is a thorough review of the octopus learning research conducted by researchers based in Naples using naturally occurring behaviors. Octopuses were rewarded by finding a food item and repelled from capturing hermit crabs by the sting of their accompanying anemone. Testing was based on putting octopuses within their homes in the tank and training them to stay at home for the presentation of one stimulus (S⁻) and to come out and attack another stimulus (S⁺). This led to an understanding of their visual capacity, of their ability to learn, and of the localization of the learning capacity within particular lobes of the brain (Wells, 1965). Packard (1963) described octopuses' behavior in these situations as a dichotomy between approach and withdrawal, with displacement activity representing the conflict between the two. While this dichotomy may accurately describe behavior within the limited testing situation, Packard's conclusion that "the majority of its behavior is to be

Figure 15.2. Octopus with sonic tag observing diver. Such tagging is one way of "observing" the animal while minimizing the influence of the experimenter. (Photo courtesy of Jim Cosgrove.)

described and understood in terms of the approach–withdrawal dimension" seems unwarranted. More recent work (Mather, 1988) indicates that most of the natural behavior of octopuses within the ocean involves quiet, alert waiting or chemotactile exploration (Figure 15.2). Bitterman (1975) also pointed out problems with this paradigm resulting simply from the fact that octopuses learn so quickly. When he visited the Naples labs, he noticed that "the animal often refused to wait quietly at home, advancing to the front of the tank before the introduction of the stimulus and having to be driven back by the experimenter, who would beat on the edge of the tank with a clenched fish [*sic*] or some heavy object." Octopuses also grabbed hold of objects and were reluctant to release them. Walker et al. (1970) suggested that on these grounds "control of the stimulus is inadequate, the definition of response is uncertain, and the delay of reinforcement is considerable." Octopuses' quick learning, which made the paradigm interesting, also made it difficult.

Ironically, just as technological advances emerged that could have contributed to better controlled and more objective testing situations, interest in the octopus as a model for learning began to wane. In 1969

247

Nixon produced an apparatus with an automatic food dispenser. Walker et al. (1970) later designed a simple maze, with the reward of reentering the water. Hales, Crancher, and King (1972) designed an apparatus that rewarded an octopus for inserting its arm into a tube out of the water, and Henderson, Woodward, and Bitterman (1975) designed a food delivery system to avoid the grabbing-and-holding problem. None of these devices was ever widely used, perhaps because researchers had long since realized that the octopus was not the elusive simple model that would enable learning theorists to "find the engram."

Future research on octopuses must be designed to cope with the problems of interactions between researcher and subject. The octopus is not a stereotyped "simple" invertebrate. Researchers must recognize the octopus's ability to learn about humans and account for this ability in their experimental designs. One approach would be to mechanize and attempt to remove the experimenter as a variable. Alternatively, we can recognize the octopus as a complex animal, evaluate the two-way street of octopus–experimenter influences, and attempt to control and/or use them as a means of understanding the animal.

References

Ambrose, R. F. (1984). Food preferences, food availability and the diet of *Octopus bimaculatus* Verrill. *Journal of Experimental Marine Biology and Ecology, 77,* 29–44.

Anderson, R. C. (1987). Cephalopods at the Seattle Aquarium. *International Zoo Yearbook, 26,* 41–8.

Bitterman, M. E. (1975). Critical commentary. In W. C. Corning (Ed.), *Invertebrate Learning,* 139–45. New York: Plenum.

Boycott, B. B. (1954). Learning in *Octopus vulgaris* and other cephalopods. *Pubblicazione della Stazione Zoologica di Napoli, 25,* 67–93.

Boyle, P. R. (1981). Methods for the aquarium maintenance of the common octopus of British waters, *Eledone cirrhosa. Laboratory Animals, 15,* 327–31.

Dews, P. B. (1959). Some observations on an operant in the octopus. *Journal of the Experimental Analysis of Behavior, 2,* 57–63.

Dilly, N., Nixon, M., & Packard, A. (1964). Forces exerted by *Octopus vulgaris. Pubblicazione della Stazione Zoologica di Napoli, 34,* 86–97.

Forsythe, J. W., & Hanlon, R. T. (1980). A closed marine culture system for rearing *Octopus joubini* and other large-egged benthic octopods. *Laboratory Animals, 14,* 137–42.

Ghiretti, F. (1960). Toxicity of octopus saliva against crustacea. *Annals of the New York Academy of Sciences, 90,* 726–41.

Griffin, D. R. (1982). *Animal Mind–Human Mind.* Berlin: Springer.

Hales, R. S., Crancher, P., & King, M. G. (1972). An apparatus for operant conditioning of *Octopus oyaneus* Gray. *Behavior Research Methods and Instrumentation, 4,* 145–6.

Henderson, T. B., Woodward, W. T., & Bitterman, M. E. (1975). Measurement of consummatory behaviour in octopuses. *Behavior Research Methods and Instrumentation, 7,* 265–6.

Lane, F. W. (1960). *Kingdom of the Octopus.* New York: Sheridan House.

Lee, H. (1875). *The Octopus; or the "Devil-fish" of Fiction and Fact.* London: Chapman & Hill.

Lettvin, J. Y., & Maturana, H. R. (1961). Octopus vision. *MIT Quarterly Progress Report, 64,* 288–92.

Mather, J. A. (1984). Behaviour and interaction of *Eledone moschata* in the laboratory. *Animal Behaviour, 33,* 1138–44.

(1988). Daytime activity of juvenile *Octopus vulgaris* in Bermuda. *Malacologia, 29,* 69–76.

(1989). Ethical treatment of invertebrates: How do we define an animal? *Animal Care and Use in Behavioral Research: Issues and Applications.* Washington, DC: Animal Welfare Information Center, National Agricultural Library.

(1991). Foraging, feeding, and prey remains in middens of juvenile *Octopus vulgaris* (Mollusca, Cephalopoda). *Journal of Zoology, London 224,* 27–39.

(In press). Interactions of juvenile *Octopus vulgaris* with scavenging and territorial fishes. *Marine Behaviour and Physiology.*

Mather, J. A., & O'Dor, R. K. (In press). Foraging strategies and predation risk shape the natural history of juvenile *Octopus vulgaris. Bulletin of Marine Science.*

McMichael, D. F. (1971). Venomous mollusks. In W. Bucherl & E. Buckley (Eds.), *Venomous Animals and Their Venoms: Vol. 3. Venomous Invertebrates,* 384–93. New York: Academic Press.

Nixon, M. (1969). The time and frequency of responses by *Octopus vulgaris* to an automated food dispenser. *Journal of Zoology, London, 158,* 475–83.

(1987). The diets of cephalopods. In P. R. Boyle (Ed.), *Cephalopod Life Cycles,* 2: 201–19. New York: Academic Press.

Packard, A. (1983). The behaviour of *Octopus vulgaris. Bulletin of the Oceanographic Institute of Monaco,* Special No. 1D, 35–49.

Tranter, D. J., & Augustine, O. (1973). Observations on the life history of the blue ringed octopus *Haplochlaena maculosa. Marine Biology, 18,* 115–28.

Van Heukelem, W. F. (1977). Laboratory maintenance, breeding, rearing, and biomedical research potential of the Yucatan octopus (*Octopus maya*). *Laboratory Animal Science, 27,* 852–9.

Voss, G. L. (1973). *Cephalopod resources of the world,* Circular 149. Rome: Food and Agriculture Organization of the United Nations.

Walker, J. J., Longo, N., & Bitterman, M. E. (1970). The octopus in the laboratory: Handling, maintenance, training. *Behavior Research Methods and Instrumentation, 2,* 15–18.

Wells, M. J. (1962). *Brain and Behavior in Octopus.* Stanford, CA: Stanford University Press.

(1965). Learning and movement in octopuses. *Animal Behaviour,* Suppl. 1, 115–26.

(1978). *Octopus: Physiology and Behaviour of an Advanced Invertebrate.* London: Halstead Press.

Wittich, A. C. (1966). Account of an octopus bite. *Florida Academy of Science Quarterly Journal, 29,* 265–6.

Wodinsky, J. (1977). Hormonal inhibition of feeding and death in *Octopus:* Control by optic gland secretion. *Science, 198,* 948–51.

—16—

The scientist and the snake: relationships with reptiles

Bonnie B. Bowers and Gordon M. Burghardt

Editors' introduction

Like Mather, Bowers and Burghardt discuss some unexpected effects of the scientist's relationship with unorthodox research animals. Reptiles, like octopuses, are difficult for humans to interpret. "Affectionate" or affiliative behavior in reptiles is unlikely to be recognized by most humans. Nevertheless, interactions between humans and reptiles may have profound effects.

Although individual recognition of humans by animal subjects is not generally a noteworthy result, it is more surprising in animals such as Mather's octopuses or the reptiles described by Bowers and Burghardt. As a consequence, such recognition is less likely to be anticipated or considered part of the research agenda, even by some workers in the field.

Bowers and Burghardt's brief description of the iguana's response to sleep disruption underscores the fact that the effects of the scientist's actions may be far from immediate and are therefore even less likely to be identified.

There are few animals that have as emotionally charged a relationship with people as do snakes. Other reptiles are different. Lizards may not be liked by most people, but neither are they intensely hated or feared. Crocodilians are large and dangerous and are treated as such. Turtles, of all the reptiles, have a generally benign reputation and are even liked by many people. Turtle motifs are common in tourist shops throughout the world, and among those reptiles that are actually kept as pets, turtles are probably the most common. (This was especially true before keeping baby turtles as pets was discouraged for health and conservation reasons.)

Snakes figure prominently in myths and religion (Morris & Morris, 1965) and are strong fear-inducing stimuli (Mundkur, 1983, 1988). This could well be derived from the fact that many snakes are highly venomous and, in most parts of the world, are not easily distinguished from nonvenomous species. Indeed, many harmless snakes mimic venomous ones in appearance and behavior (Greene, 1987; Pough, 1987). Morris and Morris (1965) found that more than a quarter of British schoolchildren named snakes as the most disliked animal. They suggested that a fear of snakes may be innate in humans, a view echoed by Coss (1968). In any event, snakes are feared, revered, avoided, or indiscriminately slaughtered. Paradoxically, there are probably no more dedicated animal fanciers and breeders than those who collect, keep, and breed snakes. Snakes are among the most popular animals in reptile houses at zoos (Marcellini & Jenssen, 1988).

Thus, there is a marked ambivalence about snakes and human relationships to them. The issues considered here are (1) whether the nature of this relationship is merely one of fear and fascination on the part of people, without the snake or other reptile showing any reciprocal response, and (2) how experience with humans may affect the subsequent behavior of reptiles toward people. There are few empirical data on these questions, and even anecdotes dealing with this topic are rare.

Factors limiting human–reptile bonds

One hardly expects to find the same emotional bond with reptiles as is often found between scientists and more social, "human-like" animals – for example, Dian Fossey and her mountain gorillas (Fossey, 1983). Many reptiles, including most snakes, are reputed to be asocial outside of breeding periods (Prater, 1933). Evidence is accumulating, however, that many snakes do aggregate as neonates and also as adults outside of the breeding season, and some may develop social dominance hierarchies (Barker, Murphy, & Smith, 1979; Carpenter, 1984; Gillingham, 1987; Schuett & Gillingham, 1989). Recent work in our laboratory has shown not only that neonate plains garter snakes (*Thamnophis radix*) that compete for food develop dominance relationships (as do snapping turtles; Froese & Burghardt, 1974), but that competitors preferentially aggregate with non-competitors outside feeding situations (Yeager & Burghardt, in press). In other words, relationships among conspecifics in snakes may be more extensive than previously thought and include individual recognition, as well as species and sex recognition (e.g., Froese, 1980).

Many lizards and crocodilians are highly social and gregarious (Burghardt, 1977). Postnatal parental care is most highly developed in crocodilians, absent in turtles, and exhibited in a small fraction of squamate reptiles (Shine, 1987). Behavior we typically consider affectionate seems largely absent in most reptiles. Possible exceptions are the maternal neonate licking reported in some lizard species (e.g., *Eumeces* spp.; Evans, 1959; Somma, 1987) and the social tongue touching documented in green iguanas (*Iguana iguana*; Burghardt, Allen, & Frank, 1986) and western fence lizards (*Sceloporus occidentalis*; Duvall, 1982).

Reptiles do not express the typical overt signals we interpret as affiliative behavior; for example, snakes do not wag their tails like a dog when their owner is present. Coupled with the lack of an elaborate parental–offspring communication system that has affective components readily interpreted by humans (Burghardt, 1988), responses to people may be subtle, slow, and seemingly composed largely of antipredator and avoidance responses. The controversial question of whether play occurs in reptiles is similarly compounded by these factors and the mammalian bias we inevitably bring to forming and interpreting interspecific relationships (Burghardt, 1988).

The temporal factor is supported by the following example. The position of sleeping adult green iguanas in Venezuela was noted by shining a spotlight on them in trees from tens of meters away. At most they opened their eyes but showed no other overt response. However, the following night they often had disappeared from areas where they had been noted regularly (Rodda, Bock, Burghardt, & Rand, 1988).

The typical reptile's appearance is not likely to promote bonding with humans. Most adult reptiles are scaly, with inexpressive faces and often with claws and long pointed teeth. However, our experience has been that many students who begin working with neonate snakes in our laboratory become attached to selected individuals and may even consider them attractive and certainly interesting. Neonate reptiles do have larger heads, more prominent eyes, and somewhat foreshortened snouts that are typical of the baby releasers Lorenz (1943) discussed in mammals and birds. Still, bonds are formed. Pope (1975) reared an Indian Python (*Python molurus*) named Sylvia for 14 years. He not only dedicated his excellent book on giant snakes to her, but presented many data and observations derived from her. Her gentleness with children was legendary, and they reciprocated.

The sensory abilities that reptiles possess are quite foreign to humans. Many of the stimuli reptiles respond to occur in the area where human senses are the weakest – the detection of chemical cues (Burghardt, 1970; Halpern & Kubie, 1984). Even the visual and acoustic displays of more social species of reptiles rarely overlap with familiar mammalian signals and are generally difficult to interpret. Head bobbing and push-up displays between male green anoles (*Anolis carolinensis*) do not have the same emotional impact as would the territorial howling of wolves in a similar situation. Here the lack of screams, whimpers, and songs has an effect. Neonate crocodilians do have contact (grunt) and distress (scream) vocalizations that resemble those of neonate gallinaceous birds and water-fowl (Herzog & Burghardt, 1977). Physiological measures and sensory information-gathering responses such as visual orientation and tongue flicking may be salient measures to develop in assessing reptilian responses to conspecific and interspecific stimuli.

Some rather brutal experiments have been carried out on reptiles, which illustrate our relative insensitivity to them. G. K. Noble, one of the most eminent early students of reptile behavior, performed a series of experiments on garter snakes to determine what factors may be responsible for stimulating and maintaining courtship in males (Noble, 1937). His experiments were ingenious but sometimes cruel. For example, he decapitated female Butler's garter snakes (*Thamnophis butleri*) and moved the bodies around the male's cages to assess the role of movement. He also broke some females' backs to see if copulation could occur without female assistance. Since reptiles do not engender much sympathy from the general public, these experiments might not arouse the same wrath from animal rights and welfare groups today as would such studies performed on mammals. One seldom sees reptile examples in animal welfare brochures. A major exception is the campaign to have shrimp boats use excluders to keep from drowning *endangered* sea turtles (Rudloe & Rudloe, 1989).

Individual recognition of humans by reptiles

There is some evidence (mostly anecdotal) that reptiles may be able to recognize individual humans and that they may respond differently to different humans. Romanes (1882) related a series of interesting, if dubious, anecdotes concerning this matter. His first example dealt with a young alligator that would follow its owner around the house while displaying great "affection and docility" (Romanes, 1882, p. 259). Such filial imprinting might actually be possible, since crocodilian mothers do lead their young to water and the young stay with the mother for many months (Shine, 1987). In captivity, neonate alligators (*Alligator mississippiensis*) have been observed to "grunt" in recognition of their keepers and become quite accustomed to handling (Breen, 1974).

Romanes also told of a tortoise that would come to the call of only certain people and would show its "affection" by tapping its beak on their feet. The significance of this behavior is unknown, although it may be a way of gathering olfactory information or be related to feeding behavior. Perhaps his most fantastic tale concerned a boa constrictor (or python) that, after being separated from its owners, "moped, slept, and refused to be comforted" (Romanes, 1882, p. 261). When the owners appeared, the snake "sprang upon them with delight" (p. 261). Unfortunately, this snake met an untimely end when it died of shock after seeing its master lying stricken in bed. Most of us are no doubt unwilling to grant reptiles such tender feelings and faithfulness.

Two more recent anecdotes by Carl Kauffeld (1969), director and curator of reptiles at New York's Staten Island Zoo for many years, illustrate the type of evidence typically used to try to establish recognition of individual humans by reptiles. His first story deals with rattlesnakes that would show no sign of disturbance when their regular keepers cleaned their cages but that would exhibit strong defensive reactions when an unknown keeper performed the same operation. Kauffeld rightly pointed out that this differential behavior may be due either to discrimination on the part of the snakes or ineptness on the part of the new keepers. Kauffeld's other story highlights the problems found in trying to determine whether a "bond" exists between humans and reptiles. A Mohave rattlesnake (*Crotalus scutulatus*) deemed "quiet and phlegmatic" (Kauffeld, 1969, p. 6) suddenly coiled and rattled its tail defensively when its owner failed to wear the coat he usually wore when attending the snake. When the owner put the coat on again, the snake was no longer defensive. Regardless of the relationship that existed between this snake and its owner, the coat was a highly salient cue involved in initial recognition. There is other anecdotal evidence. Klauber (1956), long the world's leading expert on rattlesnakes, also claimed that they recognized their regular attendant. Grace Olive Wiley, a fancier of venomous snakes and considered a fanatic by professional herpetologists, could, through care and patience, handle almost any snake. Individual recognition was claimed since rattlesnakes would rattle at the approach of anyone but her (Pope, 1975).

The monitor lizards, family Varanidae, are considered by many to be among the most intelligent of reptiles (Auffenberg, 1981). Auffenberg cites a series of papers claiming that Komodo monitors (*Varanus komodoensis*) can recognize and respond to individual keepers. For example, he cites Lederer (1942), who reported that a captive Komodo monitor would come when people familiar to her would call her name. Similarly, the slow worm *Anguis fragilis*, a legless lizard, is said to tame easily, recognize its owner, and take food from the owner's hand (Breen, 1974). But generally all these examples lack experimental support. Even Pope (1975) could not affirm categorically, although he wanted to, that Sylvia learned to respond to him.

John Groves, reptile curator at the Philadelphia Zoo, reports two instances of apparent individual recognition in long-term captive reptiles

Figure 16.1. An elderly Aldabran tortoise that has slowly approached long-time attendant John Groves from many meters and has risen up on all four legs and solicited neck scratching. Groves is selectively sought out regardless of his clothing.

(personal communication). In the first example, Groves had to hold and manually induce a neonatal king cobra (*Ophiophagus hannah*) to eat for the first 8 months of its life. Now more than 13 years old and several meters long, the snake growls (low hiss) and moves away when Groves works with it, and this behavior is displayed only toward Groves. Visual cues seem salient for this recognition, since the recognition is immediate and made through glass. Also, if the snake is left in the exhibit area, it will return immediately to its cage if Groves appears.

A sibling was similarly reared by Groves and then sent to another zoo. When Groves visited the snake after about two years, it showed no differential response to him. The continued contact with Groves in the absence of any continued aversive interaction, that is, assist feeding, may be an important factor in the apparent discrimination by the first snake. In Groves's 18 years of experience with reptiles, no other snake has responded in a similar manner.

The Philadelphia Zoo has several Aldabran and Galapagos tortoises (*Geochelone* spp.). They rise up on their legs after learning to accept, and then solicit, being scratched on the neck by Groves, allowing the underside of their shells to be inspected (Figure 16.1). Some naïve tortoises have learned through observation to solicit neck scratching by Groves and their keepers. However, if Groves is present, other people are approached and solicited if Groves ignores their approach (personal observation). One male Aldabran tortoise will also approach and nip Groves's leg if its solicitations are ignored. These tortoises give every indication that the

255

attention and scratching are reinforcing. The biological roots of this behavior may be found in the fact that wild birds perch on tortoises and remove ectoparasites. We have observed similar responses to neck scratching in the older iguanas in our colony. When scratched, these animals cock their heads and close their eyes while extending their front legs and dewlaps.

In our laboratory, we have had similar experiences with snakes and lizards. As mentioned earlier, student caretakers originally fearful of snakes often come to find them attractive and develop individual favorites. While this attachment is most likely due to changes in the student's attitude rather than changes in the snake's behavior, more docile snakes do generally become student favorites, and many snakes tend to become less defensive after prolonged handling. In addition, constricting snakes (such as boas and rat snakes) that cling to one's arms and coil about them are often viewed more positively, especially if the snakes are not too large, and only if they do not bite, hiss, or show other signs of "disenchantment." These snakes are also more deliberate and controlled in their movements than other snakes commonly encountered in North America (such as garter and water snakes) and thus are less likely to provoke jerks and other rapid, often fearful behaviors in people. Such snakes may develop different ways of responding to experienced and nonexperienced snake handlers that may serve as the basis for some sort of individual recognition and even a form of bond. Snakes, being ectotherms, often rest in warm places for hours at a time, and quiet, warm people could serve this function as well. Many reptiles seek out areas warmer than the ambient temperature and show considerable ingenuity, persistence, and learning in doing so. For example, an adult Nile monitor lizard (*Varanus niloticus*) would rest quietly on its owner's lap for hours while its owner watched television (P. Andreadis, personal communication).

We have also maintained a colony of green iguanas (*Iguana iguana*) in a large room for several years (Figure 16.2). Usually, the lizards showed little reaction to their daily feeding and spraying, which was typically performed by the same person for a long period of time (Figure 16.3). However, when a new person fed the iguanas, there was increased defensive behavior (lateral compression, tail slapping) as well as a decrease in the amount of food eaten. After several feedings by the new person, the frequency of these behaviors would decline. Again, it is unknown whether the lizards discriminated between the old and new caretakers or whether their behavior was a consequence of the new caretaker's inexperience.

Certain iguanas were more likely to exhibit defensive behaviors than others. Such differences in "personality" appear to be quite stable and have been found in other reptile species, for example, garter snakes (Herzog &

Figure 16.2. Gordon Burghardt and Emmett, a captive-bred *Iguana iguana*.

Figure 16.3. Hand feeding a green iguana.

Burghardt, 1988). The defensive temperament of the Mexican aquatic garter snake (*Thamnophis melanogaster*), in response to the human hand differed greatly across individuals and litters. These differences remained stable for more than a year. It is unclear whether such personality differences have any effect on how reptiles respond to individual humans. We have found that there is no difference in the defensive behavior (number of strikes directed at a human hand) of the same snakes tested by two experienced experimenters (Herzog, Bowers, & Burghardt, 1989). Yearling eastern garter snakes (*Thamnophis s. sirtalis*) were used in this experiment, with half being tested by each experimenter on Day 1. On the following day, each experimenter tested the other half. We found that snakes that struck readily for one experimenter would do so when tested by the other and that unresponsive snakes would not strike readily for either experimenter.

While these accounts are suggestive, there is a great need for empirical evidence of recognition of individual humans by reptiles. Unfortunately, this is a difficult problem to study. Responses to humans are generally subtle, and recognition of an individual human may involve nothing more than "a good-natured tolerance for handling and manipulation" (Kauffeld, 1969, p. 5). For example, a juvenile eastern garter snake that had been cared for exclusively by one caretaker would sit quietly in her hand but flattened and struck violently when handled by an unfamiliar veterinarian (L. Lyman, personal communication).

One possible approach to the study of recognition of individual humans by snakes may be to measure tongue flick rates of snakes handled by a familiar and an unfamiliar person. Tongue flicking is the means by which snakes gather chemical information processed by the vomeronasal system and is their primary sensory system in many areas of behavior, including sexual, social, and feeding behavior. Tongue flick rates have been used to determine relative "interest" exhibited by snakes (Chiszar et al., 1976), and such an approach may help determine whether snakes can discriminate among humans. For example, familiar environments may produce less frequent tongue extrusions. Differences in response to familiar and unfamiliar people would indicate that discrimination of human beings is at least possible.

Effects of early experience

How does early interaction with humans affect a reptile's later behavior? We discovered, quite by accident, one instance of early rearing experience that affected later defensive behavior in *Thamnophis melanogaster* (Herzog, Bowers, & Burghardt, 1989). We were interested in the effects of stimulus movement on defensive striking in these snakes. We tested the response of a group of 1-year-olds to either a moving or a nonmoving human hand and counted the number of strikes elicited. Half of these animals had been used in a longitudinal feeding study when neonates (Halloy & Burghardt, 1990). This group had then been housed in a different laboratory room and were given meticulous care by a single caretaker. While we have no quantification of handling time, it is likely that these snakes had more direct human contact, but were more isolated than the other snakes from normal laboratory activity. It is important to note, however, that both groups of snakes had been housed in the same room for several weeks before this experiment and had been treated similarly.

258

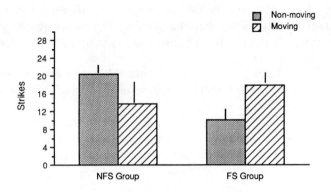

Figure 16.4. Mean number of strikes directed at a moving and nonmoving hand by the feeding-study subjects (FS) and the non-feeding-study subjects (NFS) of yearling *Thamnophis melanogaster*. (Vertical bars indicate S.E.) Snakes in the feeding-study group were used in a longitudinal feeding study when neonates. These snakes probably received more direct human contact but had been kept more isolated from general laboratory activity than the snakes in the non-feeding-study group. The pattern of striking at moving and nonmoving stimuli seen in the feeding-study snakes is very similar to that seen in neonates of this species. (From Herzog et al., 1989, © 1989 by the American Psychological Association. Reprinted by permission of the publisher.)

We found that, unlike neonate *T. melanogaster*, the 1-year-olds as a group did not respond differentially to moving and nonmoving stimuli. Herzog and Burghardt (1988) found a similar lack of differential response to moving and nonmoving stimuli in 1-year-old *T. melanogaster* that had been repeatedly tested over the course of a year. However, there was a significant movement × rearing condition interaction (Figure 16.4). In fact, the feeding study group showed a pattern of responses very similar to neonates of the same species. Unfortunately, since we were not expecting to find the difference between the two groups, we cannot pinpoint the factors responsible for it.

Herzog (1990) conducted an experiment to assess the effects of differential experience on defensive behaviors in juvenile eastern garter snakes. Three groups of snakes were used, with one group being picked up and shaken briefly by a model predator (harassed group), one group being gently handled by the experimenter (handled group), and one group being given no treatment (control group). When given a defensive test after 3 weeks of treatment, the snakes in the harassed group struck significantly more often than did the snakes that had received no treatment, with the handled snakes falling midway between the two other groups. Herzog then reversed the conditions for the harassed and control group snakes. He found that the newly harassed snakes showed a significant increase in striking and that the snakes that were harassed in the first treatment continued to strike at a high rate. This experiment shows that experience can indeed alter a snake's behavior and that the effects of such experience last at least 3 weeks. Even snakes that were gently handled had elevated strike scores. These findings have implications for experiments using laboratory-reared snakes to study defensive behavior. Care must be used to ensure that all snakes receive equal treatment. How differential treatment affects other areas of a snake's behavior – for example, species identification (Froese, 1980) – is simply not known, but these studies should alert anyone studying reptile behavior to potential problems.

The human side of the relationship

Certainly, experience with reptiles aids considerably in doing research on them. For example, some untrained experimenters have a great deal of trouble performing defensive tests on snakes, since their first reaction is to jerk their hand back when the snakes strike at it. These experimenters may also fail to elicit as many strikes from a snake as would a trained experimenter.

Similarly, there seems to be variation in the level of skill exhibited by those in our laboratory who perform extract testing on snakes. In this test a cotton swab impregnated with prey odor is placed gently in front of the snake; the number of tongue flicks to the swab is counted and the time before the swab is attacked is measured. One of the authors (G.M.B.) has been performing this test for more than 20 years and is often able to elicit prey attacks from snakes that may be unresponsive or flighty with other experimenters. After performing many such tests, an experimenter becomes attuned to the most subtle behavioral cues that the snake emits and learns which variations in his or her behavior will most likely elicit the desired behavior from the snake. Such abilities also emphasize the need for experimenter-blind testing of either stimuli or subjects in many circumstances.

Conclusions

Those of us who study reptiles do not really expect affectionate behavior from them. This is another way of saying that our interest in them is not influenced by some of the anthropomorphic and social companion factors so common among those who study or keep birds or mammals. Our interest is more often based on a fascination with a life-style, morphology, and sensory world so alien to ours in an animal with which we share a common vertebrate heritage. An emotional factor, usually kept hidden by the scientific herpetologist, may yet be involved. Looking into the eyes of an iguana, and being moved in a profound but unarticulated manner, for example, might represent a yearning to understand the vanished world of dinosaurs, arguably the most successful and dramatic occupants of our planet. Or perhaps our behavioral fascination with snakes somehow taps into being able to understand and predict the behavior of an inscrutable, mysterious, potentially dangerous being.

Thus, if it is possible that reptiles are able to recognize us, and even form a bond, it is a bonus, an additional link to an archaic life-style. But it is not one sought out by scientists studying reptiles, particularly in these days of understanding natural behavior in the field and physiological processes in the laboratory. The reentry of cognitive approaches to nonhuman animals in general has been controversial and application to reptiles rare indeed (Burghardt, 1991).

Any behavioral changes in reptiles due to individual recognition and bonding are likely to be extremely subtle or at least not readily interpreted with uncritical anthropomorphism. It is on the level of the very subtle interchange between experimenter and subject that reptile–scientist interactions might most profitably be studied. The reciprocal responses of humans and reptiles is a complex and rich area (Mundkur, 1988) deserving of further study.

Acknowledgments

Support for preparing this chapter was provided by a research grant from the National Science Foundation (BNS-8709629) to G. M. Burghardt and an NICHD training grant (HD-07303) to the Graduate Program in Ethology. We thank Paul Andreadis, Carla Breidenbach, John Groves, Hal Herzog Jr., Mark Milostan, Lani Lyman-Henley, Carey Yeager, Donna Layne, and the editors for valuable suggestions or comments on earlier drafts of this chapter.

References

Auffenberg, W. (1981). *The behavioral ecology of the Komodo monitor.* Gainesville: University Presses of Florida.

Barker, D. G., Murphy, J. B., & Smith, K. W. (1979). Social behavior in a captive group of Indian pythons, *Python molurus* (Serpentes Boidae) with formation of a linear hierarchy. *Copeia, 1979,* 466–77.

Breen, J. F. (1974). *Encyclopedia of reptiles and amphibians.* Neptune City, NJ: T.F.H. Publications.

Burghardt, G. M. (1970). Intraspecific geographical variation in chemical food cue preferences of newborn garter snakes (*Thamnophis sirtalis*). *Behaviour, 36,* 246–57.

(1977). Of iguanas and dinosaurs: Social behavior and communication in neonate reptiles. *American Zoologist, 17,* 177–90.

(1988). Precocity, play, and the ectotherm–endotherm transition: Profound reorganization or superficial adaptation. In E. M. Blass (Ed.), *Handbook of behavioral neurobiology* (Vol. 9, pp. 107–48). New York: Plenum.

(1991). Cognitive ethology and critical anthropomorphism: A snake with two heads and hognose snakes that play dead. In C. A. Ristau (Ed.), *Cognitive ethology: The minds of other animals* (pp. 53–90). Hillsdale, NJ: Erlbaum.

Burghardt, G. M., Allen, B. A., & Frank, H. (1986). Exploratory tongue flicking by green iguanas in laboratory and field. In D. Duvall, D. Müller-Schwarze, & R. M. Silverstein (Eds.), *Chemical signals in vertebrates* (Vol. 4, pp. 305–21). New York: Plenum.

Carpenter, C. C. (1984). Dominance in snakes. In R. A. Siegel, L. E. Hunt, J. L. Maleret, & N. L. Zuschiag (Eds.), *Vertebrate ecology and systematics: A tribute to Henry S. Fitch* (pp. 195–202). Lawrence: Museum of Natural History, University of Kansas Publication.

Chiszar, D., Carter, T., Knight, L., Simonsen L., & Taylor, S. (1976). Investigatory behavior in the plains garter snake (*Thamnophis radix*) and several additional species. *Animal Learning and Behavior, 4,* 273–8.

Coss, R. (1968). The ethological command in art. *Leonardo, 1,* 273–87.

Duvall, D. (1982). Western fence lizard (*Sceloporus occidentalis*) chemical signals: III. An experimental ethogram of conspecific body licking. *Journal of Experimental Zoology, 221,* 23–6.

Evans, L. T. (1959). A motion picture study of maternal behavior in the lizard *Eumeces obsoletus* Baird and Baird. *Copeia, 1959,* 103–10.

Fossey, D. (1983). *Gorillas in the mist.* Boston, MA: Houghton Mifflin.

Froese, A. D. (1980). Reptiles. In M. A. Roy (Ed.), *Species identity and attachment: A phylogenetic evaluation* (pp. 39–68). New York: Garland STPM Press.

Froese, A. D., & Burghardt, G. M. (1974). Food competition in captive juvenile snapping turtles, *Chelydra serpentina. Animal Behaviour, 22,* 734–9.

Gillingham, J. C. (1987). Social behavior. In R. A. Siegel, J. T. Collins, & S. S. Novak (Eds.), *Snakes: Ecology and evolutionary biology* (pp. 184–209). New York: Macmillan.

Greene, H. W. (1987). Antipredator mechanisms in reptiles. In C. Gans & R. B. Huey (Eds.), *Biology of the Reptilia* (Vol. 16, pp. 1–152). New York: Wiley.

Halloy, M., & Burghardt, G. M. (1990). Ontogeny of fish capture and ingestion in four species of garter snakes (*Thamnophis*). *Behaviour, 112,* 299–318.

Halpern, M., & Kubie, J. L. (1984). The role of the ophidian vomeronasal system in species-typical behavior. *Trends in Neuroscience, 7,* 472–7.

Herzog, H. A., Jr. (1990). Experiential modification of defensive behaviors in garter snakes (*Thamnophis sirtalis*). *Journal of Comparative Psychology, 104,* 334–9.

Herzog, H. A., Jr., Bowers, B. B., & Burghardt, G. M. (1989). Stimulus control of antipredator behavior in newborn and juvenile garter snakes (*Thamnophis*). *Journal of Comparative Psychology, 103,* 233–42.

Herzog, H. A., Jr., and Burghardt, G. M. (1977) Vocal communication signals in juvenile crocodilians. *Zeitschrift für Tierpsychologie, 44,* 294–304.

(1988). Development of antipredator responses in snakes: III. Long-term stability of litter and individual differences. *Ethology, 77,* 250–8.

The scientist and the snake

Kauffeld, C. (1969). *Snakes: The keeper and the kept.* New York: Academic Press.

Klaober, L. M. (1956). *Rattlesnakes.* Berkeley & Los Angeles: University of California Press.

Lederer, G. (1942). Der Drachenwaren (*Varanus komodoensis* Ouwens). *Zoologiste Garton* (Leipzig), *14*, 227–44.

Lorenz, K. (1943). Die angeborenen Formen möglicher Erfahrung. *Zeitschrift Für Tierpsychologie, 5*, 235–409.

Marcellini, D. L., & Jenssen, T. A. (1988). Visitor behavior in the National Zoo's reptile house. *Zoo Biology, 7*, 329–38.

Morris, R., & Morris, D. (1965). *Men and snakes.* London: Hutchinson.

Mundkur, B. (1983). *The cult of the serpent.* Albany: State University of New York Press.

(1988). Human animality, the mental imagery of fear, and religiosity. In T. Ingold (Ed.), *What is an animal?* (pp. 139–84). London: Unwin Hyman.

Noble, G. K. (1937). The sense organs involved in the courtship of *Storeria, Thamnophis* and other snakes. *Bulletin of the American Museum of Natural History, 73*, 673–725.

Pope, C. H. (1975). *The giant snakes.* New York: Knopf.

Pough, F. H. (1987). Mimicry and related phenomena. In C. Gans & R. B. Huey (Eds.), *Biology of the Reptilia* (Vol. 16, pp. 153–234). New York: Wiley.

Prater, S. H. (1933). The social life of snakes. *Journal of the Bombay Natural History Society, 36*, 469–75.

Rodda, G. H., Bock, B. C., Burghardt, G. M., & Rand, A. S. (1988). New techniques for identifying lizard individuals at a distance reveal influences of handling. *Copeia, 1988*, 905–13.

Romanes, G. J. (1882). *Animal intelligence.* London: Kegan Paul, Trench, Trubner.

Rudloe, J., & Rudloe, A. (1989). Shrimpers and sea turtles: Conservation impasse. *Smithsonian, 20*, 44–55.

Schuett, G. W., & Gillingham, J. C. (1989). Male–male agonistic behaviour of the copperhead, *Agkistrodon contortrix. Amphibia–Reptilia, 10*, 243–66.

Shine, R. (1987). Parental care in reptiles. In C. Gans & R. B. Huey (Eds.), *Biology of the Reptilia* (Vol. 16, pp. 275–329). New York: Wiley.

Somma, L. A. (1987). Maternal care of neonates in the prairie skink, *Eumeces septentrionalis. Great Basin Naturalist, 47*, 536–7.

Yeager, C. P., & Burghardt, G. M. (In press). The effect of food competition on aggregation: Evidence for social recognition in the plains garter snake (*Thamnophis radix*). *Journal of Comparative Psychology.*

—17—

Fear of humans and its consequences for the domestic pig

P. H. Hemsworth, J. L. Barnett, and G. J. Coleman

Editors' introduction

Humans and animals interact in a variety of settings other than conventional laboratories. The chapter by Hemsworth et al. deals with handler–animal effects in an applied setting: a pig farm. More than any other contributors, Hemsworth et al. are concerned with the effects of negative interactions between humans and animals. They examine the somewhat surprising array of human behaviors that are likely to cause fear and stress in pigs and other farm animals, as well as their physiological and behavioral effects, both immediate and chronic. It is an inescapable conclusion that were there more ethological knowledge about the pig, it would be easier to avoid the deleterious effects on behavior and productivity of negative human–animal interactions.

Introduction

Considerable research has been conducted on the behavior of farm animals in order to improve their growth, reproductive performance, and welfare. However, it is surprising that there is little objective information on the level of farm animals' fear of humans and the consequences of this for the animals. This topic may have been ignored or overlooked by behavioral scientists because the process of domestication involves adaptation to humans (Hemsworth & Barnett, 1987; Price, 1984) and many of the modern forms of animal farming involve intense contact between humans and animals. Thus, high levels of fear of humans may not have been expected in domestic animals in these farming systems. Nevertheless, there is recent evidence that farm animals may be very fearful of humans, with adverse consequences for the animals.

Pig production today is an example of the intensification of farm animal production in which humans exert significant control over animals (e.g., humans manipulate the nutritional, social, physical, and climatic environments) to the extent where the animals are totally dependent on humans for their survival in these production units. With their frequent and often intense contact, it is not surprising that humans and farm animals form extended and complex social relationships. Our recent research has shown that in some circumstances pigs may be highly fearful of humans and that this poor relationship with them may adversely affect the behavior, growth, reproductive performance, and welfare of the pigs. The main objectives of this chapter are to review the literature on the factors affecting, and the consequences of, farm animals' fear of humans, particularly that of the pig, and to consider the implications of this research for the experimental animal.

Measuring pigs' fear of humans

Considerable controversy and confusion surround the use of the word "fear," and problems such as those of terminology and definition, and whether fear really constitutes an example of motivation or drive, have been well reviewed by Hinde (1970), Murphy (1978), and Toates (1980). To avoid confusion and for the purpose of this chapter, fear will be considered an underlying motivational state that is linked, on the one hand, to particular preceding circumstances or treatments (stimuli) and, on the other hand, to a limited number of types of behavior (responses). This is similar to the approach taken by Murphy (1978), Toates (1980), and McFarland (1981).

Although the level of fear in animals has been inferred from the occurrence of behavioral changes, several authors have questioned whether an animal's behavior should be used for such an assessment (Murphy, 1978; Toates, 1980). When an animal is confronted with a fear-provoking stimulus, there are likely to be a number of behavioral responses available to the animal, and the specific response(s) elicited will depend on a number of factors, including the stimulus and its immediate environment and the past experience, physical condition, and evolutionary history of the animal. As argued by both Murphy (1978) and Toates (1980), it is difficult to determine which of several possible responses represents the greater level of fear. For example, can the following behavioral responses be defined as fear responses by chickens, and can they be ranked in order of increasing level of fear: orientation, head shaking, vocalization, crouching, and freezing? In addition, Murphy (1978) suggested that the absence of overt fear responses is not always indicative of a low level of fear. Another difficulty discussed by Murphy (1978) is that an artificial environment may prevent the animal from making behavioral responses that in a natural habitat would be functionally defined as fear responses. Therefore, there are difficulties in measuring fear and, in particular, in equating specific behavioral responses with the level of fear.

In studying pigs' level of fear of humans over the past 10 years, we have adopted a functional approach that classifies fear responses by their appropriateness in a given situation. Since it is generally accepted that fear responses function to protect the animal from harmful stimuli, we have proposed that the amount of avoidance of the experimenter or, conversely, the amount of approach to the experimenter is a useful measure of the pig's fear of humans. In our research we have equated the latency to and the amount of approach to a stationary experimenter in a standard test with the level of fear of humans in pigs. After a 2-minute familiarization period in which an individual pig is exposed to an arena that is similar to its home pen, an experimenter quietly enters and during the next 3 minutes, while the experimenter remains stationary, the approach behavior of the pig to the experimenter is recorded. The main chains of behavior of the domestic pig that may occur when it is exposed to a stationary human in a familiar environment are presented in Figure 17.1. Whether the animal directly moves from Stage 1 to Stage 2, 3, or 4 will depend on such factors as the characteristics of the human stimulus and experiential and genetic factors. The rate and extent of habituation of these fear responses will also be determined by these factors. The assumption is made that given sufficient time all pigs will experience habituation of their fear responses to the stationary human to the extent that they will move to Stage 5 and investigate the human (visual and perhaps olfactory and auditory investigation at a distance). Most animals will explore a stimulus that was initially fear provoking once the fear responses have waned (Hinde, 1970; McFarland, 1981), and thus progression to Stages 5 and 6 involves the motivation of pigs to familiarize themselves with their total environment. Another assumption is that differences among pigs in their motivation to explore the familiar arena are minor relative to the differences that exist among pigs in their fear of humans, and thus the predominant response is the response to the experimenter. This assumption is probably quite reasonable given the familiar environment and the pretest familiarization period.

It is generally accepted that exposure to fear-provoking stimuli results in a range of physiological responses by the animal, one of the most consistent being elevated corticosteroid concentrations (Mason, 1975; Selye,

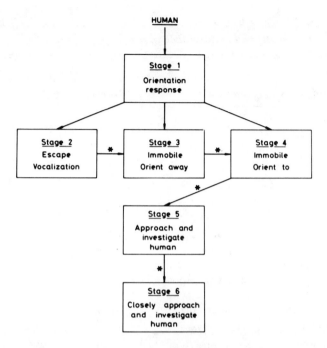

Figure 17.1. Chains of behavior of the domestic pig that may occur when the animal is exposed to a stationary human in a familiar arena. Asterisk refers to habituation of the fear response.

1976). Therefore, data on the corticosteroid response of pigs to humans would be useful in validating the use of behavioral data to assess pigs' level of fear of humans. We have recently examined data collected in a number of our experiments and have found significant correlations between the behavioral and corticosteroid response of pigs to humans. For example, the change in free corticosteroid concentrations from immediately before to 10 minutes after human exposure in the pig's pen was significantly correlated with the time taken for the pig to interact physically with the experimenter in the standard test ($r = .55$, d.f. = 48, $p < .01$). It should be noted that it was most unlikely that the blood-sampling procedure influenced the corticosteroid profile, since an extension was attached to the indwelling venous cannula so that the pigs could be blood-sampled from behind a blind without any apparent disturbance to the pigs. Therefore, these correlated behavioral and physiological changes in the presence of the experimenter provide good evidence that the approach behavior of a pig to the experimenter in the standard test is a useful measure of the pig's level of fear of humans.

Table 17.1. *The effects of handling treatments on the level of fear of humans and performance of pigs in five experiments*

Experiment and parameters	Mean for handling treatment*		
	Pleasant	Minimal**	Aversive
Hemsworth, Barnett, and Nansan (1981)			
Time to interact with experimenter (sec)[†]	119	—	157
Growth rate from 11 to 22 weeks (g/day)	709^b	—	669^a
Free corticosteroid concentrations (ng/ml)[‡]	2.1^x	—	3.1^y
Gonyou, Hemsworth, and Barnett (1986)			
Time to interact with experimenter (sec)[†]	73^a	81^{ab}	147^b
Growth rate from 8 to 18 weeks (g/day)	897^b	881^{ab}	837^a
Hemsworth, Barnett, and Hansen (1986)			
Time to interact with experimenter (sec)[†]	48^x	96^y	120^z
Pregnancy rate of gilts (%)	88^b	57^{ab}	33^a
Age of a fully coordinated mating response by			
boars (days)	161^x	176^{xy}	193^y
Free corticosteroid concentrations (ng/ml)[‡]	1.7^a	1.8^{ab}	2.4^b
Hemsworth, Barnett, and Hansen (1987)			
Time to interact with experimenter (sec)[†]	10^x	92^y	160^z
Growth rate from 7 to 13 weeks (g/day)	455^b	458^b	404^a
Free corticosteroid concentrations (ng/ml)	1.6^x	1.7^b	2.5^y
Hemsworth and Barnett (1991)			
Time to interact with experimenter (sec)[†]	55^a	—	165^y
Growth rate from 15 kg for 10 wk (g/day)	656	—	641
Free corticosteroid concentrations (ng/ml)[‡]	1.5	—	1.1

* Means in same row with different superscripts differ significantly ($^{a,b,c} p < .05$; $^{x,y,z} p < .01$).

** Treatments involving minimal human contact.

[†] Standard test to assess level of fear of humans in pigs.

[‡] Blood samples remotely collected at hourly intervals from 0800 to 1700 hours.

Further evidence to support our procedure is the effect of a number of handling treatments on the approach behavior of pigs to the experimenter in the standard test (details of the handling treatments are presented in the section "Fear of humans and its consequences for pigs"). In comparison with handling treatments that were aversive to pigs, positive handling resulted in pigs rapidly approaching and physically interacting with the experimenter in subsequent standard tests (Table 17.1).

The consequences of farm animals' fear of humans

Over the past 10 years we have been studying the effects of handling pigs. This research has consistently shown that pigs displaying high levels of fear of humans, measured on the basis of their approach behavior to an experimenter in the standard test, show evidence of a chronic stress response (i.e., sustained elevation of plasma concentrations of free corticosteroids) with detrimental effects on their growth and reproductive performance (Gonyou, Hemsworth, & Barnett, 1986; Hemsworth, Barnett, & Hansen, 1981, 1986, 1987; Hemsworth & Barnett, 1991). If the pig is highly fearful of humans and if a behavioral response such as fleeing is ineffective in avoiding this fear-provoking stimulus, the animal may have to resort to physiological coping mechanisms (i.e., a chronic stress response). The possible consequences of this chronic stress response are adverse effects on growth, reproduction, and health. Thus, while the potential magnitude of the production losses is large (Table 17.1), the nature of the response should not be unexpected, since fear is known to be a potent stressor of the pituitary–adrenal axis in animals (Selye, 1976).

The stress response and its consequences

The stress response is central to an animal's ability to adapt to short- and long-term changes in its environment, and components of this response are involved in biological processes relating to health, cardiovascular responses, tissue metabolism, and other physiological and psychological processes. Although there is some argument over the best biological indicator of the stress response (Mason, 1975; Moberg, 1985), there is an enormous body of literature on corticosteroid concentration as an indicator of stress response (Selye, 1976). There is general agreement that elevated corticosteroid concentrations indicate a physiological adaptation by animals and that a sustained elevation of corticosteroid concentrations is potentially detrimental to animals' welfare because of their actions on protein metabolism, reproduction, and the immune system.

There are two types of stress response in animals: acute and chronic. Animals display acute responses to such stressors as a new environment, mixing of unfamiliar animals, introduction of a female to a male, or transportation (Barnett, Cronin, Hemsworth, & Winfield, 1984). These responses are short term (lasting minutes to hours), and one of their major functions is to trigger a mechanism whereby glucose is provided from noncarbohydrate sources, particularly protein (from muscle or food), because the glycogen available in the liver for an immediate conversion to glucose is limited. Energy is thus mobilized for immediate response

269

to stressors. During this stage, provided that the intensity or duration of action of the stressors is not excessive, a steady state is achieved in which the increased metabolic demands are met by an increased metabolic performance.

Acute stress is essentially a physiological state that disappears on removal of the stressor; it leaves the animal depleted of energy reserves, but otherwise normal. This is an effective mechanism whereby animals can cope with or adapt to changes in their environment. As such, it is a normal event that occurs on a regular basis in all higher vertebrates and is of obvious benefit, although at the very least there is an energetic cost involved (Cahill, 1971). At this level of response, stress should be considered an adaptive mechanism.

If the stressor continues, the response proceeds to another series of events that is long term or chronic (days, months, or longer). In this phase, although still allowing the animal to cope, the response can have long-term detrimental consequences on nitrogen balance, secondary metabolism, and the cardiovascular, digestive, and immune systems. Prima facie evidence of a chronic stress response is a sustained elevation of corticosteroid concentrations (Burchfield, Woods, & Elich, 1980), and the basis of the methodology that we use to determine this has been described (Barnett, Winfield, Cronin, & Makin, 1981). Further discussion of the stress concept can be found in papers by Barnett and Hemsworth (1985), Barnett and Hutson (1987), Dantzer and Mormede (1983), McDonald (1979), and Moberg (1985).

Fear of humans and its consequences for the pig

In two studies conducted on commercial pig farms in the Netherlands (Hemsworth, Brand, & Willems, 1981) and in Australia (Hemsworth, Barnett, Coleman, & Hansen, 1989), the average level of fear of humans in sows was significantly and negatively correlated with the reproductive performance of the farm: Sows showed increased avoidance of the experimenter in the standard test at farms in which reproductive performance was low. For example, the correlation coefficient between the average time to interact physically with the experimenter in the standard test and the number of piglets weaned per sow per year at the farm was -0.52 (d.f. = 15, $p < .05$) in the latter study.

In conjunction with these on-farm studies, we conducted a number of experiments to identify the factors affecting fear of humans and the mechanisms by which fear may affect production. In these experiments (Gonyou et al., 1986; Hemsworth, Barnett, & Hansen, 1981, 1986, 1987; Hemsworth & Barnett, 1991), a number of handling treatments were imposed that resulted in a substantial range in pigs' fear of humans (data on the average time to first physical interaction with the experimenter in the standard test are presented in Table 17.1). It is noteworthy that the upper limits of this range are similar to those seen in commercial farms (Hemsworth & Barnett, 1987). The handling treatments in these experiments varied from treatments that discouraged approach to humans, by either slapping or briefly shocking the pigs with a battery-operated prodder whenever the pigs approached the experimenter, to treatments that encouraged approach to humans, by either patting or stroking the pigs whenever the pigs approached. The handling treatments were briefly conducted either three or five times per week, but an important feature was that the treatments were imposed only whenever the pig approached or remained close to the experimenter. As shown in Table 17.1, most of the aversive handling treatments increased the level of fear of humans and either depressed growth rate in young pigs or depressed reproductive performance in adult pigs.

In the experiment by Hemsworth and Barnett (1991), although there was no significant difference between treatments in growth rate over the 10-week period, pigs in the aversive treatment had a lower growth rate in the first 5 weeks of the experiment than pigs in the pleasant treatment (509 and 533 g/day, respectively, $p < .01$). It is of interest that Paterson and Pearce (1989) and Pearce, Paterson, and Pearce (1989), in experiments on the effects of aversive handling, found no significant effects of aversive handling on the growth rate of pigs, in contrast to our results. Bassett and Cairncross (1977) reported that rats regularly exposed to electric shock displayed adaptation of the corticosteroid response, and thus adaptation by some pigs to the aversive but predictable handling treatment may have been responsible for the failure of the aversive treatments to affect growth rate in the studies by Paterson and Pearce (1989) and Pearce et al. (1989). In the experiment by Hemsworth and Barnett (1991), the failure to affect growth rate in the latter part of the study suggests that adaptation to the aversive treatment may have occured.

A finding that is important for the interpretation of the results presented in Table 17.1 is that in three of these experiments (Hemsworth, Barnett, & Hansen, 1981, 1986, 1987), the pigs in the aversive handling treatments experienced a chronic physiological stress response. (In the experiment by Gonyou et al. [1986], corticosteroid concentrations were not studied, and in the experiment by Hemsworth and Barnett [1991] there was no significant difference between treatments in corticosteroid concentrations at the end of the experiment.) In the absence of humans, the mean daytime free corticosteroid concentrations of these pigs were higher than those of pigs in the other treatment. Presumably, a chronic stress response is responsible

271

for the observed depressions in growth and reproduction. For instance, it is probable that the reduction in growth rate was due to the actions of elevated corticosteroid concentrations decreasing the incorporation of amino acids into tissues and mobilizing energy reserves, particularly proteins, to provide glucose as an energy source for the central nervous system (Burchfield et al., 1980; Imms, 1967). This contention is supported by data from the aversively handled pigs used by Hemsworth, Barnett, and Hansen (1981); in addition to elevated corticosteroid concentrations, there were increases in plasma glucose and protein concentrations and a decrease in plasma urea concentration (Barnett, Hemsworth, & Hand, 1983). A chronic physiological stress response also indicates that the welfare of these pigs in the aversive handling treatment may also be at risk. A sustained elevation of corticosteroid concentrations, in part because of its detrimental effects on nitrogen balance and the immune system (Selye, 1976), suggests reduced welfare. We have evidence from other studies of a concomitant change in corticosteroid concentrations and the immune system (Barnett, Hemsworth, Winfield, & Fahy, 1987); in pigs housed in neck tethers in partial stalls, there was a significant elevation of free corticosteroid concentrations and a suppression of the immunoglobulin reactivity (IgM fraction) to an antigen (*Escherichia coli* K99).

On the basis of this research, it is proposed that the growth and reproductive performance and welfare of pigs that are highly fearful of humans may be at risk in situations where there is frequent and at times intense contact with humans. The mechanism responsible for the depressions in performance and welfare is likely to be a chronic stress response.

Fear of humans and its consequences for other farm animals

There is evidence from studies with other farm animals that contact with humans may affect the animals' performance. In a recent experiment with laying hens, we found that the level of fear of humans was significantly and inversely related to the hens' productivity (Hemsworth & Barnett, 1989). For example, the amount of avoidance by the bird of an approaching experimenter accounted for 21% of the variance in egg production. In addition, there was a significant negative relationship between corticosteroid response to handling and productivity, indicating that birds that had the highest corticosteroid increase in response to handling, and thus presumably were the most fearful of humans, had the lowest egg production. These data provide support for the extrapolation of our hypothesis to the laying hen: Birds that are highly fearful of humans are likely to experience a chronic stress response or a series of acute stress responses in a typical hen production unit in which there is frequent and at times close contact between humans and birds. The productivity and even the welfare of these birds may be at risk as a consequence of this stress response.

Gross and Siegel (1979) found that chickens that experienced frequent human contact from 5 weeks of age had higher growth rates, a higher antibody response to foreign red blood cells, and greater resistance to *Mycoplasma gallisepticum* challenge than birds that experienced minimal human contact. In the former case birds were housed in groups of 6 to 10, and the amount of human contact per cage varied from 90 to 120 seconds per day. This contact initially involved gently touching and stroking the birds' backs, and presumably these birds became less fearful of humans. Birds in both treatments were fed and cleaned twice a week, and this was the main opportunity for birds in the minimal-contact condition to receive human contact. Although the behavioral response of the birds to humans was not quantified, the authors stated that the handled birds were easy to handle during weighing and blood sampling. In another experiment with chickens, Gross and Siegel (1980) studied the effects of frequent human contact from 6 to 8 weeks of age on the subsequent performance of birds to 18 weeks of age. The handled birds had better feed-conversion efficiency than birds that experienced minimal human contact. In addition, handling decreased the deleterious effects of fasting on the antibody response to foreign red blood cells.

There are several possible explanations for these results. First, differences in the level of fear of humans may be responsible for the differences in performance, as observed in the pig. The handled birds are likely to have been less fearful of humans, perhaps because the handling was of a positive nature or the handling occurred during a sensitive period for socialization. In contrast, with minimal human contact, there may not have been an opportunity for habituation of the initial fear responses of the nonhandled birds. Second, some of the behavioral patterns of humans displayed during routine husbandry procedures may have been threatening to these nonhandled birds, which were initially quite fearful, and this may have exacerbated their fear of humans. The possible threatening nature of some of the behavioral patterns of humans used in routine husbandry will be further discussed later in this section. Third, the improved performance of handled birds may have been due to the phenomenon that is often called "infantile stimulation." Infantile stimulation of young rats and mice, presumably producing an acute stress response, results in increased avoidance learning and increased weight at weaning and, in adulthood, in decreased "general fearfulness" and an earlier maturation of the hypothalamic–hypophyseal–adrenal system (Hinde, 1970; Schaefer, 1968). It is possible that a similar phenomenon may have occurred with the handled birds. Obviously, further research is warranted to identify the mechanism(s) responsible for the improved performance of the handled birds.

The results of a number of other experiments support those of Gross and Siegel (1979, 1980). For example, Thompson (1976) and Jones and Hughes (1981) found, in general, an increased growth rate following handling of young broilers and layers. In contrast, Reichmann, Barram, Brock, and Standfast (1978) found no effects of handling on the growth of young broilers and layers, whereas Freeman and Manning (1979) found that regular handling decreased growth in young layers. A better understanding of the nature of the handling procedures and its effects on the animals' levels of fear of humans is likely to be useful in explaining these contradictory results. An experiment by Buckland, Goldrosen, and Bernon (1974) demonstrated the adverse effects of very aversive handling on the growth performance of broilers; a significant decrease in the 8-week body weight of broilers was observed with blood sampling using cardiac puncture at 4 and 8 weeks of age.

Therefore, although the results of studies on chickens appear contradictory, general handling treatments that appeared to involve considerable human contact, to be of a positive nature, and to reduce the bird's fear of humans resulted in improved growth rates in chickens. The mechanism(s) responsible for this effect is unclear. A different result might occur if the handling were minimal, irregular, inconsistent, or of a negative nature.

The dairy cow is another farm animal that has received some research attention in this area. Seabrook (1972) reported some significant associations between the personality of the stockperson and milk production of cows. In 28 one-person herds, the highest-yielding herds had stockpersons who were introverted and had a high level of confidence. Although the behavior of the cows was not quantified, Seabrook suggested that cows in the highest-yielding herds tended to be the most willing both to return from pasture and to enter the milking parlor, and to be less restless in the presence of the stockperson. Although it is possible that an underlying psychological factor of the stockperson, by affecting the stockperson's behavior, may affect cows' level of fear of humans, it is surprising that the relationships between these factors and productivity have not been further investigated in dairy cows.

Two experiments at our laboratory (Hemsworth, Hansen, & Barnett, 1987; Hemsworth, Barnett, Tilbrook, & Hansen, 1989) examined the effects of human presence at the time of calving of primiparous dairy cows. Cows that calved in the presence of humans were less restless during milking and required less assistance from the stockperson during milking. The productivity of these cows was not affected. On the basis of the approach behavior of the cows to a stationary experimenter and cortisol concentrations in the cows' milk, it is proposed that the improved milking behavior of cows that calved in the presence of humans might have been due to the these cows being less fearful of humans. It is possible that since the parturient cow is extremely sensitive to her environment, the close presence of humans at this time might result in the cow forming an attachment to humans that may be similar to, but less specific and weaker than, that which occurs between the cow and her newborn calf.

A number of studies have shown that dairy calves reared in visual and tactile isolation from conspecifics produce more milk than herd mates raised with visual and tactile contact with conspecifics (Arave, Mickelson, & Walters, 1985; Warwick, Arave, & Mickelson, 1977). It has been proposed that in the former case cows may have imprinted on the stockperson and thus may have adapted more easily to the milking procedure, which involves intense human contact. Creel and Albright (1988) rejected this hypothesis on the basis of the approach behavior of calves to a stationary experimenter; however, they found that the isolated calves had a shorter flight distance to an experimenter than control calves. The behavioral response of mature cows to humans was not examined.

Fordyce, Wythes, Shorthose, Underwood, and Shepherd (1988) examined the influence of "temperament" on the meat quality of beef cattle. The authors defined temperament as the behavioral response to handling by humans and assessed temperament by scoring the behavior of cattle that were restrained in a weighing scale. They found that cattle that were most active and vocal when restrained had the most carcass bruising and tended to have tougher meat. Although some components of the behavioral response of the bulls observed in this test would have been responses to restraint and the novel environment, a significant component of the behavioural response would have been a response to humans.

Therefore, although the data for other farm animals are not as rigorous as for the pig, there is some limited evidence that high levels of fear of humans may adversely affect the performance of animals. Clearly, further research is warranted in these species.

Human factors regulating the level of fear of humans in farm animals

We have examined some human factors that may affect pigs' level of fear of humans (Hemsworth, Barnett, Coleman, & Hansen 1989). An obvious factor was the behavior of the stockperson toward pigs. Ajzen and Fishbein (1980) have developed the theory of reasoned action, which they propose can be used to predict human behavioral intent. According to this theory,

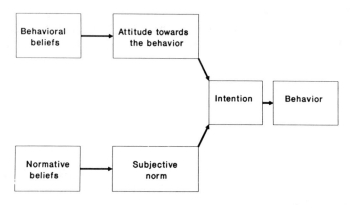

Figure 17.2. Factors determining a person's behavior according to Ajzem and Fishbein (1980). Arrows indicate the direction of influence.

the intention to perform a behavior is a function of a person's attitude toward the behavior to be performed and the perceived social pressures on the person to perform the behavior (referred to as the subjective norm). In addition, it is proposed that both attitudes and subjective norms are a function of beliefs. This approach is summarized in Figure 17.2. Ajzen and Fishbein (1980) have found that a person's attitude toward a behavior can be measured by the use of questions relating to that behavior. In most situations, normative beliefs do not contribute significantly to the prediction of behavioral intent (Ajzen & Fishbein, 1980). Therefore, in addition to the stockperson's behavior, the stockperson's attitudes and subjective norms were examined in this study. So that the farmers would not be too sensitized to the objectives of the study, attitude questions were confined to a stockperson's beliefs about pigs and behaviors toward pigs. In this study of 19 farms significant relationships were found between the attitude and behavior of the stockperson, level of sows' fear of humans, and reproductive performance of the farm.

An important finding was that the behavior of the stockperson toward pigs was a strong predictor of sows' level of fear of humans. A high proportion of negative physical actions by the stockperson was consistently and strongly associated with a high level of fear of humans in the sows. Most of these negative actions were hits, slaps, and kicks; however, many were applied with only slight or moderate force. It was also surprising that although the proportion of negative actions observed during the handling of pigs in routine mating activities (estrus detection, assistance at mating, etc.) varied from 38 to 100% across the 19 farms, the time spent handling pigs in these activities over a 7-week period was small and varied from

8 to 57 minutes per pig. The positive actions were mainly pats, strokes, and hand resting on the back of the pig. This relationship suggests that sows' fear of humans may be reduced by the stockperson reducing the proportion of negative behaviors.

Another significant finding was that a number of the attitudinal variables showed moderate to strong relationships with the sows' level of fear of humans. These relationships are probably due to the fact that the stockperson's behavior was highly correlated with his or her attitude toward pigs or to the stockperson's own behavior toward pigs. Since both the attitude and behavior of the stockperson were related to the level of fear of humans, it is not surprising that both these aspects were also related to the reproductive performance of the farm. These relationships demonstrate that there may be substantial potential to reduce pigs' fear of humans in order to improve reproductive performance by employing stockpersons that have desirable attitudinal and behavioral profiles. It is well established that human subjects show a low tolerance for discrepancies between beliefs and behavior (Festinger & Carlsmith, 1959) and will attempt to alter either attitude or behavior to reduce the discrepancy. When one is attempting to change attitude and behavior, to avoid increasing the attitude–behavior discrepancy, both attitude and behavior should be targeted.

While the study by Hemsworth, Barnett, Coleman, and Hansen (1989) examined the main physical behaviors of the stockperson, further research is required to identify the effects of more subtle behavioral patterns. Our research with the pig indicates how sensitive the animal is to a very limited amount of aversive handling, even if the animal has the opportunity to avoid the handling (Table 17.1). However, behaviors that are obviously aversive to animals may not be the only ones that affect their level of fear of humans. Some human behavioral patterns that intuitively appear to be harmless and inoffensive to pigs can markedly reduce the approach behavior of pigs. For example, a human standing in an erect posture, wearing gloves, and closely approaching pigs is more threatening to pigs than a human in a squatting posture, showing bare hands, and remaining stationary (Gonyou et al., 1986). Surprisingly, the regular display of the former behavioral patterns by an experimenter significantly reduced the growth rate of young pigs to levels similar to those found with a clearly aversive handling treatment. Therefore, behaviors that may not appear to be aversive to farm animals may in fact be quite threatening and, if displayed regularly or frequently, may markedly increase the level of fear of humans and place the animals' productivity and welfare at risk.

In the study by Hemsworth and Barnett (1989), in which a significant negative relationship was found between fear of humans and productivity of laying hens, the birds housed in the top tier of a battery-cage system had higher levels of fear of humans (and lower productivity) than birds in the lower tier. It is interesting to speculate on the human factors that may regulate laying hens' fear of humans in a production system. There was a marked tendency for birds housed in the two middle rows to show lower levels of fear than those in the outer rows. Stockpersons used the middle corridor to service birds in these two rows, while the two outer corridors serviced only one row each; therefore, the amount of time that humans spent in the middle corridor was likely to be greater than that spent in the two outer corridors. The increased human contact, perhaps by facilitating habituation of fear responses that are initially high, may be responsible for making birds in the middle rows less fearful of humans. Similarly, the birds in the bottom tier, which have a less restricted view of humans in the corridor than birds in the top, may receive more human contact and consequently display less fear of humans. Another possibility is that birds in the top tier or outer rows may perceive human behavior as more threatening. For example, for birds in the top tier, the hand of the stockperson suddenly appearing near the trough (to add food or collect eggs) may be highly threatening. Because the corridors in the outer rows are narrower, the birds are likely to have less warning of the stockperson's close approach, and so the distance at which the birds are first aware of the human's imminent approach may be less. Human behavior that startles birds could lead to an increase in fear of humans. Such factors and others may act either individually or in combination to produce these differences.

Obviously, further research is required to examine the factors that regulate animals' fear of humans. Physical interaction with humans may be most important in animals like pigs; however, with other farm animals, such as poultry, which have little human tactile contact, visual interaction with humans may be more important.

The influence of experimenter behavior on the performance of experimental animals

The data presented so far suggest that either the mere presence of a human or the behavior of a human toward a farm animal can affect the animal's performance. While the mere presence of a person may have an aversive effect on animals not habituated to the novel stimulus of that person's presence, at least in some cases the mere presence of a person may be beneficial to the animal. An example of this, cited earlier, is the improvement in milking behavior observed in dairy cattle when the stockperson is present during parturition (Hemsworth, Barnett, Tilbrook, & Hansen, 1989; Hemsworth, Hansen, & Barnett, 1989).

Most likely it is the *interaction* between the stockperson and the animal that determines the stimulus properties of the human for the animal. The characteristics of these stimulus properties include the familiarity or novelty of the human as a stimulus (i.e., the extent to which the animal has become habituated to the mere presence of the person) and the rewarding or aversive properties the person has for the animal (i.e., the extent to which the person has been associated with rewarding or aversive events).

This proposed nature of the stockperson's stimulus properties is firmly based in learning theory. The fear response elicited by the mere presence of a stranger can be regarded as a form of neophobia (fear of novel stimuli), which is quite common in a number of species (Russell, 1979). However, this response is normally short-lived when it occurs in a novel area (such as the arena used in our research) and is soon replaced by exploratory behavior (Russell, 1979). The variability in approach behavior that we observe in pigs when using the arena described earlier can therefore best be understood in terms of learned associations between a specific human (the stockperson) and aversive events followed by stimulus generalization.

Although there is some controversy about the mechanism by which avoidance behavior becomes conditioned by punishment (Walker, 1987), it is well established that animals learn to avoid unconditioned stimuli that are paired with aversive events. Through the process of stimulus generalization, the pig's response to a single human can extend to all humans (Hemsworth, Brand, & Willems, 1981).

If this is the mechanism by which the behavior of the stockperson influences the pig's performance, a number of predictions about the possible influences of the experimenter's behavior can be made. First, it can be explicitly tested whether the pig's avoidance of a person decreases as the differences increase between the stimulus person and the original person from whom the aversive stimulation originated. This is an explicit test of the stimulus generalization aspects of avoiding humans as a consequence of previous aversive experience with the stockperson. Second, it would be predicted that the pig's avoidance of the stockperson should decrease if, on repeated exposure, no punishment occurred. The prediction is based on the expectation that extinction of the avoidance response should occur in the absence of reinforcement. If the stockperson rewards the pig, this process should be accelerated as the approach response becomes reinforced at the same time as the avoidance response is becoming extinguished. Such a procedure suggests a method of intervention on farms where poor production can be attributed to poor stockperson behavior.

Implicit in this discussion is the idea that we know which human interactions are aversive and which are rewarding for farm animals. Ultimately, this is an empirical question. Our observations suggest, however, that even mild slaps, if repeated often enough and in the absence of patting or stroking, are aversive stimuli to pigs. Experimental testing of this by investigating the changes in corticosteroid concentrations and avoidance behavior associated with these quite specific human activities should be possible using procedures similar to those in our earlier studies (Hemsworth, Barnett, & Hansen, 1981). In other farm animals the relevant correlational data are not available to determine which behaviors of the stockperson are rewarding or punishing. In dairy cattle the fact that the stockperson's personality is associated with milk production (Seabrook, 1972) suggests that the mechanism by which production is affected may be similar to that in pigs. Specifically, the stockperson's personality may be related to the attitude and behavior of the stockperson, leading to conditioning of the dairy cow. In the case of chickens, however, the dominant mechanism producing fear of humans may be neophobia rather than conditioned avoidance, because there is virtually no contact between the stockperson and the bird. The findings by Gross and Siegel (1979) are consistent with this view. However, if the chickens are neophobic, intermittent exposure to a human would be aversive and would lead, in time, to a conditioned fear response in the birds, while prolonged exposure of a neutral or positive nature would reduce the birds' fear of humans.

Conclusion

In summary, it is proposed that neophobia and conditioned fear responses in farm animals provide a basis for understanding the mechanism by which the behavior of stockpersons affects production. If this is the case, there are clear areas in which selection and training of stockpersons can be improved and fear responses in farm animals can be extinguished.

This proposal also has implications for experimenters and their experimental animals. Variation among experimental animals in either the amount or nature of human contact may result in different levels of fear toward humans. This, in turn, may create within-treatment variation or confound the effects of treatment on the behavior or performance of the animals. This possibility should be recognized by the experimenter and should be controlled in the experimental method. In order to avoid the possibility of high levels of fear compromising experimental objectives when fear of humans is not being studied, the experimenter may find it most useful to manipulate the animals' fear of humans before and during experimentation so that the experimental animals consistently display low levels of fear. The experimenter can gain an approximation of the animals' fear of humans by monitoring their approach–withdrawal behavior in the presence of humans, and in situations where there is large variation in level of fear or where there is a high level of fear, the experimenter can attempt to reduce the animals' fear of humans to low levels. Techniques that the experimenter could use to reduce fear of humans include increasing the amount of human contact to enable the animals to habituate themselves to the presence of humans, replacing negative human behaviors toward the animals (e.g., those behaviors that increase withdrawal from humans) with positive human behavior (e.g., those behaviors that encourage approach to humans), avoiding situations in which the animals may associate aversive events with humans, and providing situations in which the animals may associate rewarding events with humans. Therefore, fear of humans should be considered a factor, similar to temperature variation or disease, that can influence the behavior or performance of animals and thus should be controlled in an experimental design.

References

Ajzen, I., & Fishbein, M. (1980). *Understanding attitudes and predicting social behavior*. Prentice-Hall, Englewood Cliffs, NJ.

Arave, C. W., Mickelson, C. H. & Walters, J. L. (1985). Effect of early rearing experience on subsequent behavior and production of Holstein heifers. *Journal of Dairy Science, 68*: 923–9.

Barnett, J. L., Cronin, G. M., Hemsworth, P. H., & Winfield, C. G. (1984). The welfare of confined sows: Physiological, behavioural and production responses to contrasting housing systems and handler attitudes. *Annates de Recherches Veterinaires, 15*: 217–26.

Barnett, J. L., & Hemsworth, P. H. (1985). Stress and its links with stockmanship. *Pig International*, April, pp. 20–1.

Barnett, J. L., Hemsworth, P. H., & Hand, A. M. (1983). The effect of chronic stress on some blood parameters in the pig. *Applied Animal Behaviour Sciences, 9*: 273–7.

Barnett, J. L., Hemsworth, P. H., Winfield, C. G., & Fahy, V. A. (1987). The effects of pregnancy and parity number on behavioural and physiological responses related to the welfare status of individual and group housed pigs. *Applied Animal Behaviour Sciences, 17*: 229–43.

Barnett, J. L., & Hutson, G. D. (1987). Objective assessment of welfare in the pig: Contributions from physiology and behaviour. In J. L. Barnett et al. (Eds.), *Manipulating Pig Production*. Australasian Pig Science Association, Werribee, Australia, pp. 1–22.

Barnett, J. L., Winfield, C. G., Cronin, G. M., & Makin, A. W. (1981). Effects of photoperiod and feeding on plasma corticosteroid concentrations and maximum corticosteroid binding capacity in pigs. *Australian Journal of Biological Sciences, 34*: 577–85.

Basset, J. R., & Cairncross, K. D. (1977). Changes in the coronary vascular system following prolonged exposure to stress. *Pharmacology, Biochemistry, & Behavior, 6*: 311–18.

P. Hemsworth, J. Barnett, and G. Coleman

Buckland, R. B., Goldrosen, A., & Bernon, D. E. (1974). Effect of blood sampling by cardiac puncture on subsequent body weight of broilers and S. C. White Leghorn replacement pullets. *Poultry Science, 53*: 1256-8.
Burchfield, S. R., Woods, S. C., & Elich, M. S. (1980). Pituitary adrenocortical response to chronic intermittent stress. *Physiology and Behavior, 24*: 297-302.
Cahill, G. F. (1971). Action of adrenal cortical steroids on carbohydrate metabolism. In N. P. Christy (Ed.), *The Human Adrenal Cortex*. Harper & Row, New York, pp. 205-39.
Creel, S. R., & Albright, J. L. (1988). The effects of neonatal social isolation on the behavior and endocrine function of Holstein calves. *Applied Animal Behaviour Sciences, 21*: 293-306.
Dantzer, R., & Mormede, P. (1983). Stress in farm animals: A need for re-evaluation. *Journal of Animal Sciences, 67*: 6-18.
Festinger, L., & Carlsmith, J. M. (1959). Cognitive consequences of forced compliance. *Journal of Abnormal Social Psychology, 58*: 203-10.
Fordyce, G., Wythes, J. R., Shorthose, W. R., Underwood, D. W., & Shepherd, R. K. (1988). Cattle temperaments in extensive beef herds in northern Queensland: 2. Effect of temperament on carcass and meat quality. *Australian Journal of Experimental Agriculture, 28*: 689-93.
Freeman, B. M., & Manning, A. C. C. (1979). Stressor effects of handing on the immature fowl. *Research in Veterinary Science, 26*: 223-6.
Gonyou, H. W., Hemsworth, P. H., & Barnett, J. L. (1986). Effects of frequent interactions with humans on growing pigs. *Applied Animal Behaviour Science, 16*: 269-78.
Gross, W. B., & Siegel, P. B. (1979). Adaptation of chickens to their handlers and experimental results. *Avian Diseases, 23*: 708-14.
(1980). Effects of early environmental stresses on chicken body weight, antibody response to RBC antigens, feed efficiency and response to fasting. *Avian Diseases, 24*: 549-79.
Hemsworth, P. H., & Barnett, J. L. (1987). Human-animal interactions. In E. O. Price (Ed.), *The Veterinary Clinics of North America, Food Animal Practice*. Saunders, Philadelphia, Vol. 3, pp. 339-56.
(1989). Relationships between fear of humans, productivity and cage position of laying hens. *British Poultry Science, 30*: 505-18.
(1991). The effects of aversively handling pigs either individually or in groups on their behaviour, growth and corticosteroids. *Applied Animal Behaviour Science, 30*: 61-72.
Hemsworth, P. H., Barnett, J. L., Coleman, G. J., & Hansen, C. (1989). A study of the relationships between the attitudinal and behavioural profile of stockpersons and level of fear of humans and reproductive performance of commercial pigs. *Applied Animal Behaviour Science, 23*: 301-14.
Hemsworth, P. H., Barnett, J. L., & Hansen, C. (1981). The influence of handling by humans on the behaviour, growth and corticosteroids in the juvenile female pig. *Hormones and Behavior 15*: 396-403.
(1986). The influence of handling by humans on the behaviour, reproduction and corticosteroids of male and female pigs. *Applied Animal Behaviour Science, 15*: 303-14.
(1987). The influence of inconsistent handling on the behaviour, growth and corticosteroids of young pigs. *Applied Animal Behaviour Science, 17*: 245-52.
Hemsworth, P. H., Barnett, J. L., Tilbrook, A. J., & Hansen, C. (1989). The effects

282

of handling by humans at calving and during milking on the behaviour and milk cortisol concentrations of primiparous dairy cows. *Applied Animal Behaviour Science, 22*: 313–26.

Hemsworth, P. H., Brand, A., & Willems, P. J. (1981). The behavioural response of sows to the presence of human beings and their productivity. *Livestock Production Science, 8*: 67–74.

Hemsworth, P. H., Hansen, C., & Barnett, J. L. (1987). The effects of human presence at the time of calving of primiparous cows on their subsequent behavioural response to milking. *Applied Animal Behaviour Science, 18*: 247–55.

Hinde, R. A. (1970). *Animal Behaviour: A Synthesis of Ethology and Comparative Psychology*, 2d ed. McGraw-Hill, New York.

Imms, F. J. (1967). The effects of stress on the growth rate and food and water intake of rats. *Journal of Endocrinology, 37*: 1–18.

Jones, R. B., & Hughes, B. O. (1981). Effects of regular handling on growth in male and female chicks of broiler and layer strains. *British Poultry Science, 22*: 461–5.

Mason, J. W. (1975). Emotion as reflected in patterns of endocrine integrations. In L. Levi (Ed.), *Emotions: Their Parameters and Measurements*. Raven Press, New York, pp. 143–81.

McDonald, I. R. (1979). The concept of stress as a physiological state. In *Deer Farming in Victoria*. Proceedings of a Symposium, Agricultural Note Series No. 15. Department of Agriculture, Melbourne, pp. 37–49.

McFarland, D. (1981). *The Oxford Companion to Animal Behaviour*. Oxford University Press, Oxford.

Moberg, G. P. (1985). Biological response to stress: Key to assessment of animal well-being? In G. P. Moberg (Ed.), *Animal Stress*. American Physiology Society, Bethesda, MD, pp. 27–49.

Murphy, L. B. (1978). The practical problems of recognizing and measuring fear and exploration behavour in the domestic fowl. *Animal Behaviour, 26*: 422–31.

Paterson, A. M., & Pearce, G. P. (1989). Boar-induced puberty in gilts handled pleasantly or unpleasantly during rearing. *Applied Animal Behaviour Science, 22*: 225–33.

Pearce, G. P., Paterson, A. M., & Pearce, A. N. (1989). The influence of pleasant and unpleasant handling and the provision of toys on the growth and behaviour of male pigs. *Applied Animal Behaviour Science, 23*: 27–37.

Price, E. O. (1984). Behavioural aspects of animal domestication. *Quarterly Review of Biology, 59*: 1–32.

Reichmann, K. G., Barram, K. M., Brock, I. J., & Standfast, N. F. (1978). Effects of regular handling and blood sampling by wing vein puncture on the performance of broilers and pullets. *British Poultry Science, 19*: 97–9.

Russell, P. A. (1979). Fear evoking stimuli. In W. Slukin (Ed.), *Fear in Animal and Man*. Van Nostrand Reinhold, New York, pp. 86–124.

Schaefer, T. (1968). Some methodological implications of the research on early handling in the rat. In G. Newton & S. Levine (Eds.), *Psychobiology of Development*. Thomas, Springfield, IL, pp. 102–41.

Seabrook, M. F. (1972). A study to determine the influence of the herdsman's personality on milk yield. *Journal of Agricultural Laboratory Science, 1*: 45–59.

Selye, H. (1976). *Stress in Health and Disease*. Butterworths, London.

Thompson, C. I. (1976). Growth in the Hubbard broiler: Increased size following early handling. *Developmental Psychobiology, 9*: 459–64.

Toates, F. M. (1980). *Animal Behavior: A Systems Approach*. Wiley, New York.
Walker, S. (1987). *Animal Learning: An Introduction*. Routledge & Kegan Paul, London.
Warwick, V. D., Arave, C. W., & Mickelson, C. H. (1977). Effects of group, individual and isolated rearing of calves on weight gain and behavior. *Journal of Dairy Science, 60*: 947–53.

—18—
The effect of the researcher on the behavior of poultry

Ian J. H. Duncan

Editors' introduction

Duncan is concerned with the effects of the scientist on the behavior of poultry and, like Hemsworth et al., reveals a surprising degree of fearful responses in these animals. Duncan makes the interesting point that, despite domestication, poultry have not become so inured to human presence that their behavior is free of fear. In general, Duncan offers a valuable summary of the precautions one must take to avoid "researcher effects" in conducting ethological research with a rather wary animal subject.

My first lesson on the effect of the researcher on poultry behavior occurred on the day I started graduate studies at the Poultry Research Centre in Edinburgh. My supervisor told me to sit in a chair in the middle of a large pen containing about 20 hens and record their behavior. What the hens did seemed to me to be completely unremarkable; they carried out typical hen activities, preening and pecking and scratching. What *was* remarkable was *where* they performed these activities. The birds formed a ring around my chair at a distance of about a meter. They were obviously in conflict, with a tendency to approach and investigate counterbalanced by a tendency to avoid the strange new object. Much later in my studies I came to realize that the "unremarkable" behavior patterns I had witnessed were, in fact, "displacement activities" (McFarland, 1981), symptomatic of the ambivalence associated with an approach–avoidance conflict and just as much due to my presence as was the ring formed by the birds around my chair. The lesson served its purpose well. From that day, I have attempted to reduce or eliminate the "human observer effect" in my studies. However, it

285

is not always possible to take these precautions. Moreover, in carrying out behavioral research on a captive domestic species, making observations is only one step of many in which the researcher might have an influence on the test animal. The rest of this chapter is devoted to speculating on what these influences might be and how they might have affected research results.

Evolutionary considerations

An understanding of the influences that could be exerted by a human researcher on the behavior of poultry requires some consideration of the evolutionary history of the species involved. In fact, little information is available on the early history of any of the poultry species – namely, the domestic fowl, turkey, quail, duck, goose, and guinea fowl. However, the modern wild species that appear to have a common ancestry with the domestic poultry species – that is, junglefowl (*Gallus gallus* spp.), wild turkey (*Meleagris gallopavo*), Japanese quail (*Coturnix coturnix japonica*), mallard duck (*Anas platyrhynchos*), muscovy duck (*Cairina moschata*), grey-lag goose (*Anser anser*), swan goose (*Anser cygnoides*), and guinea fowl (*Numida meleagris*) – are all hunted by human beings in various parts of the world today. It therefore seems most likely that the original relationship between human beings and the progenitors of poultry was a predator–prey one. The prevailing reaction of prey species to predators is usually a fearful one (Russell, 1979), and although the domestication process may have reduced the intensity of this response (Hale, 1969), evidence suggests that it has not removed it completely (Duncan, 1985, 1990).

We might therefore expect that the most likely "unwanted" or "uncontrolled" influence of the experimenter on poultry species would be connected with fear. In general, in experiments on the behavior of poultry, there has been an awareness of the dangers of recording observations within sight of the birds, and appropriate precautions, such as using hides with peepholes, one-way glass, and closed-circuit television, have usually been adopted. Another approach has been to isolate the bird in an operant conditioning chamber (see Figure 18.1). However, these precautions have not always been taken, and the exceptions will be discussed later.

Experiments involving human contact

There have been a small number of controlled experiments on the reactions of poultry to contact with human beings. Perhaps the most well-known of these are the studies on imprinting in domestic avian species that were probably inspired by the descriptions of domestic chicks by Spalding

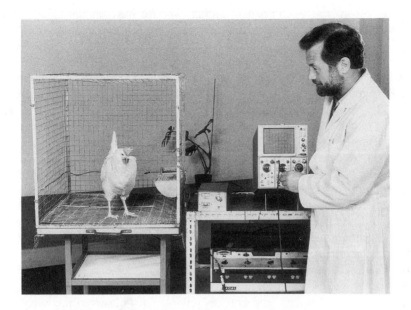

Figure 18.1. The author and his experimental subject.

(1873) and of ducklings and goslings by Heinroth (1911) and Lorenz (1935), which showed a following response to human beings. Many ethology textbooks contain photographs of Konrad Lorenz with goslings in train. However, only occasionally have human beings been used as imprinting objects in systematic laboratory studies with domestic fowl (e.g., Gray & Howard, 1957). Much more common is the anecdotal evidence of the effects of imprinting such as the case of a turkey cock described by Räber (1948). When adult, this hand-reared domestic turkey cock courted any men who walked through the park in which it lived, but attacked women. It appeared that as a result of imprinting, this bird regarded human beings as conspecifics and courted men. However, it regarded the hanging and flapping skirts of women as a sign of maleness (a displaying male turkey has a hanging snood and drooping wings) and therefore to be attacked! Of course, one observation does not make an experiment, and it would be desirable to repeat this in a controlled way to find out exactly what are the stimulus properties to which an imprinted turkey will respond.

There is a great deal of evidence that domestic poultry chicks can imprint to human beings. However, this is such a large and obvious response that it is doubtful whether it could occur in an "unwanted" way without being noticed.

287

The recent rise in interest in animal welfare has stimulated a number of investigations into the possible adverse effects of human contact with various poultry species, since it has been suggested that the interaction between human beings and animals is one of the main causes of reduced welfare (Duncan, 1974; Wood-Gush, Duncan, & Fraser, 1975; Hemsworth, & Barnett, 1987). In a review of these studies, Duncan (1990) concluded that the dominant response of poultry to human beings is a fearful one. For example, there are several reports that domestic fowl, which have had a "normal" amount of contact with human beings, are frightened and stressed by being handled (Hughes & Black, 1976; Murphy & Duncan, 1977; Beuving, 1980). In a comprehensive study of fear in domestic fowl, Murphy (1976) investigated two stocks, one with the reputation of being "flighty" and one considered to be "docile." She revealed that, as might be expected, the flighty stock showed much more avoidance of human beings than did the docile stock. However, this difference could not be explained entirely in terms of a difference in general fearfulness, since the flighty birds showed less fear of novel foods and novel objects (Murphy, 1977). This difference between the two stocks of bird was confirmed by an experiment of Jones, Duncan, and Hughes (1981). They observed the behavioral responses and recorded the heart rates of individual hens as they were slowly approached by a human being. At the start of each observation, food was presented in a novel way, and the docile birds showed more fear of this than did the flighty birds. Moreover, when the human being was far from the birds (25 to 30 m) and perhaps seen as a strange object rather than as a human being, the docile birds showed more signs of low-level fear than did the flighty birds. However, as the human being came closer, the flighty birds showed much higher indices of fear.

Results from more applied studies have also shown that domestic fowl are frightened by the close approach of a human being. For example, the catching, crating, and trucking procedures to which broilers are normally exposed before slaughter were simulated in the laboratory in a series of experiments (Duncan, 1989). The most stressful aspect of this whole procedure was the initial approach, catching, and restraint by a human being. In another experiment it was shown that birds were actually less stressed by being caught and picked up by a specially designed machine than they were by being manually caught (Duncan, Slee, Kettlewell, Berry, & Carlisle, 1986).

Attempts to reduce fear of human beings

Because of the general tendency for poultry to react fearfully to human beings and because fear is a state of suffering that reduces welfare

(Duncan, 1987), there has been interest in procedures that might reduce this fear. Imprinting has not been considered a practical solution to the problem of fearfulness in poultry because of the labor required and because it might result in too much attraction to human beings; that is, birds following human beings closely could be as big a problem as birds reacting fearfully. Yet techniques that involve habituation have been used with poultry, although they have often been called something else. The procedure called "handling," which has been shown to increase growth rates in laboratory rats when performed early in life (Ruegamer, Bernstein, & Benjamin, 1954; Weininger, 1956), has been tried with chickens with mixed results. In some studies, handling resulted in increased growth rates and decreased fearfulness (Thompson, 1976; Gross & Siegel, 1979; Jones & Faure, 1981; Jones & Hughes, 1981), whereas in others, the results were negative (McPherson Gyles, & Kan, 1961; Reichman, Barram, Brock, & Standfast, 1978; Freeman & Manning, 1979). These discrepant results can probably be accounted for by small procedural differences; for example, in all of these studies the amount of human contact before the experiment started was rather uncontrolled. There appears to have been only one experiment in which human contact was carefully controlled before the experimental treatment started. Murphy and Duncan (1978) reared chicks of a flighty and docile strain for the first 6 weeks of their lives with no human contact. When these birds were first exposed to human beings, they showed more withdrawal than control birds that had experienced "normal" contact with human beings, and the flighty chicks showed more withdrawal than the docile chicks. However, with repeated human contact, the docile chicks quickly became habituated and after 5 days were responding in the same way as control docile chicks. The flighty chicks, in contrast, were still showing more withdrawal than the controls after 21 days of exposure to human beings. This suggests that the big difference in response to human beings between adult birds of these two strains is partly a genetically determined difference in withdrawal response that is present from hatching and partly a difference in learning, birds of the docile strain showing more habituation to human beings than those of the flighty strain.

There is evidence, therefore, that the amount of previous human contact experienced by domestic fowl could have a profound effect on experimental results in experiments involving human contact as a treatment.

Experiments not involving human contact

There have been a large number of behavioral experiments on domestic fowl that have not involved human contact as a treatment, and in many of these the observer has been within sight of the birds. A typical example is the long-running series of experiments on nesting behavior that have been carried out over the past 35 years in Edinburgh (e.g., Wood-Gush, 1954; Wood-Gush & Gilbert, 1969; Duncan & Kite, 1989). In these experiments, the hens were observed in 2.4 × 2.4 m pens with solid sides to a height of about 1 m with wire mesh above, through which observations were made. The observer sat outside with body hidden but head completely exposed to the nesting birds. Did the observer's presence have any effect in these experiments? Although this was not examined in a methodical way, the answer would seem to be that they probably did not. The most obvious effect of being frightened during the prelaying period is that hens retain their eggs beyond the expected oviposition times, and this leads to a very characteristic deposition of extra calcium on the shell (Hughes, Gilbert, & Brown, 1986). Eggs of this type were not seen in these experiments. Moreover, timing of oviposition was an important component in many of the experiments, and any discrepancies in this timing due to an observer effect would probably have been detected. This conclusion, that the presence of an observer had little effect on nesting behavior (or at least on the timing of oviposition), is surprising, since descriptions of nesting behavior in feral fowl suggest that the prelaying phase is characterized by extreme wariness, particularly of human beings (Duncan, Savory, & Wood-Gush, 1978).

In fact, there is some evidence that nesting hens can be influenced by the presence of a researcher and that the effect can be most surprising. In a series of experiments designed to investigate the effect of various nest modifications on nest site selection, hens were housed communally in holding pens with no nests. Each hen was then caught and carried manually and placed on its own in a test pen about an hour before it was due to lay (Appleby, McRae, & Peitz, 1984; Appleby, Duncan, & McRae, 1988; Duncan & Kite, 1989). Quite often in these experiments a hen would enter the prelaying phase in the holding pen before a test pen was available. Typically, instead of showing normal nesting behavior, the hen would stand at the door of the holding pen. If more than one hen was in the prelaying phase, the hens would give the impression of "lining up" at the door. Moreover, if someone entered the pen, instead of showing normal avoidance, the hen would follow the person as if soliciting contact until it was picked up and carried to a test pen. It seemed as though these hens had made a positive association between human handling and gaining access to a nest site. If this was the explanation, it is quite surprising, considering that human approach and contact have been shown to be among the most frightening things that domestic fowl experience (Jones et al., 1981; Duncan, 1989) and that they tend to be particularly wary of human beings during the prelaying phase (Duncan et al., 1978). As far as we know, the nesting behavior of these birds in the test pens was unaffected; that is, it looked similar to nesting behavior in unhandled birds in other experiments, and the treatments we were examining gave consistent and explainable results. However, we had no control observations of unhandled birds.

In my experience, the studies that have been described in this chapter are typical of the approach taken by applied ethologists when working with poultry. In general, they take reasonable precautions to minimize the "researcher effect." They try to ensure that test animals have had equivalent pretesting experience of human beings and that the observer is not influencing the behavior they are interested in. If human contact is one of the experimental variables, then even more care tends to be taken. However, as pointed out earlier, even these last-mentioned experiments yielded discrepant results, which suggests that there may have been uncontrolled human effects.

In contrast to the approach of applied ethologists is the methodology employed by some psychologists working with domestic fowl. Zayan and others in Belgium were interested in social interactions among hens and used a Murchison tunnel (Murchison, 1935), and various modifications of it, in their studies (Zayan, 1987). The general form of this equipment was a straight tunnel or runway with a startbox at each end to take each of the two test birds. The test birds were transported from their holding pens in carrying boxes and transferred automatically to the start boxes lest they see the outside of the tunnel and from it make deductions about the inside of the tunnel. Although I am convinced that such deductions are way beyond the cognitive abilities of domestic fowl, nevertheless Zayan is correct in his approach. Why not err on the side of caution? Even if hens cannot make such deductions, being carried in a box is likely to be a more neutral experience for a bird than being carried in someone's hands, and it costs nothing, or very little, in effort. In addition, it is much easier to standardize the procedure by carrying the birds in a box.

Conclusions

The evidence from the studies that have been reviewed here suggests that the researcher certainly can have an "unwanted" effect on poultry. This effect is most likely to be initiated with uncontrolled human contact sometime before the experiment, and particularly at an early age, when habituation of the birds' normal fearful responses to human beings may take place. There is also evidence that birds can learn to modify their natural prelaying behavior in a surprising way in response to human contact. In general, there seems to be an awareness of the obvious "observer effect," and it is not thought that this constitutes a major source of error in poultry studies.

References

Appleby, M. C., Duncan, I. J. H., & McRae, H. E. (1988). Perching and floor laying by domestic hens: Experimental results and their commercial application. *British Poultry Science, 29*, 351–7.

Appleby, M.C., McRae, H. E., & Peitz, B. E. (1984). The effect of light on the choice of nests by domestic hens. *Applied Animal Ethology, 11*, 249–54.

Beuving, G. (1980). Corticosteroids in laying hens. In *The Laying Hen and Its Environment* (Ed. R. Moss). The Hague, Martinus Nijhoff, pp. 65–82.

Duncan, I. J. H. (1974). A scientific assessment of welfare. *Proceedings of the British Society of Animal Production, 3*, 9–19.

(1985). How do fearful birds respond? In *Second European Symposium on Poultry Welfare* (Ed. R.-M. Wegner). Celle, German Branch of the WPSA, pp. 96–106.

(1987). The welfare of farm animals: An ethological approach. *Science Progress, Oxford, 71*, 317–26.

(1989). The assessment of welfare during the handling and transport of broilers. In *Third European Symposium on Poultry Welfare* (Eds. J. M. Faure & A. D. Mills). Tours, French Branch of the WPSA, pp. 93–107.

(1990). Reactions of poultry to human beings. In *Social Stress in Domestic Animals* (Eds. R. Zayan & R. Dantzer). Dordrecht, Kluwer Academic, pp. 121–31.

Duncan, I. J. H., & Kite, V. G. (1989). Nest-site selection and nest-building behaviour in domestic fowl. *Animal Behaviour, 37*, 215–31.

Duncan, I. J. H., Savory, C. J., & Wood-Gush, D. G. M. (1978). Observations on the reproductive behaviour of domestic fowl in the wild. *Applied Animal Ethology, 4*, 29–42.

Duncan, I. J. H., Slee, G. S., Kettlewell, P., Berry, P., & Carlisle, A. J. (1986). Comparison of the stressfulness of harvesting broiler chickens by machine and by hand. *British Poultry Science, 27*, 109–14.

Freeman, B. M., & Manning, A. C. C. (1979). Stressor effects of handling on the immature fowl. *Research in Veterinary Science, 26*, 223–6.

Gray, P. H., & Howard, K. I. (1957). Specific recognition of humans in imprinted chicks. *Perceptual and Motor Skills, 7*, 301–4.

Gross, W. B., & Siegel, P. B. (1979). Adaptation of chickens to their handler, and experimental results. *Avian Diseases, 23*, 708–14.

Hale, E. B. (1969). Domestication and the evolution of behaviour. In *The Behaviour of Domestic Animals*, 2d ed. (Ed. E. S. E. Hafez). London, Bailliere, Tindall, & Cassell, pp. 22–42.

Heinroth, O. 1911. Beiträge zur Biologie, nahmentlich Ethologie und Psychologie der Anatiden. *Verhandlungen der 5 Internationalen Ornithologie Kongress Berlin, 1910*, 589–702.

Hemsworth, P. H. & Barnett, J. L. (1987). Human–animal interactions. In *Farm Animal Behavior* (Ed. E. O. Price). Philadelphia, Saunders, pp. 339–56.

Hughes, B. O., & Black, A. J. (1976). The influence of handling on egg production, egg shell quality and avoidance behaviour of hens. *British Poultry Science, 17*, 135–44.

Hughes, B. O., Gilbert, A. B., & Brown, M. F. (1986). Categorisation and causes of abnormal egg shells: Relationship with stress. *British Poultry Science, 27*, 325–37.

Jones, R. B., Duncan, I. J. H., & Hughes, B. O. (1981). The assessment of fear in domestic hens exposed to a looming human stimulus. *Behavioural Processes, 6,* 121–33.

Jones, R. B., & Faure, J. M. (1981). The effects of regular handling on fear responses in the domestic chick. *Behavioural Processes, 6,* 135–43.

Jones, R. B., & Hughes, B. O. (1981). Effects of regular handling on growth in male and female chicks of broiler and layer strains. *British Poultry Science, 22,* 461–5.

Lorenz, K. (1935). Der Kumpan in der Umwelt des Vogels. *Journal Für Ornithologie 83,* 137–213, 289–413.

McFarland, D. (1981). *The Oxford Companion to Animal Behaviour.* Oxford, Oxford University Press, pp. 132–3.

McPherson, B. N., Gyles, N. R., & Kan, J. 1961. The effects of handling frequency on 8-week body weight, feed conversion and mortality. *Poultry Science, 40,* 1526–7.

Murchison, C. (1935). The experimental measurement of a social hierarchy in *Gallus domesticus*: I. The direct identification and direct measurement of Social Reflex No. 1 and Social Reflex No. 2. *Journal of General Psychology, 12,* 3–39.

Murphy, L. B. (1976). *A Study of the Behavioural Expression of Fear and Exploration in Two Stocks of Domestic Fowl.* Unpublished Ph. D. dissertation, Edinburgh University.

(1977). Responses of domestic fowl to novel food and objects. *Applied Animal Ethology, 3,* 335–49.

Murphy, L. B., & Duncan, I. J. H. (1977). Attempts to modify the responses of domestic fowl towards human beings: 1. The association of human contact with a food reward. *Applied Animal Ethology, 3,* 321–34.

(1978). Attempts to modify the responses of domestic fowl towards human beings: 2. The effect of early experience. *Applied Animal Ethology, 4,* 5–12.

Räber, H. (1948). Analyse des Balzverhaltens eines domestizieren Truthahns (*Meleagris*). *Behaviour, 1,* 237–66.

Reichman, K. G., Barram, K. M., Brock, I. J., & Standfast, N. F. (1978). Effects of regular handling and blood sampling by wing vein puncture on the performance of broilers and pullets. *British Poultry Science, 19,* 97–9.

Ruegamer, W. R., Bernstein, L., & Benjamin, J. D. (1954). Growth, food utilization, and thyroid activity in the albino rat as a function of extra handling. *Science, 120,* 184–5.

Russell, P. A. (1979). Fear-evoking stimuli. In *Fear in Animals and Man* (Ed. W. Sluckin). New York, Van Nostrand Reinhold, pp. 86–124.

Spalding, D. A. (1873). Instinct, with original observations on young animals. *Macmillan's Magazine, 27,* 282–93. Reprinted in 1954 in *British Journal of Animal Behaviour, 2,* 2–11.

Thompson, C. I. (1976). Growth in the Hubbard broiler: Increased size following early handling. *Developmental Psychobiology, 9,* 459–64.

Weininger, O. (1956). The effects of early experience on behaviour and growth characteristics. *Journal of Comparative and Physiological Psychology, 49,* 1–9.

Wood-Gush, D. G. M. (1954). Observations of the nesting habits of Brown Leghorn hens. *10th World's Poultry Congress in Edinburgh,* pp. 187–92.

Wood-Gush, D. G. M., Duncan, I. J. H., & Fraser, D. (1975). Social stress and welfare problems in agricultural animals. In *The Behaviour of Domestic Animals,* 3d ed. (Ed. E. S. E. Hafez). London, Bailliere Tindall, pp. 182–200.

Wood-Gush, D. G. M., & Gilbert, A. B. (1969). Oestrogen and prelaying behaviour of the domestic hen. *Animal Behaviour, 17,* 586–9.

Zayan, R. (1987). Recognition between individuals indicated by aggression and dominance in pairs of domestic fowl. In *Cognitive Aspects of Social Behaviour in the Domestic Fowl* (Eds. R. Zayan & I. J. H. Duncan). Amsterdam, Elsevier, pp. 321–438.

—19—

Early human–animal relationships and temperament differences among domestic dairy goats

David M. Lyons

Editors' introduction

Like the previous chapters by Hemsworth et al. and Duncan, Lyons's contribution focuses on a species that functions in human culture primarily as a stock animal. His chapter touches on several themes developed elsewhere in this book: the role of early social-ization in establishing a pattern of interactions with the scientist and the use of physiological data to assess these responses.

Lyons's longitudinal research program is unusually thorough and reveals a correspondingly rich picture of the dairy goat's temperament and how this often underestimated variable (distin-guished from the construct "personality") determines behavior in a variety of settings.

On the occasion of a committee meeting for the International Ethology Symposium many years ago, committee members realized that they had all grown up in households that kept a variety of animals (Lorenz, 1983). Professor H. Hediger, who as director of a zoological garden was particu-larly interested in human–animal relationships, formulated a questionnaire, which was sent to 25 scientists unanimously considered by the committee members to be great biologists. Among these great scientists there was not one who, even at preschool age, had not spent time interacting with animals. From the findings of this decidedly unscientific survey, one might reasonably conclude that these early human–animal relationships had profound, long-term effects on the lives of these 25 scientists. In keeping with a principal theme of this volume, this chapter addresses the counterpart of these reciprocal relationships: Early interactions between animals and scientists may substantially affect the animals as well.

As a general proposition, this idea is neither new nor particularly surprising. Developmental psychologists and behavioral biologists have compiled hundreds of observations demonstrating early-experience effects on a wide range of subsequent behaviors (Immelman, Barlow, Petrinovich, & Main, 1981; Newton & Levine, 1968; Simmel, 1980), and the known effects of early human–animal interactions on later behavioral responses toward people generally corroborate this literature. Historians, anthropologists, psychologists, and biologists alike have long referred to this idea in their accounts of the initial processes of animal domestication (Price, 1984; Reed, 1980; Zeuner, 1963) and the taming of domesticated (Freedman, King, & Elliot, 1961; Hemsworth, Barnett, Hansen, & Gongov, 1986; Karsh, 1983; Murphy & Duncan, 1978) and nondomesticated animals (Csermely, Mainardy, & Spano, 1983; Gilbert, 1971; Friedman, 1964; Savishinsky, 1983; Woolpy & Ginsburg, 1967). At least for birds and mammals, the demonstration that early experiences with people have an impact on subsequent behavioral responses toward people is an expected outcome. My primary concern in this chapter is to demonstrate that such experiences may have more extensive effects on developmental outcomes, and as a consequence, the issue of early scientist–animal interactions is relevant to the conduct of psychobiological research that is not specifically concerned with behavioral responses toward people.

This report summarizes our longitudinal studies describing how domestic dairy goats provided with different opportunities to interact with people early in life consistently differ on behavioral and physiological measures of responsiveness toward a variety of environmental circumstances across distinctly different stages of life. These interrelated behavioral and physiological aspects of responsiveness or reactivity are what we have in mind when we refer to the notion of temperament. The first section briefly reviews the concept of temperament with the aim of clarifying its meaning. Behavioral and physiological assessments of temperamental differences among goats are presented in subsequent sections, followed by concluding comments on their implications for scientist–animal interactions in psychobiological research.

A psychobiological view of temperament

Temperament is an ancient concept used to establish conceptual links between behavior and the constitution of the individual (Diamond, 1974). As is the case with so many terms that are common in both scientific and nontechnical vocabularies, temperament means different things to different people. We begin then with an attempt to specify its meaning.

Temperament represents a psychobiological level of organization that modulates the functioning of behavioral and physiological systems through its effects on an individual's patterns of responsiveness. The basic idea is most clearly conveyed in the simple definition offered by Rothbart and Derryberry (1981): Temperament consists of "constitutional differences in reactivity and self-regulation." The term "constitutional" refers to the relatively enduring psychobiological makeup of an organism influenced over time by heredity, maturation, and experience; "reactivity" is the propensity toward physiological, behavioral, and psychological responsiveness to salient environmental events; "self-regulation" refers to the capacity to increase, maintain, or decrease initial levels of reactivity through regulatory control mechanisms. An individual's temperament is thus inferred from general and abiding patterns of responsiveness that are fairly consistent across a range of environmental changes and challenges.

This formulation of temperament emphasizes the temporal and dynamic (or intensive) aspects of behavioral and physiological response systems and offers a number of distinct advantages. Chief among these is its generality. The concepts of reactivity and regulation can be used to describe temperament at the neural level, at the level of integrated physiological systems, and at the behavioral level (see Rothbart & Posner, 1985). To maintain their existence, all living systems must *react* to potentially disruptive events in an ever-changing environment, while at the same time *regulate* or control these reactions to maintain some degree of stability and organization (Cannon, 1935). The concepts of reactivity and regulation thus provide an integrative view of physiological and behavioral aspects of temperament grounded in the fundamental organizational structure of living systems.

Another advantage lies in the conceptual distinction between temperament and the more inclusive concept of personality (Rothbart & Derryberry, 1981). This issue has undoubtedly contributed to the belief among some animal scientists that temperament is an ambiguous, loosely applied term that is difficult to measure objectively (Hemsworth & Barnett, 1987; Kilgour, 1975). Temperament can be viewed as the energetic aspects of personality in that it refers to patterns of responsiveness that may be empirically characterized by relatively simple and direct measures of response thresholds, latencies, amplitudes and rise times to peak intensities, and recovery times. As indicated by Rothbart and Derryberry (1981), these measures reflect the idea that most responses have a specific onset, they rise in intensity, and they peak and subside across time. Personality represents a broader, more integrated pattern of collective character in that it incorporates emotional, motivational, cognitive, and temperamental aspects of an individual's behavior. While most assessments of personality traits inevitably incorporate some aspects of temperament, it is clearly possible to restrict one's investigations to the temporal and dynamic aspects of response systems without reference to their broader contribution to personality. Indeed, the possibility of this distinction offers a conceptual basis for empirically investigating the interrelationships between temperament and personality.

Differences in temperament among domestic dairy goats

The clearest difference in the behavior of domestic animals relative to their wild counterparts is reduced responsiveness to environmental change (Price, 1984). This characteristic is evident in virtually all populations of domestic animals and pervades a wide variety of environmental circumstances and behavioral traits. In many animal-management systems, domestic dairy animals are routinely separated from their mothers and raised by people in order to maximize the availability of milk for human consumption. A principal aim of our research with domestic dairy goats has been to evaluate the effects of this typical early experience as a source of temperamental differences among goats.

The subjects for much of this research included 30 female dairy goats (Alpine breed) randomly assigned at birth to one of two rearing conditions: human-reared and dam-reared. The 15 human-reared goats were separated from their dams and fed goat milk from a multinippled bucket two or three times daily by one of four human caretakers. During each 10-minute feeding session, a caretaker gently stroked these kids. The 15 dam-reared goats were raised by their mothers as single offspring. Except for health-management procedures administered to all animals, dam-reared goats experienced little direct contact with people. Goats in each rearing condition lived together as a group in large, adjacent outdoor pens separated by wooden panels to prevent interactions between goats from different rearing conditions. All goats were weaned from the milk diet at 10 weeks of age (mothers removed from the dam-reared kids) and subsequently received identical treatment after this time. Testing procedures began at 3.5 months of age and continued intermittently for the next 22.0 months. Major normative events covered during this period of lifespan development included the transition from adolescence to adulthood and two subsequent breeding and birthing seasons.

Behavioral modes of responsiveness

Group differences and time-related changes in behavioral responsiveness in three standard types of goat–human encounters were first assessed at 14,

Table 19.1. *Behavioral responsiveness measures (mean ± SE) of goats in encounters with a person at 14, 22, and 30 weeks*

Measure	Rearing condition	Age		
		14 weeks	22 weeks	30 weeks
Latency to proximity	Human-reared	16 ± 3	6 ± 2	11 ± 6
(seconds)	Dam-reared	550 ± 30	192 ± 59	55 ± 30
Time in proximity	Human-reared	505 ± 27	466 ± 29	472 ± 26
(seconds)	Dam-reared	22 ± 14	145 ± 42	240 ± 39
Number of avoidances	Human-reared	0.7 ± 0.2	0.5 ± 0.2	0.2 ± 0.1
	Dam-reared	8.3 ± 1.0	4.9 ± 0.6	3.6 ± 0.6
Flight distance	Human-reared	2.2 ± 0.3	1.8 ± 0.3	1.6 ± 0.4
(meters)	Dam-reared	9.4 ± 0.5	6.6 ± 0.6	6.2 ± 0.6
Approach–withdrawal	Human-reared	24 ± 5	37 ± 6	36 ± 5
(squares crossed)	Dam-reared	44 ± 8	60 ± 6	61 ± 6

Source: Lyons, Price, and Moberg (1988a).

22, and 30 weeks of age in a novel arena (Lyons, Price & Moberg, 1986, 1988a). Human-reared goats consistently differed from dam-reared goats across all five measures of behavioral responsiveness, but the magnitude of these differences changed over time (Table 19.1). From 14 to 30 weeks, human-reared goats showed no significant changes in responsiveness in encounters with an unfamiliar stationary or moving person, whereas four of the five measures for dam-reared goats increasingly resembled those of human-reared goats. As a result of habituation, or some other regulatory changes in the way older, dam-reared goats perceived and responded in repeated goat–human encounters, between-group differences in behavioral responsiveness decreased over time.

While absolute scores changed over time, relative between-individual differences persisted. Rank-order correlations calculated for each measure of responsiveness across consecutive age intervals ranged from .36 to .92 (Table 19.2) and 9 of 10 coefficients were statistically significant. On the basis of this temporal consistency in between-individual differences, we calculated the mean of each goat's scores across ages for the five behavioral measures and conducted a principal-components analysis to determine whether a smaller set of variables could adequately summarize the interrelationships between these multiple measures of responsiveness toward people. All five measures were highly correlated (Table 19.3) and all loaded onto Factor 1, which accounted for 76% of the total variance. Based on an examination of the factor loadings, factor scores for this

299

Table 19.2. *Correlations over consecutive age intervals for behavioral responsiveness in goat–human encounters*

	Age interval	
Measure	14–22 weeks	22–30 weeks
Latency to proximity	.73	.36
Time in proximity	.92	.75
Avoidance	.80	.69
Flight distance	.90	.83
Approach–withdrawal	.68	.71

Note: If $r_s < .39$, $p < .05$.
Source: Lyons, Press, and Moberg (1988a).

Table 19.3. *Intercorrelations between 14- to 30-week summary measures of behavioral responsiveness in goat–human encounters*

Measure	1	2	3	4	5
1. Avoidance	—	.40	.78	.82	−.84
2. Approach-withdrawal		—	.53	.30	−.53
3. Flight distance			—	.82	−.85
4. Latency to proximity				—	−.89
5. Time in proximity					—

Note: If $r_s < .39$, $p < .05$.
Source: Lyons, Press, and Moberg (1988a).

dimension were interpreted as collectively representing a general mode of behavioral reactivity toward people. Goats with higher factor scores were those that, on average, scored consistently larger latencies to approach a person closely, larger flight distances, larger approach–withdrawal scores, larger avoidance scores, and smaller time in proximity scores.

In terms of this general measure of reactivity, factor scores of all dam-reared goats exceeded those of all human-reared goats. As expected, different postnatal experiences with people significantly affected subsequent modes of responsiveness toward people. A follow-up study extended the generality of these between-group differences and revealed some informative constraints on their expression.

When the goats were 55 weeks of age, we conducted a series of home pen tests to determine each goat's latency to proximity scores in encounters with a number of objects or events. This simple measure of be-

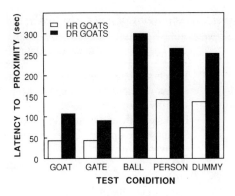

Figure 19.1. Mean latency to proximity scores of human-reared (HR) and dam-reared (DR) goats in test encounters with five different objects at 55 weeks.

havioral responsiveness was easily recorded and, as indicated in Table 19.3, was previously shown to be significantly correlated with a number of more detailed measures of responsivity. While individual differences in the 14- to 30-week behavioral reactivity factor scores reliably predicted individual differences for nearly all latency to proximity scores in the five test conditions at 55 weeks (Lyons, Price, & Moberg, 1988a), between-group similarities in certain situations highlighted some limitations on the expression of these temperament differences. Across all five test conditions, human-reared goats rather quickly approached and engaged the salient features of each situation. In keeping with earlier findings, significantly greater latency scores were evident in dam-reared goats in encounters with an unfamiliar person, and between-group differences were equally apparent in encounters with a large beach ball and a stuffed cloth dummy (Figure 19.1). Significant between-group differences were not evident, however, in encounters with an unfamiliar same-age goat and an open gate.

Both human-reared and dam-reared goats had been raised with peers, and all animals were familiar with open gates as a consequence of routine management procedures. None of the goats had any previous experience with beach balls or stuffed dummies. For human-reared goats this distinction made little difference. These animals consistently displayed a rather uninhibited style of coping with environmental challenge and change. In contrast, heightened levels of reactivity in dam-reared goats were reflected in a more inhibited behavioral style toward engaging certain kinds of environmental events. For dam-reared goats, cross-situational comparisons suggested that encounters with people represent a small subset of a much larger class of situations with the distinguishing characteristics of novelty

301

Table 19.4. *Behavioral definitions of items assessed in the milking parlor*

Item	Behavioral definition
Excitable	Reacts strongly to a change in the environment
Tense	Shows restraint in movement and posture; carries body stiffly alert
Watchful	Looks readily at a change in the environment
Apprehensive	Seems to be anxious about everything; fears and avoids any kind of risk
Confident	Behaves in a positive, assured manner; not restrained or tentative
Friendly toward people	Initiates proximity and/or contact with people
Fearful of People	Retreats readily from people

Source: Lyons (1989).

or uncertainty. Latent temperament differences not observed in more common day-to-day events were most clearly expressed in those situations that represented a substantial challenge or change in the status quo.

In terms of personality traits, these contrasts in behavioral responsiveness seem most closely associated with the familiar concept of timidity or shyness. While directly recorded measures of responsiveness provide useful information on temperament aspects of behavior, they generally fail to capture many other qualities of behavior that are important in describing personality traits. Such qualities are most clearly reflected in an individual's behavioral style, emotional tone, and apparent intentions and motivations. Goats may, for example, approach a person in a confident as opposed to a tentative manner, and avoidance can occur in seemingly fearful contexts or as a simple lack of interest. Behavioral recording methods based on observer's ratings explicitly recognize these qualities of behavior that are poorly suited for most direct recording techniques. Based on humans' powerful abilities to accumulate, filter, and integrate many complemental sources of information (Block, 1977), these methods provide a formal description of the everyday judgments people make about personality traits.

During routine milking procedures in a modern dairy, we used an observer rating method to assess the timidity of half the goats in each rearing condition. All goats were approximately 23 months of age, and those animals from previous research not examined in this study were omitted for reasons unrelated to this investigation. A list of seven behaviorally defined items (Table 19.4) was selected for assessment by myself and one other person. This person had not worked with the subjects before the study began, but had extensive experience milking other dairy goats. Of course, I was quite familiar with the goats from previous research. Before starting, we discussed the standard milking protocol, the list of items and their behavioral definitions, and the written instructions for the questionnaire to be completed at the end of the study. While the study was in progress, we generally worked alone during different milking sessions and did not discuss the behavior of the goats until we completed the questionnaire. By taking these precautions we could act as independent assessors.

The method of rating individual goats on the seven behavioral items was identical to that used by Feaver, Mendl, and Bateson (1986). This method has the advantage over other methods (e.g., Dickson, Barr, Johnson, & Weickert, 1970; Kerr & Wood-Gush, 1987; Kilgour, 1975; Stevenson-Hinde, Stillwell-Barnes, Zunz, 1980; Tulloh, 1961) of not requiring observers to impose their judgments onto an abbreviated, discontinuous scale. One coding form with 16 calibrated horizontal lines was used for each behavioral item (for an example, see Feaver et al., 1986). Each line corresponded to an individual goat and represented a continuum across a particular behavioral item. The left-hand end of the scale represented the minimum expression and the right-hand end the maximum expression of a behavioral item for all individuals in the study.

After each animal received a minimum of 21 days of twice-daily milking sessions, both observers independently completed a coding form for each behavioral item by marking a cross on the calibrated line corresponding to our own assessment of each goat. We based assessments on overall impressions of a goat's behavior gained during all interactions with that individual during milking procedures and assigned ratings relative to other goats in this study alone. Assessments were converted to a numerical score by measuring the distance of the marked cross from the left-hand end of the scale. Numerical scores were subsequently ordered by rank for statistical analysis.

Interobserver Spearman correlations for all seven behavioral items were statistically significant; the convergence of ratings from two independent raters speaks well for people's ability to perform observer rating tasks. To maintain at least three times as many subjects as measured behavioral items in the factor analysis, only the five most reliable items were retained. The mean of the observers' ratings was computed for each goat on the five retained items. Mean ratings for the five items were highly correlated, and all five items loaded onto Factor 1, which explained 97% of the total variance. Factor loadings indicated that Factor 1 could be interpreted as a dimension of timidity. Observers reliably rated timid goats as being relatively more tense, watchful, excitable, fearful of people, and less friendly toward people. Each goat's factor score served as an operational measure of timidity.

Individual differences in timidity scores evaluated at 92 weeks of age in the milking parlor predicted individual differences in more direct measures of behavioral responsiveness assessed (1) at 14 to 30 weeks in goat–human encounters in the novel arena (r_s = .75; p < .01), (2) at 55 weeks in encounters with an unfamiliar goat (r_s = .55, p < .05), novel inanimate objects (ball, r_s = .75, p < .01; dummy, r_s = .70, p < .01), and an open gate (r_s = .44, p < .05), and (3) at 100 weeks in goat–human encounters in the home pen (r_s = .84, p < .01; see Lyons, 1989). Taken together, these results clearly demonstrate that behavioral aspects of responsiveness are an enduring and rather general characteristic that at a broader level are reflected in what is commonly recognized as timidity or shyness.

Physiological modes of responsiveness

Our first indication that these general and abiding behavioral differences were associated with differences in physiology came from measurements of pituitary–adrenal responsiveness in the 14- to 30-week goat–human test encounters (Lyons, Price, & Moberg, 1988a). While mean posttest corticosteroid concentrations in plasma increased significantly from basal values in both dam-reared and human-reared goats, the magnitude of change was considerably greater among dam-reared goats (Figure 19.2). To examine the relationship between behavioral reactivity toward people and pituitary–adrenal responsiveness, we subtracted each goat's basal corticosteroid value from its posttest value so that pituitary-adrenal responsiveness was expressed as the deviation from basal corticosteroid levels. Individual differences in 14- to 30-week behavioral reactivity factor scores reliably predicted differences on this 14- to 30-week measure of pituitary–adrenal responsiveness (r_s = .45, p < .05). Preliminary results suggested, however, that physiological differences could be attenuated when human-reared and dam-reared goats were tested in goat–human encounters in the presence of their pen mates (Lyons, Price, & Moberg, 1986).

In a subsequent study of social modulatory effects on pituitary–adrenal responsiveness (Lyons, Price, & Moberg, 1988b), we examined the extent to which these persistent differences in responsiveness of human-reared and dam-reared goats were reflected in the behavior and physiology of their offspring. The 10-week-old kids of 12 human-reared and 12 dam-reared goats served as subjects. These 32 kid goats participated in a counterbalanced series of standard goat–human test encounters conducted in each of four social settings: (1) alone, (2) with their mother, (3) with a

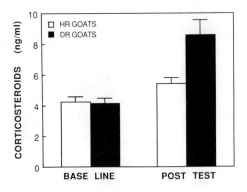

Figure 19.2. Basal and posttest corticosteroid concentrations (mean ±SE) in human-reared (HR) and dam-reared (DR) goats (14- to 30-week summary measures from goat–human test encounters).

familiar adult goat raised like their mother, and (4) with a strange (unfamiliar) adult goat raised like their mother.

Kid goat behavior changed significantly with changes in the person's behavior (stationary, moving, or pursuing) and the social context (alone, mother, adult penmate, or stranger). At the same time, a high level of individual consistency was maintained across goat–human encounter types (Kendall's coefficient $W = .69$, $p < .001$) and social contexts ($W = .70$, $p < .001$). Some kids were consistently inhibited from approaching and closely interacting with a person, whereas others were not. To determine whether these persistent differences in kid goat timidity were associated with differences in pituitary–adrenal responsiveness, we used a post hoc classification approach (Simpson, 1985) to identify kids within our sample that scored at specified levels in time spent in proximity to the person. On the basis of this cross-situational measure, kids were classified into one of two groups. Kids with a score below the mean were considered timid, and those with scores above the mean were considered bold.

The timid kid group consisted of 18 individuals; 8 and 10 kids were raised and tested with human-reared and dam-reared adult goats, respectively. The bold kid group included 14 kids; 8 and 6 kids were raised and tested with human-reared and dam-reared adult goats, respectively. When observed alone in goat–human test encounters, posttest corticosteroid concentrations in timid kids were significantly greater than their control values, whereas corticosteroid concentrations in bold kids were not (Lyons et al., 1988b). The fact that neither physiological differences in this situation nor behavioral differences across situations (as indicated by the classification results) were significantly associated with their mother's rearing

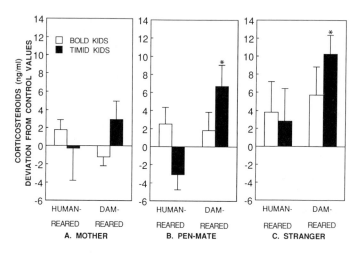

Figure 19.3. Posttest deviations from control corticosteroid concentrations (mean ± SE) in bold kids and timid kids accompanied by an adult female goat (mother, pen mate, or stanger) in goat–human test encounters. Dam-reared adult goats were fearful of people, and human-reared adults were friendly toward people. Asterisk indicates significantly different ($p < .05$) from control (Lyons, Price, & Moberg, 1988b).

condition strongly suggests the absence of pervasive transgenerational influences on kid goat behavioral and physiological responsiveness. Similar findings have been obtained in controlled cross-fostering studies of rats (Galef, 1970; Hughes, 1975).

In agreement with results observed when tested alone, corticosteroid concentrations in bold kids remained no different from control values whether mothers, pen mates, or strangers participated in test encounters (Figure 19.3). In contrast, pituitary–adrenal responsiveness in timid kids was influenced by the adult goat's behavior toward the person and the nature of the social relationship between the kid and the adult goat. The *immediate* presence of the mother unconditionally reduced kid pituitary–adrenal responsiveness (Figure 19.3A), whereas the modulatory effects of other adult goats on kid pituitary–adrenal responsiveness depended on the behavior of these adults. Posttest corticosteroid concentrations in timid kids were significantly greater than control values when accompanied by timid dam-reared adult pen mates, whereas corticosteroids in timid kids accompanied by bold human-reared adult pen mates were not (Figure 19.3B). This same pattern was accentuated in the company of strangers (Figure 19.3C).

In sum, posttest corticosteroid concentrations in bold kids who spent above average amounts of time near the person were never significantly different from control values regardless of the social context or the adult goat's behavior in goat–human encounters. The robust nature of this finding in bold kid goats indicates that an individual's characteristic mode of behavioral responsiveness may be reflected in a rather consistent pattern of pituitary–adrenal responsiveness. Moreover, cross-situational comparisons of pituitary–adrenal responsiveness among timid kids demonstrate how the expression of responsivity also depends on specifiable features in the environment. Of the three types of adult goat companions we examined, only mothers unconditionally reduced kid pituitary–adrenal responsiveness. In undisturbed social settings, mothers, siblings, and age mates are most often the closest animals to a kid between 9 and 12 weeks of age, and adult female herd members are generally avoided (Lickliter, 1987). Strange adult females can be actively excluded from integrated herds for up to 4 weeks (Addison & Baker, 1982). While mother goats occupy a position of privileged influence in modulating the expression of kid pituitary–adrenal responsiveness, long-term maternal effects do not appear to be as pervasive as one might expect.

These data also indicate a need for further research on the intensive and temporal aspects of the pituitary–adrenal response itself. Based on previous studies of sheep (Basset & Hinks, 1969; Moberg & Wood, 1981; Pearson & Mellor, 1976), the 9- to 10-minute posttest samples used in our research represent the minimal time required to detect a rise in circulating corticosteroid concentrations in response to provocative environmental events. Consequently, our measure of pituitary–adrenal responsiveness may better reflect an aspect of rise time to peak intensity, rather than peak intensity itself. The generally consistent elevations in corticosteroid concentrations in kids accompanied by strange adult goats (Figure 19.3C) might therefore be indicative of a pituitary–adrenal response that occurs more rapidly (without necessarily being of greater peak intensity) than those observed in other social settings. In comparison with diminutive laboratory rodents, large domestic farm animals are well suited for repeated blood-sampling procedures required by this kind of neuroendocrine research (Katz, 1987).

By their very nature, domestic dairy goats are also particularly well suited for investigations of the psychoneuroendocrine aspects of lactation. In all lactating mammals, stored milk is distributed in various proportions between alveolar and sinus compartments in the mammary glands. Alveolar milk constitutes 70 to 90% of the stored volume before suckling in most mammalian species (Cross, 1977) and is accessible only by neuroendocrine-mediated milk ejection processes. Ruminants like domestic goats have capacious udder cisterns that hold approximately 80% of their stored milk; this can readily be extracted without oxytocin release. Nevertheless, significant amounts of alveolar milk are evident in domestic dairy goats due to the exceedingly large volumes of milk these animals routinely produce. In conjunction with our assessments of timidity in human–reared and dam-reared goats during routine milking procedures in a modern dairy, we measured differences in the efficacy of (alveolar) milk ejection (Lyons, 1989).

Residual milk not extracted by routine milking procedures was monitored every third evening milking session for the first 20 days postpartum in the dairy barn. Standard milking procedures were used to collect milk first by the teat vacuum cups of a milking machine and then by hand. Each goat subsequently received 5 IU of oxytocin injected intravenously into the right jugular vein. Residual milk not extracted by routine milking procedures was then collected immediately by hand milking. Quantities extracted by machine milking, preinjection hand milking, and postinjection hand milking were weighed separately. The sum of these measures represented the total milk yield (grams) for an individual goat. The weight of residual milk collected after the oxytocin injection was divided by an individual's total milk yield to provide a measure of milk ejection impairment in routine milking procedures. This standardized, dimensionless measure, expressed in terms of percent residual milk, was computed to eliminate individual differences in absolute milk yields. (Individual total milk yields summed across the seven sessions ranged from 9.6 to 14.9 kg.) Higher percent residual milk values indicated greater inhibition of milk ejection processes during routine milking procedures.

Human-reared and dam-reared goats never differed significantly in total milk yields, but initially greater percentages of total yields were obtained as residual milk in dam-reared goats (Figure 19.4). Dam-reared goat residual milk scores decreased significantly over successive milking sessions, whereas scores for human-reared goats consistently remained at low levels – a temporal pattern that replicates early measures of behavioral responsiveness (Table 19.1). For a direct comparison with behavioral measures, a summary residual milk score was computed by taking each goat's mean residual milk score over all seven milking sessions. Mean percent residual milk scores of dam-reared goats (9.7 ± 0.8) were significantly greater than those of human-reared goats (5.9 ± 0.7). Individual differences in residual milk scores were correlated significantly with differences in the behavioral measure of timidity assessed concurrently in the milking parlor ($r_s = .72$, $p < .01$) and a more direct, composite measure of behavioral responsiveness toward people conducted 1 month later in the goats home pens ($r_s = .58$, $p < .05$; see Lyons, 1989).

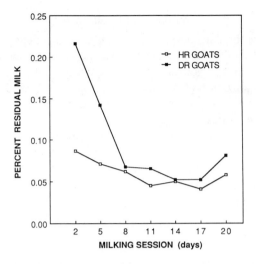

Figure 19.4. Mean residual milk scores of human-reared (HR) and dam-reared (DR) goats over a 20-day period (Lyons, 1989).

The greater residual milk volumes extracted from dam-reared goats when oxytocin was injected following routine milking procedures suggest that central inhibition of oxytocin release was responsible for the impaired ejection of alveolar milk in dam-reared goats. While the ability of exogenous oxytocin to restore milk yields to normal has been advanced as evidence for a central mechanism in the inhibition of milk ejection in cows (Ely & Petersen, 1941), humans (Newton & Newton, 1948), rabbits (Cross, 1955), guinea pigs (Chaudhury, Chaudhury, & Lu, 1961), and rats (Grosvenor & Mena, 1967), the participation of peripheral inhibitory mechanisms cannot be ruled out. Activation of the sympathetic–adrenal system can impair milk ejection through (1) constriction of mammary blood vessels carrying oxytocin to the myoepithelium, (2) inhibition of the binding of oxytocin directly at the mammary myoepithelium, and (3) increased resistance of the ducts within the mammary gland impeding milk flow to the exterior (Goodman & Grosvenor, 1983; Grosvenor & Mena, 1974). Although we did not find significant differences between human-reared and dam-reared goats using heart rate as a crude index of sympathetic activity (Lyons & Price, 1987), studies using pharmacological agents that block the peripheral actions of the sympathetic–adrenal system are needed to determine their importance in the inhibition of milk ejection in domestic dairy goats.

Summary and implications

The importance of considering an individual's distinctive capacities and propensities as significant sources of variation in psychobiological systems has been recognized at neurophysiological, behavioral, social, and ecological levels of organization (Denenberg, 1979; Hinkle, 1974; Lomnicki, 1982; Porges, 1985; Rose, 1980), but little empirical research has systematically addressed these issues. During the most recent epoch of investigative psychobiology in which many of the more obvious questions have been considered, it seems that individual differences could be safely ignored without seriously affecting experimental results. With increasing maturation of knowledge and the application of sophisticated technical advances, more precise questions are now being addressed. As questions concerned with the functioning of individual organisms grow more subtle, the answers may increasingly become obscured by a kind of nonrandom "noise" representing the range of natural variation observed among individuals (Martin & Kraemer, 1987). One theoretical framework for relating stable individual differences in behavior and physiology is the concept of temperament.

The early experiences imposed in our studies contributed significantly to differences in the way goats responded in a wide variety of environmental settings from 3.5 to 25.0 months of age. In comparison with the more "normal" experiences of dam-reared goats, human-reared goats were separated from their mothers and raised by people following husbandry procedures not unlike those used in laboratory nurseries at certain animal research facilities. In response to environmental challenge and change, human-reared goats consistently appeared less tense, alert, watchful, and hesitant to engage closely in novel situations. Differences in behavioral reactivity and timidity coincided with differences in pituitary–adrenal responsiveness and the inhibition of neuroendocrine-mediated milk ejection processes.

This range of behavioral and physiological outcomes is taken as evidence that certain early experiences may significantly alter those processes involved in the development of a fairly well integrated, superordinate level of psychobiological organization that we refer to as temperament. The consistent differences found in measures of reactivity across a two-year period of life-span development clearly demonstrate some degree of constancy or continuity in the extent to which temperament exerts a controlling influence on the reactivity of specific (subordinate) behavioral and physiological response systems. Equally evident is the fact that individuals adapt and change. Regulatory processes operating at physiological (i.e., circadian and circannual cycles in hormonal states), cognitive (i.e., habituation, associative learning, etc.), and emotional (i.e., fluctuations in mood) levels of organization modify the significance of environmental events and thereby alter the expression of temperamental differences in organism–environment transactions.

With regard to temperament, what is not at all clear is the nature of the developmental processes involved at this level of psychobiological organization and the specific experiential and genetic (see Lyons et al., 1988a) factors that influence the course of these developmental processes. In the research reported here, the experiential effects of routine goat–human interactions early in life are confounded with maternal separation. We attempted to separate these factors in a pilot study of five goat kids that were removed from their mothers at birth and fed through a milk delivery device that required no exposure to people after a brief training period entailing 2 to 6 hours of interactions with a caretaker over the first 2 to 4 days of life. On the basis of informal judgments, these motherless, artificially reared goats appeared more timid than motherless, human-reared goats, but most of the 11 direct measures of behavioral responsiveness in test encounters with a person, a goat, or social isolation more clearly resembled human-reared than dam-reared goats (unpublished data). These results suggest the somewhat surprising possibility that extensive early human–animal interactions may have a minimal effect beyond that produced by maternal separation and very limited early interactions with people.

The implications of these findings for scientist–animal interactions are by now fairly obvious. Many testing paradigms in psychobiological research necessarily entail some sort of environmental challenge or change that is often accompanied by the presence of the scientist. These are the very situations that seem to maximize the expression of individual differences in temperament. When such differences are not taken into account in experimental designs, they will contribute to variation in behavioral and physiological research results and thereby decrease the probability of detecting significant experimental effects of interest. This problem is particularly relevant to research examining more subtle influences of particular experimental manipulations, as illustrated by our study of social modulatory effects on pituitary–adrenal responses in goat kids (Lyons et al., 1988b). Because of their consistently low levels of responsiveness across all conditions (Figure 19.3), we were able to identify bold kids as inappropriate subjects for this aspect of the study. An alternative for dealing with problems of this nature may lie in appropriate environmental modifications of the experimental situation (i.e., Barlow, 1968).

One source of differences in temperament may be attributed to animals' early experiences with people. Whether such procedures are required because a particularly valuable animal subject is not adequately cared for by its biological mother, or an experimental protocol requires repeated testing of an infant without its mother (a procedure that is clearly facilitated by nursery rearing in certain nonhuman primates), the experiences of being raised by people may substantially alter certain aspects of the behavioral and/or physiological systems being studied. Such effects are by no means limited to domestic goats, which have never been popular research subjects, nor are they completely predictable on an *a priori* basis. For example, Singh (1966, 1968) compared the behavioral responsiveness of rhesus monkeys that had been captured in Indian cities with those living in jungle areas not frequented by people. The urban monkeys showed a greater interest in complex visual stimulus displays, and were more active, manipulative, and not at all timid around people, much like human-reared goats. In contrast, motherless female rhesus monkeys and domestic cats raised by humans in peer groups similar to our human-reared goats exhibit atypical levels of behavioral responsiveness toward adult males that impair reproductive success (Mellen, 1989), and they are often abusive toward their infants (Suomi, 1978). Neither of these effects could have been predicted from our research with goats. Both human-reared and dam-reared goats were quite prolific in bearing offspring, and minimal differences were observed across a number of measures of mother–kid relationships (unpublished data). While our findings demonstrate a range of developmental outcomes that are influenced by early experiences with people, the extent to which these experiences affect specific psychobiological systems in goats and other animals remains an issue that can be addressed only through empirical investigation.

Acknowledgments

Much of this research was supported by a Jastro-Shields Research Scholarship, University of California, Davis. While writing this chapter, I was supported by a National Research Service Award HD07293, National Institute of Child Health and Human Development, and Grant RR00169, National Institutes of Health, Division of Research Resources.

References

Addison, W. E., & Baker, E. (1982). Agonistic behavior and social organization in a herd of goats as affected by the introduction of nonmembers. *Applied Animal Ethology, 8*: 527–35.
Barlow, G. W. (1968). Dither: A way to reduce undesirable fright behavior in ethological studies. *Zeitschrift für Tierpsychologie, 25*: 315–18.

Bassett, J. M., & Hinks, N. T. (1969). Microdetermination of corticosteroids in ovine peripheral plasma: Effects of venipuncture, corticotrophin, insulin, and glucose. *Journal of Endocrinology, 44*: 387–403.

Block, J. (1977). Advancing the psychology of personality: Paradigmatic shift or improving the quality of research? In *Personality at the Crossroads*, D. Magnusson & N. S. Endler (Eds.). New York: Wiley, pp. 37–63.

Cannon, W. B. (1935). Stress and strains of homeostasis. *American Journal of Medical Science, 189*: 1–14.

Chaudhury, R. R., Chaudhury, M. R., & Lu, F. C., (1961). Stress-induced block of milk ejection. *British Journal of Pharmacology, 17*: 305–9.

Cross, B. A. (1977). Comparative physiology of milk removal. *Symposium of the Zoological Society of London, 41*: 193–210.

(1955). Neurohormonal mechanisms in emotional inhibition of milk ejection. *Journal of Endocrinology, 12*: 29–37.

Csermely, D., Mainardi, D., & Spano, S. (1983). Escape-reaction of captive young red-legged partridges (*Alectoris rufa*) reared with or without visual contact with man. *Applied Animal Ethology, 11*: 177–82.

Denenberg, V. H. (1979). Paradigms and paradoxes in the study of behavioral development. In *Origins of the Infant's Social Responsiveness*, E. B. Thoman (Ed.). Hillsdale, NJ: Erlbaum, pp. 251–89.

Diamond, S. (Ed.). (1974). *The Roots of Psychology*. New York: Basic Books.

Dickson, D. P., Barr, G. R., Johnson, L. P., & Weickert, D. A. (1970). Social dominance and temperament of Holstein cows. *Journal of Dairy Science, 53*: 904–7.

Ely, F., & Petersen, W. E. (1941). Factors involved in the ejection of milk. *Journal of Dairy Science, 24*: 211–23.

Feaver, J., Mendl, M., & Bateson, P. (1986). A method for rating the individual distinctiveness of domestic cats. *Animal Behaviour, 34*: 1016–25.

Freedman, D. G., King, J. A., & Elliot, O. (1961). Critical period in the social development of dogs. *Science, 133*: 1016–17.

Friedman, H. (1964). Taming of the Virginia opossum. *Nature, 201*: 323–4.

Galef, B. G., Jr. (1970). Aggression and timidity: Responses to novelty in feral Norway rats. *Journal of Comparative and Physiological Psychology, 70*: 370–81.

Gilbert, B. K. (1971). The influence of foster rearing on adult social behavior in fallow deer (*Dama dama*). In *The Behavior of Ungulates and Its Relation to Management*, V. Geist & F. Walther (Eds.). Morges: IUCN Publ. No. 24, Vol. 1, pp. 247–73.

Goodman, G. T., & Grosvenor, C. E. (1983). Neuroendocrine control of the milk ejection reflex. *Journal of Dairy Science, 66*: 2226–35.

Grosvenor, C. E., & Mena, F. (1967). Effect of auditory, olfactory and optic stimuli upon milk ejection and suckling-induced release of prolactin in lactating rats. *Endocrinology, 80*: 840–6.

(1974). Neural and hormonal control of milk secretion and milk ejection. In *Lactation: A Comprehensive Treatise*, B. L. Larson & V. R. Smith (Eds.). New York: Academic Press, pp. 227–76.

Hemsworth, P. H., & Barnett, J. L. (1987). Human–animal interactions. In *Farm Animal Behavior, Veterinary Clinics of North America: Food Animal Practice*, E. O. Price (Ed.). Philadelphia: W. B. Saunders, Vol. 3, No. 2, pp. 339–56.

Hemsworth, P. H., Barnett, J. L., Hansen, C., & Gonyou, H. W. (1986). The influence of early contact with humans on subsequent behavioural response of pigs to humans. *Applied Animal Behavior Science, 15*: 55–63.

313

Hinkle, L. E., Jr. (1974). The concept of "stress" in the biological and social sciences. *International Journal of Psychiatric Medicine, 5*: 335–57.

Hughes, C. W., Jr. (1975). Early experience in domestication. *Journal of Comparative and Physiological Psychology, 88*: 407–17.

Immelmann, K., Barlow, G. W., Petrinovich, L., & Main, M. (Eds.). (1981). *Behavioral Development.* Cambridge University Press.

Karsh, E. B. (1983). The effects of early handling on the development of social bonds between cats and people. In *New Perspectives on Our Lives with Companion Animals,* A. H. Katcher & A. M. Beck (Eds.). Philadelphia: University of Pennsylvania Press, pp. 22–8.

Katz, L. S. (1987). Endocrine systems and behavior. In *Farm Animal Behavior, Veterinary Clinics of North America: Food Animal Practice,* E. O. Price (Ed.). Philadelphia: Saunders, Vol. 3. No. 2, pp. 393–404.

Kerr, S. G. C., & Wood-Gush, D. G. M. (1987). The development of behavior patterns and temperament in dairy heifers. *Behavioral Processes, 15*: 1–16.

Kilgour, R. (1975). The open-field test as an assessment of the temperament of dairy cows. *Animal Behaviour, 23*: 615–24.

Lickliter, R. E. (1987). Activity patterns and companion preferences of domestic goat kids. *Applied Animal Behavior Science, 19*: 137–45.

Lomnicki, A. (1982). Individual heterogeneity and population regulation. In *Current Problems in Sociobiology,* King's College Sociobiology Group (Ed.). Cambridge University Press, pp. 153–67.

Lorenz, K. (1983). Foreword. In *Proceedings of the International Symposium on The Human–Pet Relationship.* Vienna: Institute for Interdisciplinary Research on the Human–Pet Relationship, pp. 7–8.

Lyons, D. M. (1989). Individual differences in temperament of dairy goats and the inhibition of milk ejection. *Applied Animal Behavior Science, 22*: 269–82.

Lyons, D. M., & Price, E. O. (1987). Relationships between heart rates and behavior of goats in encounters with people. *Applied Animal Behavior Science, 18*: 363–9.

Lyons, D. M., Price, E. O., & Moberg, G. P. (1986). Adjustments in defensive behavior and plasma corticosteroids of dairy goats in interactions with people (abstract). *Journal of Animal Science, 63* (Suppl. 1): 166.

(1988a). Individual differences in temperament of domestic dairy goats: Constancy and change. *Animal Behaviour, 36*: 1323–33.

(1988b). Social modulation of pituitary-adrenal responsiveness and individual differences in behavior of young domestic goats. *Physiology and Behavior, 43*: 451–8.

Martin, P., & Kraemer, H. C. (1987). Individual differences in behaviour and their statistical consequences. *Animal Behaviour, 35*: 1366–75.

Mellen, J. D. (1989). Reproductive behavior of small captive exotic cats. Unpublished Ph.D. dissertation, University of California, Davis.

Moberg, G. P., & Wood, V. A. (1981). Neonatal stress in lambs: Behavioral and physiological responses. *Developmental psychobiology, 14*: 155–62.

Murphy, L. B., & Duncan, I. J. H. (1978). Attempts to modify the responses of domestic fowl toward human beings: II. The effect of early experience. *Applied Animal Ethology, 4*: 5–12.

Newton, G., & Levine, S. (Eds.). (1968). *Early Experience and Behavior: The Psychobiology of Development.* Springfield, IL: Thomas.

Newton, M., & Newton, N. R. (1948). The let-down reflex in human lactation. *Journal of Pediatrics, 33*: 698–704.

Pearson, R. A., & Mellor, D. J. (1976). Some behavioural and physiological changes in pregnant goats and sheep during adaptation to laboratory conditions. *Research in Veterinarian Science, 20*: 215–27.

Porges, S. W. (1985). Spontaneous oscillations in heart rate: Potential index of stress. In *Animal Stress*, G. P. Moberg (Ed.). Bethesda, MD: American Physiological Society, pp. 97–111.

Price, E. O. (1984). Behavioral aspects of animal domestication. *Quarterly Review of Biology, 59*: 1–32.

Reed, C. A. (1980). The beginnings of animal domestication. In *Animal Agriculture: The Biology, Husbandry and Use of Domestic Animals*, H. H. Cole & W. N. Garrett (Eds.). San Francisco: Freeman, pp. 3–19.

Rose, R. M. (1980). Endocrine responses to stressful psychological events. In *Advances in Psychoneuroendocrinology: Psychiatric Clinics of North America*, E. J. Sachar (Ed.). Philadelphia: Saunders, Vol. 3, No. 2, pp. 251–76.

Rothbart, M. K., & Posner, M. I. (1985). Temperament and the development of self-regulation. In *Neuropsychology of Individual Differences*, L. C. Hartlage & C. F. Telzrow (Eds.). New York: Plenum, pp. 93–123.

Rothbart, M. K., & Derryberry, D. (1981). Development of individual differences in temperament. In *Advances in Psychology*, M. E. Lamb & A. L. Brown (Eds.). Hillsdale, NJ: Erlbaum, Vol. 1, pp. 37–86.

Savishinsky, J. S. (1983). Pet ideas: The domestication of animals, human behavior, and human emotions. In *New Perspectives on Our Lives with Companion Animals*, A. H. Katcher & A. M. Beck (Eds.). Philadelphia: University of Pennsylvania Press, pp. 112–31.

Simpson, M. J. A. (1985). Effects of early experience on the behaviour of yearling rhesus monkeys (*Macaca mulatta*) in the presence of a strange object: Classification and correlation approaches. *Primates, 26*: 57–72.

Simmel, E. C. (Ed.). (1980). *Early Experiences and Early Behavior: Implications for Social Development*. New York: Academic Press.

Singh, S. D. (1966). The effects of human environment upon the reactions to novel situations in the rhesus monkey. *Behaviour, 26*: 243–50.

(1968). Effect of urban environment on visual curiosity behavior in rhesus monkeys. *Psychonomic Science, 2*: 83–4.

Stevenson-Hinde, J., Stillwell-Barnes, R., & Zunz, M. (1980). Subjective assessment of rhesus monkeys over four successive years. *Primates, 21*: 66–82.

Suomi, S. J. (1978). Maternal behavior by socially incompetent monkeys: Neglect and abuse of offspring. *Journal of Pediatric Psychology, 3*: 28–34.

Tulloh, N. M. (1961). Behaviour of cattle in yards: II. A study of temperament. *Animal Behaviour, 9*: 25–30.

Woolpy, J. H., & Ginsburg, B. E. (1967). Wolf socialization: A study of temperament in a wild species. *American Zoology, 7*: 357–63.

Zeuner, F. E. (1963). *A History of Domesticated Animals*. New York: Harper & Row.

—20—

The effect of the researcher on the behavior of horses

Sharon L. Crowell-Davis

Look back at our struggle for freedom,
  Trace our present day's strength to its source;
And you'll find that man's pathway to glory
  Is strewn with the bones of the horse.

Anonymous

Editors' introduction

As Crowell-Davis notes, given the importance of the horse in human culture as a source of both labor and pleasure, it is quite surprising to find a virtual absence of scientific information about human–horse interactions. If anything, there seems to have been a concerted attempt to avoid direct investigation of the effects of humans on horses. For example, studies of learning in horses, which, of necessity, involve human experimenters, have typically taken pains to exclude, or minimize the presence of, humans.

Crowell-Davis's chapter surveys what little is known of the effects humans have on horse behavior, both from the scientific literature and from her own ethological research program. She argues that this knowledge can be quite useful in designing ethological protocols. For example, evidence is presented that even our scant understanding of human–horse interactions may facilitate the study of social behavior in horses.

As Ginsburg and Hiestand noted with regard to canids, horses have an impressive sensitivity to minimal cues emitted by humans, an observation that would surprise few riders but continues to lie beyond the domain of scientific inquiry.

For thousands of years, humans and horses (*Equus caballus*) have lived and worked together in an intimate biological symbiosis. Horses provided labor as draft and riding animals and, in return, were able to expand their range and number through the protection provided by humans, who were carried by horses on migrations across the globe.

Living as they do in such close contact with horses, it is not surprising that humans have been making informal observations of their behavior for some time. For example, in the King James Bible, Job 39: 21–2, we learn of the horse that "he paweth in the valley, and rejoiceth in his strength: he goeth on to meet the armed men. He mocketh at fear, and is not affrighted; neither turneth he back from the sword." Apparently horses pawed before being ridden into battle.

The horse is a large ungulate. Thus, the reasons for carrying out research on this species are often different from those for studying many other species, and the methods employed differ as well. The expense of purchasing and feeding domestic horses makes it impractical to use them for most research that addresses purely theoretical issues, especially when smaller and substantially less expensive species are available. Why study the horse at all? Most research papers fall into one and, occasionally, both of two major categories.

First, in many locations throughout the world, and especially in the United States, horses have escaped human management and become feral. In some places, competition between feral horses and native wildlife has produced a need for understanding feral horse ecology. This includes an understanding of social structure and related behaviors in order to predict changes in the effect of horse herds on their environment over time, and determine ways to manage populations most effectively. Studies on the horse's wild relatives, that is, zebras and asses, are conducted primarily for similar reasons, although the animals are natural residents of the habitat rather than introduced species.

Second, while the horse has recently lost its status as the primary means of transport for human travel and labor for the ploughing of fields, it continues to form the basis of a multibillion dollar industry. Considering the fact that the horse has been replaced by various motors and motorized vehicles, one might expect that the species would have been reduced to only a few specimens kept as curiosities in zoos and that horsemanship would become a rare skill. Instead, there is a large industry based on the use of the horse in sporting events and as a companion animal. For example, there are currently 5.25 million horses in the United States, supporting a $15.2 billion industry (American Horse Council, 1987).

The individuals who breed, raise, train, and show horses are interested in any information that might assist them in improving various aspects

317

of production, such as conception rates, live births, and performance, or decrease the incidence of various problems, such as aggression, which produces injury. Because of this, much of the research on domestic horses is in the realm of applied behavior, such as determining what variables of social structure affect aggression (Houpt & Wolski, 1980), investigating the function of coprophagy (Crowell-Davis & Caudle, 1989), and investigating whether cribbing, weaving, and stall walking are inherited (Vecchiotti & Galanti, 1986).

Thus, to some degree, the study of horse behavior is comparable to the study of human behavior in that knowledge about the behavior of the species represents an end in itself, to be used for practical applications, rather than simply an example to support a general theory. Nevertheless, the results of research on applied behavior can also bear on issues of general theory.

Researcher effect: studies of feral horses

Studies of feral and semiferal horses present problems of researcher effect that are typical of most wildlife studies. Feral horses are, with a few exceptions, not habituated to the presence of human observers. Proximity sufficient to provide good observations is likely to produce flight, especially in wooded terrain (e.g., Pellegrini 1971). In some studies determination of home range, territoriality, and movement patterns has been based on the location of tracks, feces, and horsehair. Most researchers have relied on direct observation, but found the distance to which they could approach limited. For example, Feist (1971), in a study of feral horses in the Pryor Mountains, made observations in part from a truck and reported that "I found that some bands would tolerate the truck to within distances of 50 to 100 yards, while other bands took immediate flight at my approach."

Problems of the opposite kind can occur when feral animals are accustomed to tourists. For example, Tyler (1972), in a study of New Forest ponies, found that "many ponies could be approached to within 5 yd but those used to cars and tourists would often then move towards the observer for food, and others would then show signs of alarm and move away." Tyler found that if the observer remained within 10 to 50 yards, depending on the individual pony, the animals "usually looked up when the observer first approached but then showed apparent disregard and very quickly resumed the activity in which they had previously been engaged."

Avoidance of an observer effect in studies of feral horses requires that the researcher carry out a period of preliminary observations. This not only allows the observer to collect ad libitum data so as to refine his or her protocol, but also allows the animals to become habituated to the observer. The observer also learns what distances and movements will produce a minimal effect while still allowing adequate data collection.

Researcher effect: studies of domestic horses

Similar problems exist for observational studies of domestic horses. The subjects bring into the research situation a varied background of experience with humans that can affect their response. Some horses and ponies run away at the approach of human observers. In some rare cases, the animals may exhibit obvious signs of alarm, periodically stopping to look back at the approaching human with tensed or even trembling body musculature, pricked ears, and/or flaring nostrils. Presumably, these are animals that have had little handling by humans, some or much of which may have been aversive.

The simple presence or absence of humans on a regular basis can, however, affect the flight response. During research I conducted on foals living at pasture, a relative of the owner commented at the end of one research season that the foals were far more tolerant of humans approaching them in the pasture than foals of previous years had been. The only difference was that the foals had been exposed to a human standing nearby for 20 to 40 hours a week during the first 4 to 6 months of life, whereas previous foals had been exposed to humans for only a few minutes a day when someone entered the broodmare pasture to check for injuries and illnesses. The question of what effect this had on the foals' ability to tolerate handling later in life remains unanswered.

In 13 years of research on several breeds of horses at numerous types of research sites, for example, inside barns and outside in paddocks and pastures of various sizes, I have found that the opposite situation, that of the horse approaching the researcher, is the most common problem. One obvious reason is that humans are often a source of food. In addition, the development of a strong social relationship between the horse and one or more humans before the horse is used as a research subject may alter the individual's response to the researcher, even if the researcher attempts to maintain a strictly neutral relationship.

One of the most dramatic examples of the presence of a human altering social interactions occurred during research on maternal, developmental, and social behavior in which a stallion was present in the pasture. I had been studying the herd for several months, and its regular members, having become habituated to my presence and having learned that I neither fed nor petted them, typically ignored me. When a new mare was placed in the pasture, typical interactions of mutual investigation between the established herd members and the newcomer, alternating with flight by the new pony, ensued.

After a while, the other mares ceased their investigation of the newcomer and resumed normal activities, but the stallion continued the pursuit. After galloping around the pasture, the mare approached me and positioned herself so that I was between her and the stallion to the extent that she kept her head immediately behind my back. Fortunately, the stallion, which had been used as a riding stallion for several years and had been taught what horse people refer to as "manners" (i.e., attempts to bite and paw at humans had been punished), continued his attempts to reach the mare by trying to run around me. This represented considerable restraint on the stallion's part, for many stallions would simply have knocked an intervening human aside. For my part, I attempted to get out from between them while being alert to the stallion's movements. The mare, however, succeeded for some time in maintaining her position at my back. After we had been running circles around each other for several minutes, a sight which no doubt would have been amusing from the sidelines, I maneuvered close enough to the fence line to duck through. The mare remained next to me until the last moment, then turned and galloped off.

Observations had to be terminated for the rest of the day. It is interesting that not only did the mare seek out a human rather than some other physical blockade present in the pasture, such as a tree or a shed, but she kept her head close to my back. It is probable that her view of the stallion was partially to fully blocked, at least some of the time. Over the next two days, the mare was rapidly integrated into the herd to the point that there were no more chases around the pasture, and she remained close to the rest of the herd. I can only speculate that it was probably the mare's prior experience with humans that led to this unusual involvement of a researcher in a social interaction.

More routinely, when a researcher is beginning to work with a given herd, many horses initially become alert and orient themselves to the entrance of a human into the pasture; that is, they raise their heads and rotate erect ears toward the person. Their heads are positioned so that the person is within their visual field. Some subjects either approach and stop at variable distances to continue observing the human, or fully approach and initiate contact. It is then necessary to spend several days habituating the horses to the observer and extinguishing the approach behavior that results from previous positive reinforcement by other humans.

It is possible for the horses to learn to distinguish among humans and/or situations and adjust their responses accordingly. For example, in 1988 several observational studies were carried out on a herd of 15 Belgian mares and 10 foals at the Snyder Foundation Equine Reproduction

Research Center in Athens, Georgia. The subjects were maintained on a pasture of approximately 10 hectares. There were two gates into the pasture, one of which (Gate A) was used daily by the horses in order to enter a lane to the barn to be fed supplemental grain. One technician did most of the feeding and was usually the person who opened the gate for the horses. The researchers did not participate in taking the mares to the barn for supplemental feeding.

When it was noticed that a researcher entering by Gate A produced an alerting response among the subjects, all researchers began consistently entering by the other gate (Gate B). The horses developed clearly different responses to three different situations. If a researcher or other person entered by Gate B, there was no detectable response. If the technician who usually fed the mares approached Gate A, the mares would become alert and orient toward her. If she opened it, they would immediately begin moving toward the gate at various speeds ranging from a canter to a walk. If other persons approached and opened Gate A, then entered the pasture and closed the gate, the mares would briefly become alert but would not approach.

This situation changed when some of the researchers assisted with feeding after the conclusion of the studies. For example, the first time I opened Gate A, the mares, who until now had ignored me, became alert but initially only gradually began approaching the gate. As I stood there with the gate wide open calling to them, they gradually increased their speed. The second day, however, the response was much more rapid. Almost immediately, several mares began trotting and cantering toward me. This rapid change illustrates how important it is for an observational researcher to avoid participating in the routine care and feeding of equine subjects.

Even during routine interactions with habituated herds, the individual temperament of various animals may produce different responses to researchers that can disrupt observations of intraspecies interactions. In the Belgian herd described earlier, two mares were aggressive toward humans in the pasture. Their behaviors included head threats (laying back the ears while moving the head or the body toward the threatened individual), bite threats, kick threats, and charges in which they would rush at a person while exhibiting a head threat. Our initial response, in an attempt to avoid interacting, was to retreat. However, the aggression continued and appeared to be increasing.

Observations from greater distances were not feasible. First, some of the studies being conducted required detailed observations that could be obtained only at close proximity, usually 10 to 30 m. Second, the herd was sometimes widely dispersed. Positioning oneself 30 to 60 from a focal animal subject would not prevent another horse from moving in behind to within 10 to 20 m, sufficient proximity from which the horse could launch an attack while one's attention was on the focal animal. With the obvious threat to human health if any of the 1500- to 1800-pound animals should attempt to follow a threat with an actual bite or kick, we interpreted the behavior as a dominance–subordinance interaction and concluded that our previous actions had constituted a form of submission that was serving to reinforce dominance-related aggression. Retreating from an aggressor is a common form of submissive response in horses. We decided that we should become dominant to the mares but only aggress when aggressed against; that is, instead of losing the occasional encounters by retreating, we had to win them. In this way, we hoped to eliminate the attacks, while maintaining as much neutrality as possible.

Subsequent aggressive threats were met by a researcher yelling at the horse, waving his or her clipboard at it, and if distance allowed, running at it. Any other researchers close enough to provide agonistic aiding did so. Physical contact was never made during the return aggressions. The immediate response of both mares was to retreat from the observer(s). The long-term effect of this change in researcher response was rapid and total cessation of the attacks by one mare. The rate of attack by the other mare decreased. Attacks occurred only if a researcher inattentively allowed the mare's or his or her own movements to take them to within approximately 10 m of each other. For the remainder of the study, all researchers were careful to keep one another informed of the location of the problem mare, especially when her graze walking was causing her to approach someone from behind gradually. Other than eliminating the danger of attacks, the change from submissive to aggressive responses to threatening behavior resulted in no change in the response of the mares to the observers that we were able to detect. They continued normal feeding and intraspecies social interactions whether observers were within or outside the periphery of the herd.

Foals present different problems than do adults. They engage in substantially more play behavior (Crowell-Davis, Houpt, & Kane, 1987) and engage in much investigative behavior both within and outside the context of play. During play, they may trot, canter, or gallop, repositioning themselves rapidly and making it difficult to position oneself so as to observe them without being too close. During play and investigative behavior, foals will pick up, strike, and paw at a variety of objects, including rocks, limbs, stones, and pieces of paper. Given this general pattern of play and investigation, it is not surprising that the researcher is occasionally confronted with the problem of a foal's galloping up and attempting to investigate him or her.

Much of this can be avoided by positioning oneself at the maximum distance at which accurate data can be obtained and standing so that an adult horse is between oneself and the foal. One can then make observations by looking around the head or hindquarters of the adult, or over its back if the adult is a pony. When a foal approaches in spite of these precautions, I have found that avoiding eye contact and keeping one's arms against one's body while slowly walking away usually results in a rapid loss of interest.

Occasional foals may attempt to direct behaviors against a human observer that are identical to those carried out during combat play, that is, rearing, striking, and biting. In such cases, retreat again only exacerbates the problem while immediate aggressive responses decrease and may eliminate the behavior.

While the problems just described occur occasionally, it should be emphasized that they are not common, in terms of either the number of animals involved in agonistic behavior, or the frequency and intensity of alerting/approach responses. In the course of making observations on several hundred animals over a number of years, I have encountered notable human-directed aggression from adult horses only with the two animals described here. Most domesticated horses exhibit a mild alerting response to any novel human, but rapidly habituate themselves if that human remains in frequent close proximity without directly interacting with them, and continue their normal maintenance behaviors and intraspecies social behaviors.

Interactions between humans and horses: research that has not been done

Given that the motivation for much of the research on horses is a need to provide information useful to horsepeople, there is a noticeable gap in the literature. Specifically, questions about the human–horse relationship and the effect of human behavior on horse behavior in general, and specifically on equine learning, remain virtually unasked. For example, from a collection of articles on "horse behavior" that I attempt to keep complete with all literature published to date, I pulled 50. The selection was random within the following rules: complete papers only (no abstracts), which must have been published in English in refereed journals. Only one paper was included with any given author as senior author. The papers are listed in the Appendix.

Of the 50 articles, only one addressed the issue of the effect of humans on horses, that by Blackshaw, Kirk, and Cregier (1983). However, the article is not a report of original research. Instead, it is a description of a horse-training method developed in 1914 by Kell B. Jeffery. While the article describes a number of stimuli that can be used to alter a horse's behavior, it relies on personal opinion (e.g. "Kirk (1978) believes that it is essential that a neck rope . . . be kept on the horse throughout his training") rather than any kind of testing.

Six articles were on equine learning. However, in spite of the fact that the economic significance of the horse derives substantially from its ability to respond appropriately to stimuli administered by humans, all six studies involved apparatuses that substantially decreased or eliminated the human factor. None of the situations was comparable to that which a working horse encounters. In the oldest study, a horse was conditioned to nudge a lever in a horizontal plane for half-cups of grain as positive reinforcement (Myers & Mesker, 1960). In another study, horses were presented with a three-choice test (one door different from the other two) for a food reward (Mader & Price, 1980). In another, ponies were trained to jump a small hurdle in response to a buzzer to avoid an electric shock and to move backward in response to a visual cue to avoid shock. The latter part of the experiment did involve a signal from a human. One experimenter would hold his or her forearm and hand in a vertical position and wave it from side to side. In addition, if a pony tried to turn completely around, its handler would intervene. Otherwise, the pony was allowed free movement. Any possible experimenter/handler effect that may have occurred was not treated in the discussion (Rubin, Oppegard, & Hintz, 1980). The other three studies concerned maze learning (Kratzer, Netherland, Pulse, & Baker, 1977; Haag, Rudman, & Houpt, 1980; Heird, Lokey, & Cogan, 1986).

Of the remaining 43 articles, most were observational, usually without any human intervention whatsoever, although occasionally there were minor degrees of handling that involved, for example, leading a stallion to a mare to observe his sexual response. However, the handling was not the focus of the study and was generally mentioned only briefly, if at all. The one exception in which the effect of the handler on the horse is mentioned, McDonnell, Kenney, Meckley, and Garcia (1985), is a study of shock-conditioned suppression of sexual behavior in stallions. The authors reported that "stimuli associated with handling, such as the lead shank or the experimenter, apparently became conditioned stimuli predicting a no-shock period."

One research project, not in the 50 selected randomly, involved experimental comparison of early experience. Heird, Bell, and Brazier (1987) divided 40 horses into five treatment groups, which received a different number of days of handling by humans. Subsequently, they were tested for habituation in a T-maze and their ability to learn that a bucket of feed was alternately placed in the left and right arms of the maze. Trainability ratings were provided for performance under saddle by four different trainers. The trainability ratings of the four trainers were averaged and compared among treatment groups. In the discussion, all differences among groups were considered to be solely the product of differences in the number of days of handling. The possibility of an individual handler effect was not considered. "Handling," in fact, was treated as a unitary stimulus, even though it involved a variety of specific human–horse interactions. No information was given on whether any attempt was made to balance which horses were handled by which persons, making it difficult to evaluate the experiment. Handling is also treated as a unitary stimulus in an earlier article on this topic (Heird, Lennon, & Bell, 1981).

In a study by Fiske and Potter (1979) scores obtained on a test of reversal learning in a Y-maze were compared with the subjective trainability score provided by a single trainer. Since "trainability" is not defined and the assessment was the subjective opinion of a single person, it is difficult to evaluate the usefulness of the results of this comparison.

Waring (1983) has been studying the effect of very early experience with humans (5 minutes to 15 hours postpartum) on subsequent responses to humans by young foals. However, Waring considers the results to date to be inconclusive, citing as problems such factors as heritable tendencies, maternal influences, and prior experiences resulting from management procedures inadvertently unique to each mare and foal.

From this partial literature survey, one can see a tremendous avoidance in horse behavior research of studies on the effect of humans on horses' behavior. Especially in the research on learning, which the authors often justify on the basis of understanding the horse's learning abilities because of practical applications, the researchers clearly avoid interacting with the subjects and make a point of preventing the horses from even seeing them. This contrasts sharply with the popular literature, which is replete with anecdotes and opinions on the topic. Why have scientists studying horse behavior so diligently avoided the topic?

The background for this bias in the selection of research projects may well have begun with Clever Hans. Hans was a horse that supposedly had been trained to carry out sophisticated mathematical problem solving. He would give answers by pawing the ground the appropriate number of times. Careful assessment eventually revealed that Hans had no understanding of math, but he *was* able to detect subtle and unconscious movements made by humans present whenever he had pawed the appropriate number of times. Hans responded to these movements by ceasing to paw (Pfungst, 1965).

Hans is often used to point out the potential of human cues to invalidate an experiment. Such commentary misses the point of what the studies of Hans demonstrated that horses *can* do. Horses can learn to respond in specific ways to very subtle cues provided by another species. This ability provides a significant part of the basis for the horse's position in human culture. The dressage horse is a prime example. After extensive training, the dressage horse not only changes its gait in response to cues from its rider that are almost invisible to the casual observer, but alters its length and rate of stride and makes very slight changes in the direction of limb movement, arch of the neck, and angle of the body relative to the direction of movement, again in response to subtle cues. A minuscule shift in weight and tensing of specific muscles by the rider can produce a very specific action by the horse – for example, the levade, a very controlled half-rear.

Among trainers of dressage horses and other horses that must exhibit specific behaviors in response to very subtle cues, there would be substantial interest in finding answers to some of the following questions: Can one predict at an early age which horses will, when adults, best be able to detect and be most likely to respond appropriately to certain stimuli? Do specific early social and learning experiences alter a horse's learning abilities and ability to discriminate among various intensities and types of sensory stimuli? If so, how? What specific learning experiences should be introduced at what age to produce the best performance?

Why haven't such studies been carried out? Over the course of many thousands of hours of research on horse behavior, I have periodically considered such research, and discarded it. Since virtually no research has been conducted in this area, other researchers who may have considered such projects obviously discarded the idea as well. Methodologies and protocols for observational studies are well established. One requirement presumably is simply a neutral observer who objectively records occurrences of operationally defined behaviors, then evaluates and describes them and interprets them in light of current theory. In the field of learning, there are well established and repeatedly used procedures for evaluating operant processes and assessing cognitive abilities, all without raising the ghost of Clever Hans by involving the researcher.

Experiments that might include humans and that directly assess a horse's response to and learning from humans, however, would present numerous pitfalls. The case of Clever Hans tells us that any human involved may be providing information or cues that the horse is more sensitive to than any of the humans present. The visual, auditory, and olfactory abilities of the horse are different from ours. The horse also has a chemosensory organ that we lack, the vomeronasal organ. Thus, there might be significant variables in such studies of which we are not aware.

One such variable might be a response to female menses. In a preliminary study of the possibility of this problem, which has been described anecdotally by horsepeople, especially regarding stallions, Miller (1976) interviewed eight horsewomen with an average of 17 years (range 5 to 50) of experience in handling adult breeding stallions. None had observed any difference in the stallions' responses to them that could be attributed to menstruation. Nevertheless, I am not aware of any controlled studies published to date on the effect of specific variables of human physiology and behavior on horse behavior.

In addition, one must consider the fact that horses bring into the test situation prior experience with humans that the experimenter may not be aware of. A history of having received a large amount of aversive stimulation or routine petting and feeding from a person of a given sex, size, scent, or hair color could alter a subject's reaction.

There is also a logistical problem of maintaining consistency on a single stimulus variable across several humans. Just how much pressure is applied in a "hard" push versus a "gentle" push? Again, to what unknown cues is the horse responding of which we are not aware? Perhaps how far apart the fingers are spread, or exactly which muscle the hand is placed on, is more relevant than how hard one pushes. Such problems can make the introduction of the human element into a study seem impossibly confounding.

Do the difficulties of such research mean it is impossible? No. They do make it more difficult and challenging. If such projects were initiated, no doubt the early ones would serve primarily as examples of the specific problems that can and will occur. As such, however, they would teach us how to improve and refine our methodology and terminology. Ultimately, we would develop a better understanding of human–horse relations and develop and expand theories on interspecies communications. The models developed could assist us in understanding other human–animal relations. The challenge remains for modern researchers to set aside the taboo against the direct involvement of humans in experiments and to design experiments that treat the "human effect" as a variable that not only exists, but has to be understood.

Researcher personality type and experience

Finally, in considering future research, it may be worthwhile to take into account the personality and experience of the individual researcher, especially if the subjects will be able to see the researcher, whether the study is observational or interactive. When involved in research on any animal, and perhaps with horses in particular, researchers bring substantial personal biases into the studies. They also have different degrees of experience with and understanding of horses. This can substantially affect how they behave toward their subjects, even on a level that humans cannot easily detect.

Many researchers of horse behavior whom I know personally have a major interest in the species that extends outside their professional life. Many own horses, and some would call themselves "horse lovers." Most have at least some experience with the species. Previous experience with horses probably has a significant effect on whether researchers become involved in studies of them. Research on horses is potentially dangerous, and one must be experienced in handling them and interpreting their behavior sufficiently to anticipate future movements that might cause injury. Persons who lack such experience may be less likely to enter into research on the species and may be more likely to abandon such research if they experience logistical difficulties related to special handling problems. The experience necessary to carry out the research successfully will, in almost all cases, have to have been obtained on one's own, before initiating any kind of research activity. In contrast, persons who have had extensive involvement with horses before beginning their research career, and who continue their involvement with the species as part of their scientific profession, may also be partially affected by an aesthetic attraction to the species.

There are, however, cases in which researchers have selected the horse specifically because it suited a theoretical model they were testing and/or they wanted to expand the number and type of species that had been used to test a particular model (e.g., Dr. Dan Estep, personal communication). In such cases, education in species-specific behavior began after the development of a research interest. A childhood or youth of long summer days spent with horses, which may have biased and probably altered behavior toward horses, as well as providing education in species-typical behavior, is not brought into the research situation.

The effect of differences in experience and attitude about horses has not been examined. Given the sensitivity of horses to human communication signals, it may well be that, unknown to the researchers, there are subtle effects on the behavior of their subjects.

Summary

Even in observational studies in which researchers avoid directly interacting with the subjects to any degree and habituate the horses to their presence before beginning the studies, horses may respond to the presence of researchers by directing specific behaviors toward them. Sometimes this is primarily a form of investigative behavior, especially with foals. At other times it is probably a result of prior learning, that is, aversive or pleasant experiences associated with humans before the horses became research subjects.

Scientists studying horse behavior have primarily used observational methods. When manipulations are made necessary by the topic of research, they have avoided any direct interactions with the subjects. This has resulted in an absence of controlled, objective studies on the effect of humans and specific human behaviors on the behavior of horses, in spite of the fact that such studies would be invaluable to the horse industry.

Appendix: fifty articles on horse behavior

Arnold, G. W., & Grassia A. (1982). Ethogram of agonistic behaviour for thoroughbred horses. *Applied Animal Ethology, 8*, 5–25.

Arora, R. L. (1986). Studies on breeding behaviour of mares. *Indian Veterinary Journal, 63*, 214–16.

Asa, C. S., Goldfoot, D. A., & Ginther, O. J. (1979). Sociosexual behavior and the ovulatory cycle of ponies (*Equus caballus*) observed in harem groups. *Hormones and Behavior, 13*, 49–65.

Back, D. G., Pickett, B. W., Voss, J. L., & Seidel, G. E. (1974). Observations on the sexual behavior of nonlactating mares. *Journal of the American Veterinary Medical Association, 165*, 717–20.

Berger, J., & Cunningham, C. (1987). Influence of familiarity on frequency of inbreeding in wild horses. *Evolution, 41*, 229–31.

Blackshaw, J. K., Kirk, D., & Cregier, S. E. (1983). A different approach to horse handling, based on the Jeffery method. *International Journal for the Study of Animal Problems, 4*, 117–23.

Boy, V., & Duncan, P. (1979). Time-budgets of Camargue horses: I. Developmental changes in the time-budgets of foals. *Behaviour, 71*, 13–202.

Buntain, B. J., & Coffman, J. R. (1981). Polyuria and polydypsia in a horse induced by psychogenic salt consumption. *Equine Veterinary Journal, 13*, 266–8.

Campitelli, S., Carenzi, C., & Verga, M. (1982–3). Factors which influence parturition in the mare and development of the foal. *Applied Animal Ethology, 9*, 7–14.

Carson, K., & Wood-Gush, D. G. M. (1983). Behaviour of thoroughbred foals during nursing. *Equine Veterinary Journal, 15*, 257–62.

Clutton-Brock, T. H., Greenwood, P. J., & Powell, R. P. (1976). Ranks and relationships in Highland ponies and Highland cows. *Zeitschrift für Tierpsychologie, 41*, 202–16.

Collery, L. (1974). Observations of equine animals under farm and feral conditions. *Equine Veterinary Journal, 6*, 170–3.

Crowell-Davis, S. L. (1985). Nursing behaviour and maternal aggression among Welsh ponies (*Equus caballus*). *Applied Animal Behaviour Science, 14*, 11–25.

Delpietro, H. A. (1989). Case reports on defensive behaviour in equine and bovine subjects in response to vocalization of the common vampire bat (*Desmodus rotundus*). *Applied Animal Behaviour Science, 22*, 377–80.

Dodman, N. H., Shuster, L., Court, M. H., & Dixon, (1987). Investigation into the use of narcotic antagonists in the treatment of a stereotypic behavior pattern (crib-biting) in the horse. *American Journal of Veterinary Research, 48*, 311–19.

Duncan, P., Feh, C., Gleize, J. C., Malkas, P., & Scott, A. M. (1984). Reduction of inbreeding in a natural herd of horses. *Animal Behaviour, 32*, 520–7.

Duren, S. E., Dougherty, C. T., Jackson, S. G., & Baker, J. P. (1989). Modification of ingestive behavior due to exercise in yearling horses grazing orchardgrass. *Applied Animal Behaviour Science, 22*, 334–45.

Fagen, R. M., & George, T. K. (1977). Play behavior and exercise in young ponies (*Equus caballus* L.). *Behavioral Ecology and Sociobiology, 2*, 267–9.

Francis-Smith, K., & Wood-Gush, D. G. M. (1977). Coprophagia as seen in thoroughbred foals. *Equine Veterinary Journal, 9*, 155–7.

Fretz, P. B. (1977). Behavioral virilization in a brood mare. *Applied Animal Ethology, 3*, 177–280.

Gates, S. (1979). A study of the home ranges of free-ranging Exmoor ponies. *Mammal Review, 9*, 3–18.

Ginther, O. J., Scraba, S. T., & Nuti, L. C. (1983). Pregnancy rates and sexual behavior under pasture breeding conditions in mares. *Theriogenology, 20*, 333–45.

Haag, E. L., Rudman, R., & Houpt, K. A. (1980). Avoidance, maze learning and social dominance in ponies. *Journal of Animal Science, 50*, 329–35.

Heird, J. C., Lokey, C. E., & Cogan, D. C. (1986). Repeatability and comparison of two maze tests to measure learning ability in horses. *Applied Animal Behaviour Science, 16*, 103–19.

Hillman, R. B., Olar, T. T., Squires, E. L., & Pickett, B. W. (1980). Temperature of the artificial vagina and its effect on seminal quality and behavioral characteristics of stallions. *Journal of the American Veterinary Medical Association, 177*, 720–2.

Hoffmann, R. (1985). On the development of social behaviour in immature males of a feral horse population (*Equus przewalski* f. *caballus*). *Zeitschrift Saugetierkunde, 50*, 302–14.

Jeffcott, L. B. (1972). Observations on parturition in crossbred pony mares. *Equine Veterinary Journal, 4*, 209–12.

Keiper, R. R., & Sambraus, H. H. (1986). The stability of equine dominance hierarchies and the effects of kinship, proximity and foaling status on hierarchy rank. *Applied Animal Behaviour Science, 16*, 121–30.

Kratzer, D. D., Netherland, W. M., Pulse, R. E., & Baker, J. P. (1977). Maze learning in quarter horses. *Journal of Animal Science, 46*, 896–902.

Leadon, D. P., Jeffcott, L. B., & Rossdale, P. D. (1986). Behavior and viability of the premature neonatal foal after induced parturition. *American Journal of Veterinary Research, 47*, 1870–3.

Mader, D. R., & Price, E. O. (1980). Discrimination learning in horses: Effects of breed, age and social dominance. *Journal of Animal Science, 50*, 962–5.

McCall, C. A., Potter, G. D., & Kreider, J. L. (1985). Locomotor, vocal and other behavioral responses to varying methods of weaning foals. *Applied Animal Behaviour Science, 14*, 27–35.

McCort, William D. (1984). Behavior of feral horses and ponies. *Journal of Animal Science, 58*, 493–9.

McDonnell, S. M., Kenney, R. M., Meckley, P. E., & Garcia, M. C. (1985). Conditioned suppression of sexual behavior in stallions and reversal with diazepam. *Physiology and Behavior, 34*, 951–6.

Miller, R. (1981). Male aggression, dominance and breeding behavior in Red Desert feral horses. *Zeitschrift für Tierpsychologie, 57*, 340–51.

Montgomery, G. G. (1957). Some aspects of the sociality of the domestic horse. *Transactions of the Kansas Academy of Science, 60*(4), 419–24.

Myers, R. D., & Mesker, D. C. (1960). Operant responding in a horse under several schedules of reinforcement. *Journal of Experimental Analysis of Behavior, 3*, 161–4.

Odberg, F. O. (1971). An interpretation of pawing by the horse (*Equus caballus* Linnaeus): Displacement activity and original functions. *Saugetierkunde Mitteilungen, 21*(1), 1–12.

Rubenstein, D. I. (1981). Behavioural ecology of island feral horses. *Equine Veterinary Journal, 13*, 27–34.

Rubin, L., Oppegard, C., & Hintz, H. F. (1980). The effect of varying the temporal distribution of conditioning trials on equine learning behavior. *Journal of Animal Science, 50*, 1184–7.

Salter, R. E., & Hudson, R. J. (1978). Habitat utilization by feral horses in western Alberta. *Naturaliste Canadien, 105*, 309–21.

Schoen, A. M. S., Banks, E. M., & Curtis, S. E. (1976). Behavior of young Shetland and Welsh ponies (*Equus caballus*). *Biology of Behavior, 1*, 192–216.

Shaw, E. V., Houpt, K. A., & Holmes, D. F. (1988). Body temperature and behaviour of mares during the last two weeks of pregnancy. *Equine Veterinary Journal, 20*, 199–202.

Tyler, S. J. (1972). The behaviour and social organization of the New Forest ponies. *Animal Behaviour Monographs, 5*, 85–196.

Vecchiotti, G. G., & Galanti, R. (1986). Evidence of heredity of cribbing, weaving and stall-walking in thoroughbred horses. *Livestock Production Science, 14*, 91–5.

Wallach, S. J. R., Pickett, B. W., & Nett, T. M. (1983). Sexual behavior and serum concentrations of reproductive hormones in impotent stallions. *Theriogenology, 19*, 833–40.

Wells, S. M., & Goldschmidt-Rothschild, B. V. (1979). Social behaviour and relationships in a herd of Camargue horses. *Zeitschrift für Tierpsychologie, 49*, 363–80.

Williams, M. (1974). The effect of artificial rearing on the social behaviour of foals. *Equine Veterinary Journal, 6*, 17–18.

Wolski, T. R., Houpt, K. A., & Aronson, R. (1980). The role of the senses in mare–foal recognition. *Applied Animal Ethology, 6*, 121–38.

Wood-Gush, D. G. M., & Galbraith, F. (1987). Social relationships in a herd of 11 geldings and two female ponies. *Equine Veterinary Journal, 19*, 129–32.

Acknowledgments

I thank the Snyder Foundation for Equine Reproduction Research and Mr. and Mrs. Karl Butler, who have provided the majority of the animals used in my research. Special thanks go to the colleagues and students with whom I have shared the many hours of fieldwork and discussion that were critical background for this manuscript, especially Dr. Al Caudle, Dr. Dan Estep, Dr. Hank Davis, Dr. Jane Hillsman, Dr. Katherine Houpt, Sally Ann Earl-Costello, Bridgitte DeVaughn, Mary Elizabeth Turner-Ellard, and Denise Smith-Funk. Very special thanks go to GlanNant Polonaise, a former research subject who is now a special friend. Sometimes our subjects teach us more about bonding than any textbook ever can.

References

American Horse Council. (1987). *The Economic Impact of the U. S. Horse Industry.* Washington, DC, American Horse Council.

Blackshaw, J. K., Kirk, D., & Cregier, S. E. (1983). A different approach to horse handling, based on the Jeffery method. *International Journal for the Study of Animal Problems, 4,* 117–23.

Crowell-Davis, S. L., & Caudle, A. B. (1989). Coprophagy by foals: Recognition of maternal feces. *Applied Animal Behaviour Science, 24,* 267–72.

Crowell-Davis, S. L., Houpt, K. A., & Kane, L. (1987) Play development in Welsh pony (*Equus caballus*) foals. *Applied Animal Behaviour Science, 18,* 119–31.

Feist, J. D. (1971). Behavior of feral horses in the Pryor Mountain wild horse range. Unpublished M. S. thesis, University of Michigan.

Fiske, J. C., & Potter, G. D. (1979). Discrimination reversal learning in yearling horses. *Journal of Animal Science, 49,* 583–88.

Haag, E. L., Rudman, R., & Houpt, K. A. (1980). Avoidance, maze learning and social dominance in ponies. *Journal of Animal Science, 50,* 329–35.

Heird, J. C., Bell, R. W., & Brazier, S. G. (1987). Effects of early experience upon adaptiveness of horses. In M. W. Fox & L. D. Mickley (Eds.), *Advances in Animal Welfare Science, 1986–87.* Boston: Martinus Nijhoff, pp. 163–9.

Heird, J. C., Lennon, A. M., & Bell, R. W. (1981). Effects of early experience on the learning ability of yearling horses. *Journal of Animal Science, 53,* 1204–9.

Heird, J. C., Lokey C. E., & Cogan, D. C. (1986). Repeatability and comparison of two maze tests to measure learning ability in horses. *Applied Animal Behaviour Science, 16,* 103–19.

Houpt, K. A., & Wolski, T. R. (1980). Stability of equine hierarchies and the prevention of dominance related aggression. *Equine Veterinary Journal, 12,* 18–24.

Kirk, D. H. (1978). *Horse Breaking Made Easy: Based on the Jeffery Method.* Copyright D. H. Kirk, 17 Warrache Street, Bracken Ridge, Australia, 4017. In J. K., Blackshaw, D. Kirk, & S. E. Cregier (1983). A different approach to horse handling, based on the Jeffery method. *International Journal for the Study of Animal Problems, 4,* 117–23.

Kratzer, D. D., Netherland, W. M., Pulse, R. E., & Baker, J. P. (1977). Maze learning in quarter horses. *Journal of Animal Science, 46,* 896–902.

Mader, D. R., & Price, E. O. (1980). Discrimination learning in horses: Effects of breed, age and social dominance. *Journal of Animal Science, 50,* 962–65.

McDonnell, S. M., Kenney, R. M., Meckley, P. E., & Garcia, M. C. (1985). Conditioned suppression of sexual behavior in stallions and reversal with diazepam. *Physiology and Behavior, 34,* 951–6.

Miller, R. M. (1976). Observations on the reactions of mature stallions to the presence of menstruating women. *Veterinary Medicine/Small Animal Clinician, 71,* 678–9.

Myers, R. D., & Mesker, D. C. (1960). Operant responding in a horse under several schedules of reinforcement. *Journal of the Experimental Analysis of Behavior, 3,* 161–4.

Pellegrini, S. W. (1971). Home range, territoriality and movement patterns of wild horses in the Wassuk Range of western Nevada. Unpublished M. S. thesis, University of Nevada.

Pfungst, O. (1965). *Clever Hans (the Horse of Mr. von Osten)*, ed. Robert Rosenthal. New York: Holt, Rinehart & Winston.

Rubin, L., Oppegard, C., & Hintz, H. F. (1980). The effect of varying the temporal distribution of conditioning trials on equine learning behavior. *Journal of Animal Science, 50,* 1184–7.

Tyler, S. J. (1972). The behaviour and social organization of the New Forest Ponies. *Animal Behavior Monographs, 5,* 85–196.

Vecchiotti, G. G., & Galanti, R. (1986). Evidence of heredity of cribbing, weaving and stall-walking in thoroughbred horses. *Livestock Production Science, 14,* 91–5.

Waring, G. H. (1983). *Horse Behavior: The Behavioral Traits and Adaptations of Domestic and Wild Horses, Including Ponies.* Park Ridge, NJ: Noyes.

—21—

Imprinting and other aspects of pinniped–human interactions

Ronald J. Schusterman, Robert Gisiner, and Evelyn B. Hanggi

Editors' introduction

Schusterman, Gisiner, and Hanggi offer evidence that early social interactions between California sea lions and human researchers result in major changes in the animals' behavior. These changes may be measured directly and used to considerable advantage, whether in routine handling or in specific experimental work, to explore the cognitive and perceptual abilities of the animals.

Rocky is a female California sea lion (*Zalophus californianus*) in Santa Cruz, California, who "understands" a human gestural sign language (Schusterman & Gisiner, 1988; Schusterman & Krieger, 1984, 1986). Hoover is a male harbor seal (*Phoca vitulina*) who was reared by people and used to "speak" English when he was at the New England Aquarium (Ralls, Fiorelli, & Gish, 1985). Mendicant feral California sea lions at Pier 39 in San Francisco have learned to do a variety of tricks for fish handouts from tourists (Nolte, 1990). Sea lions and seals are getting shot, shot at, or acoustically harassed following their disturbance of partyboat fishermen (Miller, Herder, & Scholl, 1983). All of these stories are examples of contemporary human–pinniped interactions and depend on skills acquired by individual animals during the course of their development. During development, the acquisition of learned skills may be facilitated by the bonding that occurs between individuals. Among the behaviors that reflect attachment and bonding are "following" and vocalization.

In this chapter we first discuss bonding between newborn *Zalophus* pups and human surrogate mothers and document the long-lasting effects that such "imprinting" has on the attachment behaviors of more mature

334

California sea lions. Next, we draw on our understanding of the natural history and behavioral ecology of feral pinnipeds to attempt to comprehend more fully the motivations of our animals and to improve control over the outcomes of pinniped–human interactions. We emphasize our work with human-raised or imprinted sea lions, since such animals tend to treat humans as social peers more than do wild-born sea lions or captive pups reared by their own mothers.

Young animals may become imprinted on certain characteristics of the environment at specific periods of development and thus learn about selective features of their parents, siblings, or habitat (McFarland, 1985). According to the originator of the concept, Konrad Lorenz, imprinting, as an irreversible learning process, is part of a "phylogenetic program determining precisely when a young organism is to learn what" (Lorenz, 1981). In filial imprinting, young animals, under natural conditions, may respond to their mother in a variety of ways, such as following, vocalizing, and nuzzling. Such behavioral patterns are genetically coded or innate. However, acquiring knowledge of the mother figure is not genetically coded but has to be acquired; that is, youngsters learn about certain features of the attachment figure. As Lorenz has delineated the phenomenon, the tendency to form a bond is innate, and the learning process or imprinting determines which stimulus configuration is selected, thus establishing the basis for the formation of filial attachments with one particular individual or class of individuals.

Imprinting, at least filial imprinting, studied intensively in birds, is probably widespread in mammals like otariid pinnipeds. These animals breed colonially in crowded groups where there is a high potential cost if social attachment by pup to mother is misdirected. There have been relatively few experimental studies of social attachment in pinniped pups (see Trillmich, 1981). The objective of our studies was to determine whether newborn captive sea lion pups formed behavioral attachments to human surrogate mothers in a manner similar to that described for pups bonding with their biological mother in the wild (e.g., see Schusterman, 1981). In the first section of this chapter, we describe several experiments that we believe do indeed demonstrate that California sea lion pups, within a narrow window of time after birth, form a relatively exclusive and long-term attachment to their original human caretakers. (Schusterman [1985, 1986] originally presented this material on videotapes. Copies of these tapes are available on request.)

335

Imprinting experiments

Methods

In all of these experiments, California sea lions were fed immediately before test observations in order to minimize the effects of food motivation, and observations were videotaped for later analysis. In these tests we were interested primarily in the amount of time a sea lion spent with a "passive person," that is, one who did not initiate contact with the animal. Animals were considered to be in "proximity" to the caretakers when they remained within 0.5 m of a person for at least 3 seconds. "Interactions" with people consisted of following and emitting the mother call and such contact behavior as climbing on the person as she or he was in a sitting or squatting position, nuzzling, resting or sleeping on or next to the person, and nonnutritive sucking of the chin or an article of the person's clothing. Sea lions made threats toward people only occasionally during these test observations. Interactions with other sea lions consisted principally of play-chase or play-fight. An animal was scored as "solitary" when it swam or locomoted without interacting with others or sat or rested alone. In the playback experiment (Experiment 3), "orientation and searching" consisted of looking at the speaker or locomoting directly toward the speaker. "Proximity" was scored when the sea lion was within about 1 to 3 m of the speaker for at least 3 seconds, depending on the location of the speaker for each test animal.

Experiment 1 was conducted at three different oceanariums: Marine World/Africa USA, Redwood City, California; Marineland of the Pacific, Palos Verdes, California; and Sea Life Park, Hawaii. All enclosures where testing occurred contained a single pool and ranged in size from about 20 × 14 to 6 × 4 m. Altogether, 20 California sea lions were observed in their home pools in the presence of two or three people, who repeatedly switched positions on haul-out or deck areas around the pools; included in three of the four tests were the current caretaker and the original surrogate "mother" or caretaker. In the small holding pool at Marineland, one person was the current caretaker and the other was a stranger. The comparison was between seven hand-reared or "experimental" sea lions who were cared for and bottle-fed by a person (their surrogate "mother") within 96 hours of birth and 13 "controls," who were cared for and nursed by their biological mothers for at least 21 days after birth, but usually for between 6 and 9 months after birth. Table 21.1 lists the rearing history of experimental and control sea lions; it can be seen that all animals were more than 9 months old when the first test observations were made and that all animals but one (Patty from Marineland's small holding pool) were weaned before testing. Additional observations were made on three of the sea lions that were considered by the oceanarium personnel to be bonded to people and that were likely to remain at the park for another year. These sea lions were Buckwheat, who was initially nursed by Jenny Montague; Auntley, who was initially nursed by Marlee Breese; and Scooter, who was nursed by Brad Andrews. All three of these sea lions (see Table 21.1) had little or no contact with their original surrogate mothers for at least 1 month before testing.

In Experiment 2, a sea lion pup named Rio, born at Marine World, Redwood City, California, was taken from her nonlactating mother about 10 hours after birth. She was then hand-raised at our laboratory in order to follow her interactions with human trainers in greater detail. Furthermore, we wanted to test the strength of a pup–surrogate mother attachment by determining the extent to which the latter might affect the pup's exploration of a fear-provoking environment (swimming in a large, deep pool it had never been in before in order to gain access to the imprinted figure). The experimental evidence for what has been called the "secure base effect" (Rajecki, Lamb, & Obmascher, 1978) suggests that the presence of the attachment figure and communicative behaviors by an attachment figure will facilitate exploration of even relatively dangerous environments by the young of a variety of animal species. All attachment tests on Rio were conducted at the Long Marine Lab, University of California, Santa Cruz. (See Schusterman & Gisiner [1988] for a description of the pool and enclosure.)

In Experiment 3, olfaction and vision were eliminated as signals from the attachment figure by using audio tape playback voices (for Auntley and Scooter) or a hidden caller (for Rio) in order to test whether the pups' "representation" of the imprinted figure could be retrieved or activated by vocal signaling alone. All three sea lions tested in this experiment had no acoustical contact with their original surrogate mother for at least 1 month before testing. Contingent on its head being above water, each sea lion was given about 2 minutes of its original caretaker's voice, followed by about 4 minutes of a control voice (either the current caretaker or a novel voice) and finally 2 minutes of its original caretaker's voice. The times between the different voices were each about 10 seconds. In each case the words used by the people calling were essentially the same and included the name given to the sea lion. Each of the three sea lions was tested this way only once. Tape recordings were used for Auntley and Scooter. Recordings and playback were done with a Sony tape deck model TC-D5M and an Aiwa SC-A5 speaker. Although we initially tried playbacks with Rio, her indifference to the recordings forced us to abandon them and use live calling. Olfactory and visual cues were controlled by positioning both calling and

Table 21.1. *History and characteristics of California sea lions tested for attachment to people*

Name	Sex	Age tested (months)	Separated from mother	Bottle feeding (months)	Tube-fed	Weaned	Captive-born	Attached to people
Tested at Marine World's large holding pool								
Buckwheat[a]	M	10	5 hr	6	No	Yes	Yes	Yes
Alfalfa	M	10	9 mo	No	No	Yes	Yes	No
Froggy	M	10	9 mo	No	No	Yes	Yes	No
Theodore	M	22	9 mo	No	No	Yes	Yes	No
Elliot	M	22	21 days	No	Yes	Yes	Yes	No
Tested at Marineland's large holding pool								
Scooter[a]	F	33	4 days	9	Yes	Yes	Yes	Yes
Cecil	M	57	2 days	4	Yes	Yes	Yes	Yes
900	F	33	12 mo	No	No	Yes	Yes	No
741	F	57	3 mo	No	No	Yes	Yes	No

589	M	69	12 mo	No	No	Yes	Yes	No
1087	F	33	12 mo	No	No	Yes	Yes	No
1138	F	57	12 mo	No	No	Yes	Yes	No
724	M	69	12 mo	No	No	Yes	Yes	No
Tested at Marineland's small holding pool								
Xavier	M	21	4 days	11	Yes	Yes	No	Yes
Patty	F	9	3 days	9	Yes	No	Yes	Yes
Lolita	F	21	2 days	11	Yes	Yes	No	Yes
1146	X	9	6 mo	No	No	Yes	Yes	No
Tested at Sea Life Park's small holding pool								
Auntley[a]	F	12	1 day	11	Yes	Yes	Yes	Yes
Gregg	M	12	25 days	9	Yes	Yes	Yes	No
Huapala	F	12	10 mo	No	Yes	Yes	Yes	No

[a] These sea lions were given a choice between the original caretaker and their current caretaker.

noncalling surrogate mother and current caretaker in the same general location behind a visually opaque barrier. Thus, any change in Rio's behavior should have been a function of the voice she heard, since visual cues were eliminated and olfactory cues were held constant. In Auntley's playback, the speaker was placed about 2 m above the ground in a palm tree, and in Scooter's, the speaker was placed on the corner of a 1.5-m-high rectangular wall surrounding Marineland's large holding pool. For Scooter's test, there were approximately eight other California sea lions in the enclosure; for Auntley's, there were four other California sea lions and several harbor seals in the enclosure; and for Rio's test, there were two other California sea lions in the enclosure.

Results and discussion

Experiment 1. Figure 21.1 shows the results of our first set of observations. All 7 experimental sea lions showed stronger and more persistent attempts to make contact with people than did the 13 controls. The effect was strongest for those sea lions (Buckwheat, Scooter, Cecil, and Auntley) that had access to their original surrogate mother. However, even those sea lions that did not have access to any of their original caretakers (Xavier, Patty, and Lolita in the small holding pool at Marineland) showed the effect to some extent. Thus, when food motivation and exploratory motivation appear well controlled and length of time in captivity is also controlled for, hand-reared California sea lions interact more extensively with people, particularly their original caretaker, than those raised by their biological mothers.

Additional observations of the three "imprinted" sea lions, ranging in age from about 1 to 3 years, are summarized in Figure 21.2. All three individuals virtually ignored other sea lions, even animals that were their own age and sex (see Table 21.1), when they were in the presence of their original caretaker. All of Scooter's interactions with other sea lions were accounted for by a single individual, Cecil, a 5-year-old male, who had also been raised by Original Caretaker 1 (OC-1). Cecil often chased Scooter away from OC-1 and Scooter then spent time interacting with OC-2.

The three focal animals typically paid no attention to their regular or current caretaker and interacted preferentially with their original surrogate mother. The very few interactions they did have with their current caretaker were usually negative, consisting of mild, open-mouth threats. Illustrative of the preference test, Figure 21.3 shows that when yearling male sea lion Buckwheat was given a choice, he preferred his original mother surrogate (A) Jenny (light hair) to his current caretaker Cindy (dark hair). After Jenny and Cindy switched positions, Buckwheat would

340

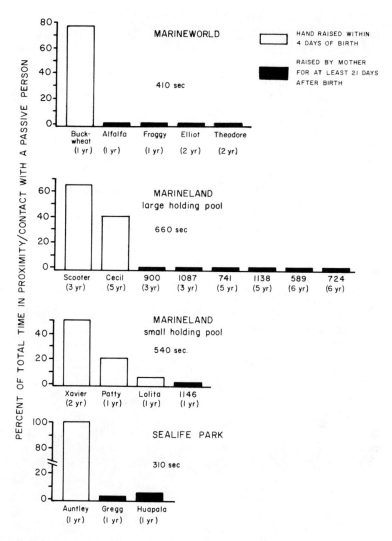

Figure 21.1. Bar graphs for each group of sea lions show the total time they were first tested with people in the enclosures. Proximity or contact time is shown as a function of whether the sea lion was raised by people within 4 days of birth or whether it was raised by its mother for at least 21 days after birth.

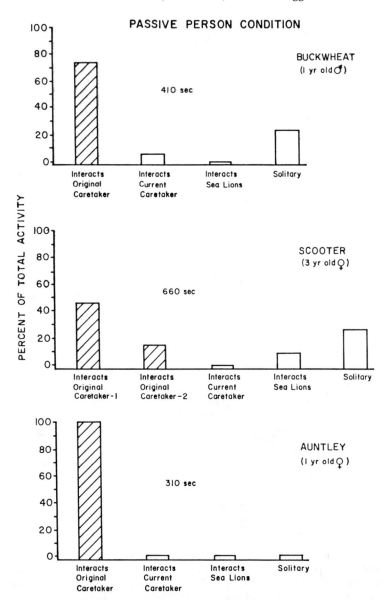

Figure 21.2. On the basis of data shown in Figure 21.1, additional time was spent observing and scoring other behavioral categories from the videotapes. Time periods shown here were matched to those previously used on Buckwheat, Auntley, and Scooter. The percentage of time spent in four mutually exclusive and exhaustive behavioral categories is given. "Passive person condition" refers to the fact that the caretakers did not initiate contact during the test.

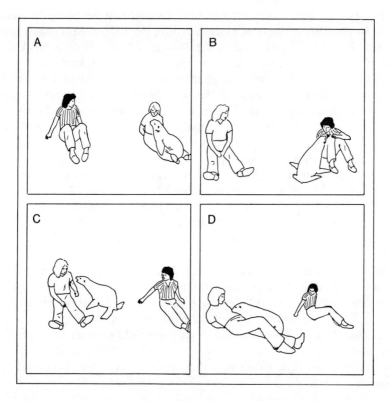

Figure 21.3. Yearling sea lion Buckwheat showing he prefers his original caretaker Jenny (the blond) to his current caretaker Cindy. After Jenny and Cindy switched positions, Buckwheat sniffed Cindy, oriented, and gave a "mother call" to Jenny, climbed on her, and nuzzled her. (This figure was drawn from slides.)

sometimes sniff Cindy (B), occasionally rebuffing her with a mild threat, then search, and give a "mother call" to Jenny (C), rapidly locomote to her while vocalizing, climb back on Jenny and nuzzle her (D). All three subjects responded to calls by their original caretakers, and frequently appeared to use olfaction to confirm identification by voice. Attachment behavior included frequent calling by the two yearlings, and all three showed nuzzling and some nonnutritive sucking of the surrogate's neck and chin or an article of clothing. Following, another attachment behavior, occurred most often when the surrogate changed positions or tried to leave the enclosure. Contact behaviors such as climbing on and resting on the legs or body of the surrogate were seen in all three focal animals. During these tests, it was unclear as to how exclusive the attachments were to the original caretakers. Additional choice tests in 1984 with Buckwheat (similar

343

to the one just described), for example, showed that he always preferred Jenny to all other trainers at Marine World, even those who began caring for him following Jenny's initial caretaking, that is, about a week after Buckwheat's birth. However, when these early caretakers were pitted against later ones, Buckwheat's attachment behaviors toward the earlier caretakers were similar, although not as intense nor as frequent as those displayed toward Jenny (the original caretaker). Also, note in Figure 21.2 that Scooter interacted by sitting in contact with OC-2 during her testing.

Additional tests on the strength of attachment were conducted on Buckwheat and Auntley. For example, in one test their original caretakers walked at a fairly brisk pace around the enclosure for 1 minute, and the extent to which these yearlings followed was noted. Buckwheat and Auntley started following their surrogate mothers immediately and continued to do so throughout the entire test with one exception; Buckwheat interrupted his following by diving into the pool half-way through the test (presumably to themoregulate) and then immediately resumed following. Both yearlings directed mother calls at the original caretakers almost continuously while they pursued them. Follow-up studies one and two years later with all three of these sea lions indicated that the attachments these animals had formed toward their original human caretakers as pups were quite durable. This occurred despite the fact that after these sea lions were weaned at about 6 or 7 months of age, the original caretakers (Jenny and Marley) had little to do with their feeding and their food-reinforced behavior.

Experiment 2. Rio's initial surrogate mother was Michelle Jeffries (Figure 21.4), who was the first person to bottle-feed the pup 10 hours after birth and who had almost continual contact with Rio for 48 hours. At that point, Evelyn Hanggi and Michelle alternated 24-hour shifts. After the first week, other caretakers also cared for and nurtured Rio several times a week, and their frequency of caretaking increased as Michelle's and Evelyn's decreased over the following months.

Within the first week, it became obvious that Rio was exclusively attached to Michelle. All attachment behaviors noted previously with Buckwheat, Auntley, and, to some extent, Scooter were directed by Rio more frequently and more intensively to Michelle than to Evelyn or any of the other caretakers. When Rio was in contact with Michelle, the pup often threatened all other caretakers, including Evelyn. Such aggressive behavior in the presence of the original mother surrogate, Michelle, toward social targets is similar to findings in precocial chicks where the presence of the chick's mother increases the likelihood that the chicks will emit aggressive pecks (Hogan & Abel, 1971).

344

Figure 21.4. Rio at 6 months of age nursing on formula from a bottle held by her original caretaker Michelle.

Rio learned to nurse from a bottle during the first 2 days with Michelle. At the time of the transition between surrogate mothers, Michelle gave Rio the bottle and fed her for a couple of minutes and then Evelyn came in and began feeding her for the first time. This was the first contact Evelyn had with Rio. The transition went smoothly, and the pup accepted food from both of the caretakers equally well. During those first days, Rio ate well, nuzzled, followed, and vocalized to both Michelle and Evelyn. Then Michelle took over caretaking duties again, and Rio was moved from Marine World/Africa USA in Redwood City to Long Marine Lab in Santa Cruz, California. Evelyn was with her the following day and noticed that her behavior had changed. Rio avoided contact and did not come to her when Evelyn entered the pen. When Evelyn tried handling her, she nipped at her hands. Rio's calling diminished, and from then on she ate less from Evelyn. This pattern was typical over the first 5 months, and only 22% of the time did she eat as much from Evelyn as she did from Michelle. In contrast, Rio ate consistently and well from Michelle. Only 2.7% of the days when Michelle cared for her did she eat less compared with what Evelyn had fed her on the previous day.

Her reactions toward both Michelle and Evelyn had changed. Rio initiated more approaches to Michelle than she did to Evelyn and showed excited greeting behaviors when Michelle went to her but not when Evelyn or anyone else did. Such behaviors included running to and climbing on

345

Michelle and sucking on her face and arms. Rio greeted Michelle with loud vocalization, sniffed her face, and did a characteristic rapid head shake at the same time. Rio had shown these same behaviors toward Evelyn, but they diminished to almost nothing after the first 2 weeks. Indeed, she stopped calling to and sucking on Evelyn's neck or chin altogether within 1 month.

During her first 5 months on formula, Rio took 70% as much from Evelyn as she did from Michelle and only about 20% as much from three additional caretakers as she did from Michelle. If Rio refused to eat from any of the caretakers other than Michelle, the mere presence of Michelle would facilitate her formula intake, either directly from Michelle or from the caretaker as soon as Michelle entered the area.

Once Rio arrived in Santa Cruz, she was housed at the lab in a fenced enclosure containing a small pool (4.8 × 2.3 m and 1.2 m deep). When Rio was 2 months old, we took her into a much larger pool (7.6 m in diameter and 1.8 m deep) for the first time. We designed a test in which Evelyn interacted affiliatively with her on land for 2 minutes, then got into the water for 2 minutes, called to Rio and tried to get her to go in, and finally sat with her on land for an additional 2 minutes. Then Evelyn left the area and Michelle went through the same procedure.

For the whole 2 minutes Evelyn was in the pool, Rio refused to go in. However, Rio showed an interest by leaning her head down to the water several times. In contrast, when Michelle went into the water, Rio immediately followed her by diving in, swam in a relaxed manner with her, and stayed in the pool for 2 minutes. We had been a little apprehensive about Rio's ability to get out of the pool on her own, so as a safety measure, we built her a ramp. However, when Michelle climbed out, Rio, not needing the ramp, jumped right out after her. Next, Evelyn went back in. Rio did not follow her until 75 seconds had elapsed. She appeared very panicky and did not stop thrashing about until she found the ramp. Evelyn called to her the whole time she was in the water, but Rio essentially ignored her. After finding the ramp, Rio did go to Evelyn and swam with her much as she did with Michelle. Her previous swim probably made this possible. Once she gained some experience with Michelle, she was more inclined to go into the water.

Rio was swimming in a fear-provoking context that was facilitated by the presence of the attachment person, and later she was able to generalize it to other people. This is a case where we were able to use the attachment person to overcome an aversive situation and promote an adaptive response.

We were able to use this bond to elicit exploratory behaviors as well. For instance, Rio followed Michelle wherever she went, including areas she

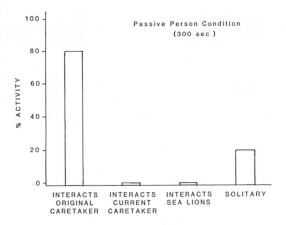

Figure 21.5. The percentage of time Rio spent in four mutually exclusive and exhaustive behavioral categories.

had never been to before. This was very useful for such tasks as moving her about and weighing her.

Following weaning at 12 months, Rio was no longer fed by Michelle. At 14 months of age, Rio was given a preference test and, like the other three hand-reared sea lion pups and as Figure 21.5 shows, she ignored her current caretaker and two other sea lions and interacted in a positive way almost exclusively with Michelle, with whom she had had no sensory contact of any kind for 2 months.

Experiment 3. As Figure 21.6 shows, voice playbacks for Auntley and Scooter and calls by individuals hidden behind a blind for Rio demonstrated that these imprinted sea lions were able to distinguish the voices of their attachment figures from those of other individuals. Both Rio and Auntley vocalized quite frequently to their mother surrogate's voice and hardly at all to the control or current caretaker's voice. Auntley's vocalizations occurred just before the onset of the current caretaker's voice, as the sea lion was beneath the tree holding the speaker that had just emitted its mother surrogate's voice. Auntley continued these vocalizations within that context for about 30 seconds before she ceased orienting to the speaker, stopped vocalizing, and left the proximity of the speaker. Scooter, at 5 years of age, showed no vocal reply to playbacks of her mother surrogate's voice. However, Scooter was quite responsive to voice playbacks of her surrogate mother as reflected by her orientation responses to the speaker. It should be noted that except for a blind adult male California sea lion in the enclosure with Auntley, no other California sea lions were

347

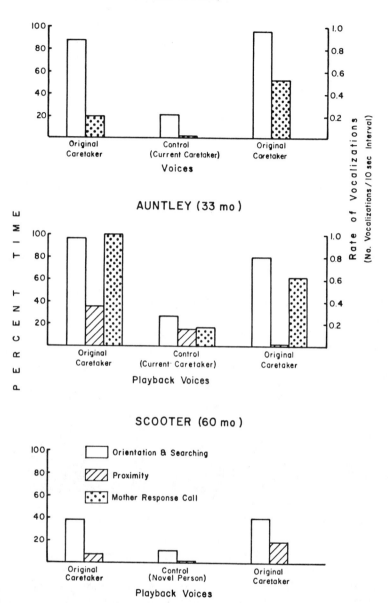

Figure 21.6. The left ordinate refers to the percentage of time sea lions spent orienting to and searching for the source of the calling voice (either live for Rio or tape playback for Auntley and Scooter) as a function of who was calling. Time spent in proximity to the speakers as a function of playback voices is also given. The right ordinate refers to rate of vocalizations (mother response call) for Rio and Auntley as a function of who was calling.

348

responsive to calls by any of the voices in any of the three different enclosures used in this experiment. Thus, the experiment confirms our initial impression that California sea lions become imprinted on the voice of their original human caretakers. Since auditory cues seem critical for individual recognition in sea lions, and because a human voice contains most of the frequencies found in a sea lion pup attraction call, perhaps it is not so surprising that a sea lion pup imprints on a human voice.

Summary and discussion

Many of the behavioral interactions that California sea lion pups had with their attachment figures in these studies were quite similar to the kinds of interactions observed in the field between pups and their biological mothers (Gisiner & Schusterman, 1991; Peterson & Bartholomew, 1967; Trillmich, 1981). Moreover, when yearling, or even older, sea lions and fur seals are seen with their mothers, their interactions are also similar to those observed in our sea lion–human surrogate mother interactions. Both inter- and intraspecies types of bonding appear to proceed in much the same way. A pup suckles from its biological mother soon after birth and begins recognizing her voice within a few days. This bonding between offspring and biological mother is usually exclusive among otariid pinnipeds and can, at least in captivity, last a lifetime (Hanggi & Schusterman, 1990). In California sea lions, when pups are bottle-fed by a human caretaker soon after birth and cared for by that person intensively for at least 2 or 3 days, there is a good probability that the pup will form an exclusive attachment to that surrogate mother and that the bond will be long lasting. The pup will recognize its surrogate mother's voice and will approach her vocal signaling under a wide variety of circumstances. Like pups with their biological mothers, human-raised pups benefit from the presence of their attachment figures. Their presence promotes exploration, swimming, and many other adaptive responses, including feeding.

Thus, a sea lion is similar to a lamb that follows the person who feeds it with a bottle, even when it is not hungry. Though it has been weaned and plays with other sheep, the lamb will still approach and follow its former keeper (McFarland, 1985). In a similar manner, a weaned yearling will stop playing with other sea lions, and will approach and attempt to interact positively with its former keeper. Continuing with the similarity between a lamb and a sea lion that have imprinted on a person, in both cases juveniles follow the person and call to her or him as if the person were its mother, and as adults both retain some attachment to the person, showing that imprinting for sheep and sea lions can have both long- and short-term aspects.

Our results suggest that sea lion pups imprint first on the voice of their caretaker (and probably on her or his smell as well). The sensitive period for this type of filial imprinting by a sea lion on a human voice, containing most of the frequencies found in sea lions pup attraction calls, appears to range from a few hours to several days after birth. In a later section of this chapter, we discuss the costs and benefits of working with a mature sea lion that has earlier formed an attachment to a person.

Using behavioral cues during interactions with pinnipeds

A student of captive pinnipeds can benefit from an understanding of the motivations and outcomes typical of certain behaviors in wild pinnipeds. We have had the good fortune to have spent time watching wild pinnipeds and also to have interacted with captive pinnipeds. The following are some of the techniques we use with captive pinnipeds that draw on an understanding of the natural history and behavioral ecology of wild pinnipeds.

Species and sex differences in aggression

Despite the apparent docility or tameness of many wild pinnipeds and those pinnipeds common in captivity such as the California sea lion, there is one obvious point that no student of wild pinnipeds is likely to forget: Pinnipeds are large carnivores. That is to say that pinnipeds survive by hunting and killing things, and they are as large or larger than the average human. Pinniped social interactions involve hard physical contact and biting. Trainers of pinnipeds, like trainers of many large animals, maintain control of the animal only by maintaining the illusion of social dominance and/or superior ability in physical confrontation. Once that illusion is lost, the animal is in control. Maintaining the illusion requires an understanding of the social cues used to indicate dominance and aggression.

To use examples from species we know best, we have chosen the northern elephant seal (*Mirounga angustirostrus*) and the California sea lion. Male northern elephant seals may exceed 5 m in length and 2,000 kg in weight. Nevertheless, they can usually be moved easily by a human holding up a stick that makes her or him "taller" than the elephant seal. (Fighting males balance themselves as high as possible to gain leverage advantages during contests, and an obviously "shorter" male usually backs down before a taller adversary.) It is also possible to move elephant seals by clapping hands or banging on metal drums, by making other banging noises that simulate the clapthreat vocalization of an aggressive male elephant seal, or by thumping on the ground to simulate the ground-shaking attack of a large male. California sea lions, although large, do not exceed 2.5 m in length, and large adult males weigh 300 to 350 kg. They also try to gain height during contests, but since they stand a little more than a meter high, a human nearly 2 meters tall can actually overdo things by crowding or leaning over a sea lion, since the sea lion may panic and lash out in "self-defense."

Sea lions, especially males and females with pups, select a site that they will defend aggressively, usually against individuals of the same approximate age and sex. In captive sea lions this site-specific aggression, or territorial behavior, often manifests as barring a doorway or an entrance to a pool from other sea lions or the trainer (Schusterman & Dawson, 1968). Even a subadult sea lion weighing 50 kg or less can be nearly impossible to dislodge from a site, almost as if it had glued itself to the spot. Usually one can dislodge or aggressive territorial sea lion only by approaching from behind (which often requires two people and enough space for the escaping sea lion) or by behaving like a social dominant, approaching without hesitation, making a lot of noise, and perhaps expanding one's profile by carrying a large piece of cloth or plywood. During drives of fur seals on the Pribilof Islands, rapidly opened umbrellas are used to move recalcitrant or straying individuals.

We know of one especially clear example of the value of knowing species differences in behavior. Trainers familiar with California sea lions know that part of their aggressive, territorial display behavior is a rapid lateral shaking of the head, like a person emphatically indicating a "no, no, no" response. Working with the much larger and less familiar Steller sea lion (*Eumatopias jubatus*), trainers were gratified to see this sea lion emphatically nodding its head and reinforced the behavior with food, thinking it might be useful in a show or as a step to some other behavior. In fact, the nodding of the Steller sea lion was the aggressive equivalent of the lateral head shaking of the California sea lion. Trainers were, in fact, positively reinforcing aggressive signals directed at them (S. Allen, personal communication).

Pinnipeds as a group tend to show great dimorphism in size, form, and behavior between the sexes. Despite the fact that male sea lions are much larger and more aggressive than females, both sexes are common in captivity and one can find trainers with preferences for working with either sex. In our experience, mature males, despite their large size and aggressive posturing and signaling, are easier to control because their social communicative behaviors tend to contain more ritualized, stereotyped behaviors. Not only are their intentions easy to read, but they respond very consistently to the trainer's use of the same signals (or an approximation). Given the behavioral ecology of polygynous pinnipeds, this is not surprising. Mistakes in interpreting social aggressive signals from other males or in the social aggressive signals a male gives can have serious effects on survival and reproductive success.

Since female sea lions are smaller and less aggressive than males, many trainers find them less intimidating than males (and cheaper to feed). However, we have found females to be more dangerous because (1) they do not provide obvious signals of aggressive intent to the same degree as males, and (2) they are less likely to inhibit direct physical aggression, like biting, because the typical social consequences of such actions are not as severe for females as males. A female sea lion that bites a male inflicts relatively little damage compared with a male rival. Furthermore, females can use submissive social behaviors to inhibit male retaliation. Contests between females are between individuals of approximately equal size (relative to the males) who also lack some of the specialized armament (large canines, heavy jaw and neck muscles) found in males. On the rookeries, contests between females are less injurious than male contests but are also much more frequent and escalate rapidly. We hear of people being bitten by females much more often than by males. Bites from males are often not as severe as female bites, because males tend to inhibit biting against an "adversary" when the adversary is high ranking, or, perhaps more appropriately, when the sea lion is unsure of the retaliatory capability of the individual it bites.

Imprinting sea lions

As techniques have improved over the years there have been more successful births of captive sea lions and, as a consequence, more successful attempts at hand rearing those pups that, for one reason or another, must be raised to weaning by bottle feeding by human "surrogate mothers." As reported earlier in this chapter, sea lion pups tend to form an exclusive social bond early in life with their primary human caretakers. Although it is unclear what effect this early social bond with a human has on subsequent social and sexual interactions with other sea lions, we have found that all hand-reared pups are later able to interact "normally" with other sea lions, mate, and give birth.

Nevertheless, all trainers and animal caretakers that we have spoken to about hand-reared sea lions report differences in the way these sea lions interact with people relative to sea lions born in the wild or raised in captivity by a female sea lion. This may also be true of harbor seals. For example, although captive, mature male harbor seals are highly vocal during the breeding season, they do not appear to incorporate human spoken words spontaneously into their repertoire. The mimicry of human speech by Hoover when he was mature may have been related to his early social ties with humans (Ralls et al., 1985).

Our direct experience comes from our interactions with Rio, now almost 6 years old, who we hand-reared as part of our imprinting study (see the preceding section) and have since incorporated into our experimental studies of complex learning and cognition. In practical terms this means that Rio has had extensive and continuous experience interacting with sea lions and a large number of people. Although Rio is more attentive to, and more interactive with, trainers than are sea lions reared by their biological mothers she has also been more difficult to handle. Her interactants are currently limited to a small number of persons with extensive field and/or captive experience of sea lion behavior. In all other circumstances, save one, she has been trained to "keep her distance."

The one exception is during play in the water. We regularly join Rio in her pool for play and social interaction, and she interacts nonaggressively even with complete strangers. Her play may get a little rough at times (inhibited biting or mouthing of hands and feet, clinging to swimmers), but she interacts without the aggressive vocalizations and open-mouth displays we see on land. The same social effect is observed in populations of wild sea lions. Groups of sea lions float together in close proximity, and the primary interactions are relaxed play nipping and chasing, nose-to-nose olfactory inspection, and gentle physical contact; there is little or no vocalization. The same group on land would be characterized by aggressive barking, pushing, biting, and jockeying for comfortable sleeping positions. The transition from land to water brings about a very rapid change in behavior from intense social competition to nonaggression and play. In some populations of California sea lions these groups, referred to as 'milling groups," become focal sites of sexual play (mock mounting, chasing) and copulation with territorial males (Health, 1985). In Steller sea lions these milling groups form at the edge of the rookery in the evening, and groups of about 3 to 10 females later split off and head to sea, apparently to feed (R. Gisiner, personal observations).

Out of the water, Rio has been harder to control and more aggressive than animals raised by their biological mothers. We attribute this difference to her social bond with humans, and not to any behavioral pathology brought about by the "abnormal" circumstances of her rearing. When frustrated by a difficult training task or made uncomfortable by the close proximity of a person or another sea lion, Rio is more likely to direct her aggression at a person than at other sea lions or inanimate objects. Under the same circumstances other sea lions might direct their aggression at another sea lion or an object near the person (a bucket, a training target, the edge of a door).

Another factor has added to Rio's tendency to engage humans in aggressive social interactions. When she has challenged or "attacked" some individuals they have flinched, fled, cried out, or given off other submissive/ subordinate signals. These types of interactions constitute the large majority of play and nonbreeding social interactions in juvenile and adult sea lions; two individuals posture or tussle, one breaks away, and the other immediately presses its advantage and chases the other. We noted earlier that sea lions in general are carnivores and have physically aggressive social interactions. Given these predispositions, it is perhaps not surprising that Rio has learned to take immediate advantage of these signs in her handlers to assert her social dominance over them. Rio, more than our other pinnipeds, tends to discriminate among individual humans and interacts with them differently. Once she has developed an ability to induce signs of fearfulness in a person, she tends to behave aggressively and dominantly and is therefore hard to control by normal food-reinforcement procedures. (In other words, social dominance sometimes outweighs food as a reinforcer.) From those individuals whom she has never successfully challenged, a word or a glance is usually sufficient to cut off aggressive vocalizations and posturing; from individuals that Rio has successfully displaced in the past, she requires prolonged and concerted efforts at control before she will "give in" to their commands. This can vary somewhat with her food motivation and the difficulty or aversiveness of the task she is being asked to perform.

Pups develop some social strategies while still under the "protection" of their mothers. It is possible, therefore, that offspring of dominant mothers take a more aggressive approach to social interactions than pups whose mothers could not protect them from the consequences of their early social interactions (in most of which they would be the smaller and therefore subordinate interactant). In Rio's case, her surrogate mothers were trainers with high social dominance who could and did protect her from the large sea lions. To some extent she is therefore a "spoiled brat," and she typically takes a very aggressive stance during her initial interactions with unfamiliar sea lions (and humans).

Conclusion

We have found that a hand-reared sea lion is capable of forming a strong, durable bond with its human caretaker. Such bonding depends on many of the same variables controlling bonding between a sea lion pup and its biological mother. We believe that an imprinted or hand-reared sea lion is more attentive to the signals of humans and more likely to respond to a human as a social "equal" than a sea lion reared by its mother either in captivity or in the wild. It is also possible that rearing by humans, who are

354

usually at the top of the dominance "hierarchy," tends to make hand-reared pups more socially aggressive than most pups raised by females of moderate to low dominance rank. This makes it extremely important that the human interacting with the imprinted sea lion understand sea lion social behavior and know the significance of sometimes subtle social cues. Since the penalties for misreading social cues in sea lions include dangerous physical contact with a large animal and injurious biting, sea lions socialized to humans should be handled carefully by experienced professionals. Under these circumstances one can enjoy the close social bonds that sea lions, normally reserve for one another (nose-to-nose contact, lying or sitting in physical contact, and play) without being aggressively confronted by a physical, and therefore social, dominant.

Acknowledgments

This research was supported by Contract N00014-85-K-0244 from the Office of Naval Research to Ronald J. Schusterman. We acknowledge the contributions of staff, volunteers, and trainers at Long Marine Lab, Marine World/Africa USA, Sea Life Park, and Marineland of the Pacific. The research we conducted could not have been done without the assistance of the sea lion caretakers. We especially thank the original "mother surrogates" Michelle Jeffries, Jenny Montague, Brad Andrews, and Marlee Breese.

References

Gisiner, R., & Schusterman, R. J. (1991). California sea lion pups play an active role in reunions with their mothers. *Animal Behaviour, 41,* 364–6.

Hanggi, E. B., & Schusterman, R. J. (1990). Kin recognition in captive California sea lions (*Zalophus californianus*). *Journal of Comparative Psychology, 104,* 368–72.

Heath, C. B. (1985). The effects of environment on the breeding system of the California sea lion (*Zalophus californianus*). Presented at the 6th Biennial Conference on the Biology of Marine Mammals, Vancouver.

Hogan, J. A., & Abel, E. L. (1971). Effects of social factors in response to unfamiliar environments in *Gallus gallus spadiceus. Animal Behaviour, 19,* 687–94.

Lorenz, K. A. (1981). *The Foundations of Ethology.* Springer, Verlag, New York.

McFarland, X. (1985). *Animal Behaviour.* Benjamin/Cummings, Menlo Park, CA.

Miller, D. J., Herder, M. J., & Scholl., J. P. (1983). *California Marine Mammal–Fishery Interaction Study,* 1979–1981. Southwest Fisheries Center, Administrative Report LJ-83-13C. National Marine Fisheries Service, La Jolla, CA.

Nolte, C. (1990). Bold, sexy sea lions delighting visitors. *San Francisco Chronicle,* Jan. 26.

Peterson, R. S., & Bartholomew, G. A. (1967). *The Natural History and Behavior of the California Sea Lion.* American Society of Mammalogists Special Publication No. 7, 1–9.

Rajecki, D. W., Lamb, M. E., & Obmascher, P. (1978). Toward a general theory of infantile attachment: A comparative review of aspects of the social bond. *Behavioral and Brain Sciences*, *3*, 414–64.

Ralls, K., Fiorelli, X. & Gish, S. (1985). Vocalizations and vocal mimicry in captive harbor seals, *Phoca vitulina. Canadian Journal of Zoology*, *63*, 1050–6.

Schusterman, R. J. (1981). Steller sea lion, *Eumetopias jubatus*. In *Handbook of Marine Mammals*, S. H. Ridgway & R. J. Harrison (Eds.). Academic Press, New York.

(1985). Imprinting in California sea lions. Presented at the 6th Biennial Conference on the Biology of Marine Mammals, Vancouver.

(1986). California sea lions imprint on human voices. Presented at the annual meeting of the International Marine Animals Trainers Association (IMATA), Vancouver,

Schusterman, R. J., & Dawson, R. G. (1968). Barking, dominance, and territoriality in male sea lions. *Science*, *160*, 434–6.

Schusterman, R. J., & Gisiner, R. (1988). Artificial language comprehension in dolphins and sea lions: The essential cognitive skills. *Psychological Record*, *38*, 311–48.

Schusterman, R. J., & Krieger, K. (1984). California sea lions are capable of semantic comprehension. *Psychological Record*, *34*, 3–23.

(1986). Artificial language comprehension and size transposition by a California sea lion (*Zalophus californianus*). *Journal of Comparative Psychology*, *100*, 348–55.

Trillmich, F. (1981). Mutual mother–pup recognition in Galagapos fur seals and sea lions: Cues used and functional significance. *Behaviour*, *78*, 21–42.

—22—

Humans as predators: observational studies and the risk of pseudohabituation

Nancy G. Caine

Editors' introduction

Caine's chapter challenges the widely held view that all aspects of an animal's behavior can be habituated to the presence of human observers. Like Hemsworth et al. and Duncan, Caine suggests that defensive aspects of scientist–animal interactions may play a large role in determining behavior. Her data suggest that confidence about uncontaminated behavioral observations on fearless animals may be unwarranted.

Caine's caveat is particularly troublesome. If true, it suggests that the human observer may be a variable in many situations where his or her presence was presumed to be neutral. Caine argues that only by knowing the full range of a species' defensive behavior can we be certain that our presence has evoked no form of defensive reaction.

The volume of research on antipredator behavior undoubtedly reflects the strength of predation-related selection pressures and their consequent adaptations. For all animals who serve as prey, antipredator tactics pervade daily life in almost innumerable ways. The point I hope to make in this chapter is that subtle, unrecognized forms of antipredator adaptations, emitted in response to human experimenters, can introduce a potentially serious bias into observational studies of animal behavior.

The admonition that I develop in the following pages is directed primarily to those who study behaviors *other* than antipredator adaptations (e.g., sexual behavior, foraging, and dominance relations). However, research on avian nest defense has identified a related concern that sets the stage for my own argument. This particular research stems from well-documented accounts of threats, lunges, and/or mobbing when a potential predator comes near the nest. These attacks can be quite intense and prolonged, and seem to vary according to type of predator and degree of risk (e.g., Buitron, 1983).

To study nest defense the experimenter must, of course, wait until a predator makes an approach. If these attacks are infrequent, a sufficient amount of data may never be collected. A common solution is to erect a model of the predator. This tactic is sometimes satisfactory, but models are often unwieldy, especially in the field, and they are usually stationary, which may reduce their credibility as predators from the prey's point of view. Some investigators avoid problems associated with models by capital-izing on the fact that many animals react to human observers with anti-predator behaviors. The observer can therefore act as experimenter and stimulus simultaneously.

According to Knight and Temple (1986), in only one of the published studies of nest defense by altricial birds (Buitron, 1983) are the data based on interactions with free-ranging predators. Our understanding of nest defense therefore relies almost exclusively on investigations that have employed either models or humans as mock predators. These studies have produced at least one very consistent and much touted result: The intensity of avian nest defense increases as the nesting cycle progresses. Parental investment theory is readily invoked to explain this phenomenon: As nestlings become older they represent greater cumulative investment by the parents, and therefore the parents are more eager to defend that investment (e.g., Barash, 1975).

Knight and Temple (1986) challenge this interpretation of the data with a disconcerting hypothesis: The positive correlation between intensity of nest defense and offspring maturation may be a by-product of a methodological bias that is common to most studies of avian nest defense. Specifically, the authors propose that "when an observer repeatedly visits or brings a potential predator to a nest and records the parent birds' responses, the nest defense behavior is gradually modified by positive reinforcement and loss of fear" (p. 318). That is, birds become more bold as they learn that there is no risk to behaving aggressively. Knight and Temple tested this hypothesis in American robins (*Turdus migratorius*) and red-winged blackbirds (*Agelaius phoeniceus*). Some of the subjects' nests were visited by a human observer every 3 days from incubation until fledging. Other nests were visited only once during this time. Multiply visited nests were associated with both greater intensity of defensive behaviors (i.e., more dives, strikes, close approaches, and calls) and greater participation in the attacks. Among nests visited only once, parents showed no more anti-predator behavior late in the nesting cycle than did parents who were tested early in the nesting cycle. Knight and Temple conclude that "increased intensity of responses to predators might be the result of repeated presentation and withdrawals of potential predators" (p. 322). Westmoreland (1989) found that multiple visits had a much smaller effect on nest defense in mourning doves (*Zenaida macroura*), but Westmoreland agrees with Knight and Temple that the behavior of human experimenters may introduce unsuspected bias into observational studies when those human are perceived as potential predators by the animals under investigation.

My own recent research involves the antipredator behavior of tamarins. Tamarins are members of the New World primate family Callitrichidae (tamarins and marmosets). They live in cohesive groups of 2 to 10 individuals who spend most of the day traveling and foraging in the midcanopy of tropical or subtropical forests. Small in size and sympatric with a variety of terrestrial and aerial predators, callitrichids probably face heavy predation pressures (Goldizen, 1987). In response to these pressures, tamarins have evolved a variety of antipredator behaviors. For example, at least three different alarm calls, given in response to stuffed owls, unfamiliar humans, and objects moved overhead, have been identified in captive saddle-back tamarins (*Saguinus fuscicollis*). Tamarins also mob predators (Bartecki & Heymann, 1987) and may display stereotypical patterns of escape maneuvers (S. Capuano, personal communication).

My reseach with captive red-bellied tamarins (*S. labiatus*) concentrates on those aspects of tamarins' antipredator behavior that promote detection and/or avoidance of predators, as opposed to those behaviors that are emitted in response to the presence of a predator. These include visual and olfactory vigilance (Caine & Marra, 1988; Caine & Weldon, 1989), use of sentinels (Zullo & Caine, 1987), and coordinated responses to danger (Caine, 1986). Many of these behaviors are particularly evident during the process of retirement to the sleeping site each evening. For example, tamarins become quiet and vigilant in the minutes preceding entrance to the site (Caine, 1987; Dawson, 1979), and rates of contact calling (a way of keeping track of one another's whereabouts) increase (Caine & Stevens, 1990). One member of the group sometimes remains out of the nest box (the laboratory equivalent of a tree hole or tangle of vines) for several minutes, scanning intently (Figure 22.1), while its group mates settle down and become quiet, at which point the "sentinel" then joins the others in the box (Zullo & Caine, 1987). These and other behaviors associated with retirement suggest that the tamarins feel particularly vulnerable at this time, and that they respond by being cautious.

Figure 22.1. Captive tamarin monkey.

The advantages of increased vigilance before entering the sleeping site are obvious if one considers that a sleeping group of tamarins is very vulnerable to nocturnal predators. The fact that at least some species of tamarins become almost torpid at night (Dawson, 1979) may render the monkeys unable to make a quick escape or muster a defensive maneuver should, in fact, their sleeping site be discovered by a predator. Insofar as callitrichids tend to enter the sleeping site at early dusk and exit the site well after dawn (Pook & Pook, 1981), the avoidance of detection by both diurnal and nocturnal predators would be facilitated by disguising the location of the site and by entering that site unnoticed.

I recently conducted an investigation of one aspect of retirement behavior that probably has an important antipredator function. As I describe later, the data from that study were consistent both with my specific hypotheses about sleeping sites and with my general argument about the importance of predation pressures in the natural history of tamarins. However, perhaps the most important (and fortuitous) contribution that the study makes is related not to tamarins' antipredator adaptations per se, but to the unanticipated and unrecognized effect that these behaviors may have on other sorts of behavior.

The investigation in question (Caine, 1990) was designed to quantify the willingness of captive red-bellied tamarins to enter their nest boxes while being observed by a human. I predicted that they would be reluctant to do so because they prefer to keep their nest sites concealed from predators, as discussed previously. The subjects were two groups of red-bellied tamarins living in large indoor enclosures. Group 1 was a family of five; Group 2 was a mated pair. The enclosures were fitted with ropes, perches, branches, and a nest box. There was no visual or olfactory contact between the groups, which were housed in different rooms.

Forty-five minutes before the usual time of retirement (about 1400 hr) I initiated one of three conditions: (1) observer present and facing the tamarins (I sat quietly in the room, watching the animals); (2) observer present but not facing the tamarins (again I sat quietly in the room but my back faced the monkeys and a video camera recorded the tamarins' behavior); (3) camera only (in this case a video camera filmed the tamarins but no observer was present). I recorded the exact time when all members of the group had entered the box. Group 1 was observed for 10 days in each of the three conditions. Group 2 was observed for 7 days in each of the three conditions. The order of the three conditions was randomly selected with the stipulation that all three conditions were represented once during each week of data collection. The study lasted for 5 months.

When being watched, the tamarins in the first group entered their nest box about 14 minutes later than they did with no observer in the room; the delay was about 18 minutes in the observer-not-facing condition. The two observer-present conditions did not differ significantly from each other. The same was true for the second group: The average time of entry to the nest box was later by about 19 minutes in the observer-facing condition and by 29 minutes in the observer-not-facing condition. Entry times showed no significant pattern of change over trials.

I then asked if the same effect would appear if the observer was unfamiliar to the tamarins. I repeated the study using six undergraduate students who had no prior exposure to these animals. Six trials each of the observer-facing and camera-only conditions were carried out with Group 1. The tamarins entered the nest box significantly later (about 14 minutes) when being watched than they did when not being watched.

These data lend support to the notion that tamarins, given their susceptibility to predators, must make careful choices about when and perhaps where to roost each night. Entering a nest site while being watched by a potential predator would undoubtedly be a risky thing to do. Indeed, my results suggest that it is not merely being watched that is disconcerting; when I sat in the room with my back to the tamarins, they delayed their entry into the box as much as when I was facing them.

The aspect of this study that called my attention to an animal–human interaction bias was the fact that I, as a very familiar stimulus, elicited the same response as did the unfamiliar observers. This led me to consider the potentially troubling methodological implication of my study: A "habituated" group of animals may continue to respond defensively in ways that are unrecognized by the investigator. The risk increases if the researcher is nonchalant about her or his criteria for habituation, which, I fear, is often the case. Statements such as "The study group was fully habituated to our presence" are commonly offered and accepted as a suitable basis for dismissing the possibility that the observers themselves were variables in the research. I myself have operated on this assumption many times, but I am no longer sure that I have reason to be cavalier.

It is disturbingly easy to contrive empirical scenarios in which unrecognized antipredator behavior could affect the data and their interpretation. First, most researchers are unfamiliar with a species' full repertoire of antipredator behavior, even if that species has been under study by the investigator for some time. Mobbing and flight and alarm calls are unmistakable, but subtler forms of defense such as those involved in surveillance and vigilance may never attract the attention of even the most diligent observer. Second, an individual or group of individuals may exhibit no particular antipredator behavior in the presence of an observer, but other behaviors may appear in altered form or reduced frequency because the individual(s) recognize the observer as a mild but real danger. For example, behavior that makes animals vulnerable to predatory attacks (e.g., grooming or copulation) may be displayed much less often, or for shorter periods of time, when an observer is present. A study of, for instance, behavioral time budgets may fail to represent accurately the proportion of time that individuals devote to various behaviors because, unbeknownst to the observer, the animals inhibit certain behaviors whenever a human is present. With regard to the sleeping habits of tamarins, an investigator whose interest and expertise were not in antipredator tactics, but in circadian rhythms, might overestimate the length of the tamarins' day based on his or her observations of retirement times. If the investigator used a blind or collected his or her data by video camera, the results could be different (see Candland, Dresdale, Leiphart, & Johnson, 1972).

Just as some kinds of behavior would likely be more affected by subtle antipredator behavior than others, some kinds of species would also be more affected than others (see Westmoreland, 1989). An additional bias enters into the situation when an *a priori* assumption has been made by the investigator that her or his study species is not very vulnerable to predation and hence is unlikely to show much in the way of antipredator behavior. I would argue that this is true in my own field of primatology, where a common assumption has been that predation pressures have played a minimal role in the evolution of most primate species (e.g., Wrangham, 1980). There are compelling reasons to disagree with this assumption (e.g., van Schaik, 1983), at least for some primate taxa, and consequently there is reason for primatologists to be more circumspect about the habituation of study groups.

We observers of behavior need to consider the possibility that subtle and unrecognized forms of antipredator behavior could influence the behavior of our research animals for the duration of a study. Even tenuous evidence of our presence, be it visual, auditory, or olfactory, may be enough to elicit defensive behaviors that compromise the validity of the data being collected. For many of us, however, there is no choice but to make observations without the aid of even a simple blind. Whereas observer presence may not create serious bias for any particular study, every effort should be made to reduce the potential for this sort of interference. The integrity of the data may depend in part on the willingness to concede that humans are probably not perceived as benign elements in the environments of many or even most species of animals.

References

Barash, D. P. (1975). Evolutionary aspects of parental behavior: Distraction behavior of the alpine accentor. *Wilson Bulletin 87*: 367–73.

Bartecki, U., & Heymann, E. W. (1987). Field observation of snake-mobbing in a group of saddle-back tamarins, *Saguinus fuscicollis nigrifrons*. *Folia Primatologica 48*: 199–202.

Buitron, D. (1983). Variability in the responses of black-billed magpies to natural predators. *Behaviour 78*: 209–36.

Caine, N. G. (1986). Visual monitoring of threatening objects by captive tamarins (*Saguinus labiatus*). *American Journal of Primatology 10*: 1–8.

(1987). Vigilance, vocalizations, and cryptic behavior at retirement in captive groups of red-bellied tamarins (*Saguinus labiatus*). *American Journal of Primatology 12*: 241–50.

(1990). Unrecognized anti-predator behaviour can bias observational data. *Animal Behaviour 39*: 195–7.

Caine, N. G., & Marra, S. L. (1988). Vigilance and social organization in two species of primates. *Animal Behaviour 36*: 897–904.

Caine, N. G., & Stevens, C. (1990). Evidence for a "monitoring call" in red-bellied tamarins. *American Journal of Primatology 22*: 251–62.

Caine, N. G., & Weldon, P. J. (1989). Responses by red-bellied tamarins to fecal scents of predatory and non-predatory neotropical mammals. *Biotropica 21*: 186–9.

Candland, D. K., Dresdale, L., Leiphart, J., & Johnson, C. (1972). Videotape as a replacement for the human observer in studies of nonhuman primate behavior. *Behavior Research Methods and Instrumentation 4*: 24–6.

Dawson, G. A. (1979). The use of time and space by the Panamanian tamarin, *Saguinus oedipus*. *Folia Primatologica 31*, 253–84.

Goldizen, A. W. (1987). Tamarins and marmosets: Communal care of offspring. In B. Smuts, D. Cheney, R. Seyfarth, R. Wrangham, & T. Struhsaker, Eds., *Primate Societies*. Chicago: University of Chicago Press, pp. 34–43.

Knight, R. L., & Temple, S. A. (1986). Why does intensity of avian nest defense increase during the nesting cycle? *Auk 103*: 318–27.

Pook, A. G., & Pook, G. (1981). A field study of the socio-ecology of the Goeldis monkey (*Callimico goeldii*) in northern Bolivia. *Folia Primatologica 35*: 288–312.

van Schaik, C. P. (1983). Why are diurnal primates living in groups? *Behaviour 87*: 120–43.

Westmoreland, D. (1989). Offspring age and nest defence in mourning doves: A test of two hypotheses. *Animal Behaviour 38*: 1062–6.

Wrangham, R. W. (1980). An ecological model of female primate groups. *Behaviour 75*: 262–300.

—23—

Human–bear bonding in research on black bear behavior

Gordon M. Burghardt

Editors' introduction

Burghardt offers an unusually moving account of the reciprocal effects of a human–bear bond. Using both historical material and examples from his own research, he paints a vivid picture of how such bonding alters the behavior of both human and animal in profound ways.

For better or worse, Burghardt is the only contributor to draw an explicit parallel between the scientist–animal bond and the relationship between human parents and children. Like Fentress, Burghardt notes that earlier forms of scientist–animal contact (e.g., dependency) often recur during periods of stress in the animal's life. Burghardt also notes a theme common to many contributors: Without the special bond between scientist and animal, a given test procedure could not have been carried out.

Whether Burghardt's conclusions stem from the special intensity of human–bear interactions or are more general in nature, Burghardt argues for a reduction in the strictures regarding clinical objectivity and anthropomorphism. He concludes that "the separation of both experimenter and emotion from the training procedures in scientific studies with bears is just not possible." No doubt a similar case might be made for research involving chimpanzees, monkeys, dogs, and wolves. In fact, the question of where or whether to draw such a line is quite legitimate. This is obviously an issue of major importance that goes to the core of how scientists are trained. Burghardt has already taken a stand with his doctrine of "critical anthropomorphism." How the rest of us respond to this challenge, be it in a more or less radical manner, remains to be seen.

Bears share with snakes the ability to beget awe, wonder, fear, and worship (Shepard & Sanders, 1985). Their large size, widespread distribution in temperate areas, humanoid appearance, intelligence, omnivorous food habits (often competitive with those of people), ecological dominance, and potential for causing physical harm are all factors that have led to ambivalent attitudes among people about how bears should be treated and valued (Pelton, Scott, & Burghardt, 1976; Burghardt & Herzog, 1989).

In this chapter I will briefly convey some accounts of individual bears in which a relationship with human beings was salient, although not necessarily for scientific endeavor. Then I will summarize the relationship my students and I had with two black bears (*Ursus americanus*) we studied. Fine lines separating subject, pet, and companion may often be impossible, not just difficult, to draw. A review of the growing ethological literature on bears will not be attempted.

Wild bears

Beginning around 1900, numerous books by hunters and naturalists about North American bears, especially the grizzly/brown bears (*Ursus arctos*), were published. Many contained descriptions of remarkable individual bears, who, even if quarry, indeed renegades or killers that needed extirpation, were seen as near equals in intelligence, bravery, and determination, warranting considerable admiration and respect. Naturalists such as Seton (1900, 1904) wrote composite portraits of the lives of individual bears based on then-current natural history knowledge. These were often presented through the *Umwelt* of the bear itself and included a good deal of mentalistic inference that was soon to become highly unfashionable in scientific circles.

One of the most objective books of the period was written by W. H. Wright (1909/1977). A famous and proficient bear hunter, Wright traded in his rifle for a camera and became a devoted bear naturalist whose observations have largely stood the test of time (Craighead, 1977). The following quote is an eloquent representation of the attitude that thoughtful hunters often developed:

In the beginning, I studied the grizzly in order to hunt him. I marked his haunts and his habits, I took notice of his likes and dislikes; I learned his indifferences and his fears; I spied upon the perfection of his senses and the limitations of his instincts,

366

simply that I might the better slay him. For many a year, and in many a fastness of the hills, I pitted my shrewdness against his, and my wariness against his, and my endurance against his; and many a time I came out the winner in the game, and many a time I owned myself the loser. And then at last my interest in my opponent overcame my interest in the game. I had studied the grizzly to hunt him. I came to hunt him in order to study him. I laid aside my rifle. It is twelve years since I have killed a grizzly. Yet in all those years there is not one but what I have spent some months in his company. (p. 11)

Hunters sometimes develop at least temporary bonds with individual grizzly bears and vice versa. Two stories, one of a rather grisly nature, indicate the stakes involved. Theodore Roosevelt – historian, naturalist, hunter, conservationist, politician, and president of the United States – wrote many books and articles concerning bears; most of his bear writings have been collected (Roosevelt, 1893/1983). In an article written in 1893 Roosevelt recounts a story told by an old hunter. He and a companion set out to follow a grizzly that had killed one of their horses. The bear stayed ahead of them and was clearly annoyed by their attention, as suggested by redirected aggression when the bear stopped to tear down saplings and claw fallen logs. The two hunters decided to split up. When his companion did not return to camp, the concerned hunter looked for him without success, finally returning to the spot where the bear had killed the horse. He followed his friend's trail, marked by distinctive boot prints, finally losing it, after 4 miles, on rocky ground. At this time he switched to tracing the bear's trail and, to his horror, found that the bear's footprints were soon on top of his companion's. Near a thicket by a streamed the bear had cut through on an oblique angle and ambushed the man. The torn and mutilated body had not been eaten. Thus, the hunter had become the quarry, killed not for food but, according to the companion, out of malice.

While such tales may be dismissed as both anecdotal and exaggerated, their frequency indicates that nettled bears do some intriguing things on a deceptive par with anecdotal reports on primates (Whiten & Byrne, 1988). Enos Mills was, like Wright, a famed hunter and guide who wrote a book on grizzly bears, and became a noted conservationist. His account of trailing a male bear in what is now Rocky Mountain National Park (reprinted in Haynes & Haynes, 1966) begins by noting the frequent reports of injured bears turning on hunters and trailing or ambushing them. He then recounts how just trailing a bear can lead to similar results:

Most animals realize that they leave a scent which enables other animals to follow them, but the grizzly is the only animal that I know who appears to be fully aware that he is leaving telltale tracks. He will make unthought-of turns and doublings to walk where his tracks will not show, and also tramples about to leave a confusion of tracks where they do show. (E. Mills in Haynes & Haynes, p. 188)

Mills had been following Old Timberline, recognizable by two missing toes, surreptitiously watching him forage and coast down snowbanks. The bear ran off after Mills rolled a rock down the mountain to within a few yards of him. The next morning Mills began trailing the bear for many miles; eventually

he discovered that I was following him. He may have seen or scented me. Anyway, instead of coming directly back and thus exposing himself, he had very nearly carried out his well-planned surprise when I discovered him. I found out afterwards that he had left his trail far ahead, turning and walking back in his own footprints for a distance, and trampling this stretch a number of times, and that he had then leaped into scrubby timber and made off on the side where his tracks did not show in passing along the trampled trail. He had confused his trail where he started to circle back, so as not to be noticed, and slipped in around me.

But after discovering the grizzly on my trail I went slowly along as though I was unaware of his near presence, turning in screened places to look back. He followed within three hundred feet of me. When I stopped he stopped. He occasionally watched me from behind bushes, a tree, or a boulder. It gave me a strange feeling to have this big beast following and watching me so closely and cautiously. But I was not alarmed. (p. 191)

Mills proceeds to tell how he, when out of sight after crossing a ridge, ran back nearly a mile to get behind the bear only to discover that the bear was waiting for him behind a boulder!

What were his intentions? Did he intend to assault me, or was he overcome with curiosity because of my unusual actions and trying to discover what they were all about? I do not know. I concluded it best not to follow him farther, nor did I wish to travel that night with this crafty, soft-footed fellow in the woods. Going a short distance down among the trees I built a rousing fire. Between it and a cliff I spent the night, satisfied that I had had enough adventure for one evening. (p. 192)

Stories about black bears do not suggest the trailing, aggression, and malicious cunning directed at people found in accounts of grizzly/brown bears. Wright (1910) followed up his writings on the grizzly with a book on the black bear. To Wright the black bear is really quite inoffensive, sort of a "happy hooligan." Powerful and potentially dangerous, it is altogether a far lesser hazard to careless and thoughtless human beings than the grizzly. Wright perhaps tries too hard to make the black bear an animal less serious about life than the grizzly and makes numerous comments we would take issue with today. Yet, until recently, even the black bear, much more numerous than any other bear in the world, was not well known in the wild.

Field biologists working with radiomarked wild bears are obtaining considerable data on black bear movements (e.g., Pelton & Burghardt, 1976) and even habituating them to the presence of humans and thus allowing observation of natural foraging, maternal care, and so on (Rogers & Wilker, 1990). This is not so surprising, since black bears, much more than browns, are likely to develop panhandling tendencies in national parks and to exist in uneasy coexistence with people. Jane Tate (1983) studied wild black bears and their interactions with tourists in the Great Smoky Mountains National Park. She reported that bears developed distinctive and often successful methods for separating tourists from their food, usually voluntarily (Tate, 1983). Many individual black bears had their own personalities, developed relationships with park personnel and researchers, and had predictable tendencies to use agonistic displays.

Captive bears

The use of bears in circus acts is well known, and food reward is frequently used during acts. Hediger (1955) discusses the variety of training procedures that may be used. Having watched many trained brown and Himalayan bears playing hockey, dancing, tight-rope walking and performing other acrobatics, driving motorcycles and cars, and so on, it is clear to me that their humanoid appearance and abilities are quite remarkable. Certainly a close relationship with trainers is necessary. Whether for historical reasons, their smaller size, or species differences in trainability, black bears are not often used, in spite of their reduced aggressiveness.

Bears are common inhabitants of zoos, and general observations are available (e.g., Hediger, 1950), although little formal research has been carried out. Outside of circuses and zoos some people have raised captive bears, especially black bears, and made observations of their behavior that would otherwise be difficult or impossible to obtain. Most of these reports are without serious scientific intent, but do not differ, except in scale, from comparable work undertaken with, for example, chimpanzees (e.g., Yerkes, 1925).

Captive bears, because of their large size and ability to maul human beings quickly, must be adapted to people so that experimental work can be carried out safely. Thus, the separation of both experimenter and emotion from the training procedures in scientific studies with bears is just not possible. Such procedures share similarities with much circus performance training (Hediger, 1955).

The first systematic study to focus on perception in bears was that of Kuckuk (1937). Two young, socialized European brown bears recognized Kuckuk at 30 m, heard sounds at 150 m that a human could detect only at

80 m, and learned a variety of tasks involving food, including locating hidden food by the use of odors. Like young human children, the bears ran away in fright when a familiar person wore a mask, but if the person wore a bear head instead, he or she was furiously attacked.

Turning to black bears, it is interesting that half of Wright's (1910) book on black bears is devoted to his experience with Ben, an orphaned bear cub he obtained at 4 to 5 months of age after the mother was shot. It is an amusing story, enlivened with photographs and interesting anecdotes that largely confirm later observations. The bear was named after Ben Franklin, the famous pet bear of James Capen Adams, better known as Grizzly Adams. The play behavior of Ben with another cub, a dog (Jim), and humans is described, as are aspects of communication. Without previous experience, the cub apparently knew how to forage and identify the herbs, roots, and fruits that Wright knew wild bears relied on. In spite of Ben's controversial reputation with townspeople, Wright was his close and affectionate companion:

But Ben was now grown so large that none but myself cared to wait on him; and when, the next spring, I found that I was going to be away in the mountains all summer, I began looking about for some way of getting him a good home. Nothing in the world would have induced me to have him killed, and I did not like to turn him loose in the hills for some trapper to catch or poison. Moreover I doubted his ability, after so sheltered a life, to shift for himself in the wilderness. But this was a problem in which the "don'ts" were more easily discovered than the "do's." (p. 49)

Ben ended up in a traveling circus after Wright received promises of kind treatment. The parting, for Wright, involved "genuine heartache."

A report on Jimmie, a Canadian black bear raised by a writer of animal books (Baynes, 1929), was less oriented to natural history. Amid many observations of locomotion, sensory abilities, feeding, sociality, and play (a favorite partner was Bingo, a dog), the love and affection, the bond between bear and human, comes through. Baynes emphasized the need for objectivity, not inferring mental states and abilities based on human psychology. Yet he found this difficult. As with Wright, the bear eventually grew too large, its play too rough, and a home at the New York Zoological Park was found. The final hours were emotionally wrenching for Baynes:

But that afternoon when [Jimmie] walked out on to the piazza, stood up on one of the posts, and with a strangely sad expression on his face looked away across those blue hills and valleys which he was never to see again, there came a chokey feeling in our throats. And when a little later he picked up a much beloved rag doll which Mrs. Baynes had made for him, sat down with it in his lap, licked its face all over for the last time and then carried it off to bed with him, we couldn't help feeling very sorry that little bears grow up into big ones. Of course our intelligence told us that he had no idea that he was going away, that his standing in the piazza post that particular afternoon was merely an interesting coincidence, and that the sadness of his expression was probably in our own imagination. (p. 137–8)

Two months later when Baynes visited the zoo and called Jimmie's name, the bear was very responsive and grabbed Baynes's hand through the bars in his paws, vocalized repeatedly, and seemed both excited to see Baynes and reluctant to let him leave. A year later Jimmie was large and healthy and, although he approached Baynes when called, seemed rather distant and puzzled.

Leyhausen (1948) was the first ethologist to describe the behavior of a black bear cub. He was in a World War II prisoner of war camp in Canada, and the inmates had a locally captured bear cub, Nelly, for a pet. He described much of her behavioral repertoire in qualitative terms, but with an ethologically trained approach (feeding, elimination, agonistic and fear responses, hibernation), described play and chase games, especially with a favored dog, and discussed the nature of bear–human relationships. Here, too, the outcome for the bear was problematic. After hibernation, she was eventually released (after Leyhausen himself) when the danger of being shot by hunters had presumably passed.

Finally, Robert Leslie (1971) reports his experiences in Canada in rearing three orphaned black bear cubs for three years. Although the report is highly anecdotal, Leslie did watch conspecifics in a natural environment, even observing his bears being courted and interacting in various ways with wild bears. Leslie's goal was to raise the bears, equip them for survival, and release them. But complications caused by, among others, hunters and developers, soured the plan. But most important was the bond:

With an almost British lack of sentimentality I had felt sorry for three cute little teddy-bear orphans, offered them temporary food and protection, trained them to meet their own responsibilities. Then, out of the finest kind of obedience, respect, trust, compatibility, and affection, there had grown a depth of mutual friendship far beyond anything I could have believed possible at the time the cubs arrived. (p. 197)

Finally, the last bear, Scratch, at 400 pounds, was too much to take care of and had to be released. A friend was to take the bear 200 miles by launch to a distant lake. Scratch had a seat in the bow.

Avoiding my eyes as he [Mark] passed the canoe, he handed me a heavy willow switch. The bear looked at me as I began to paddle south. In a bound he was over the side of Mark's boat, bellowing, and swimming toward the canoe.

Tears streaming down my cheeks, I was forced to leave the thought of betrayal in the mind of the third bear. With the switch I struck him again and again across the bridge of the nose – the nose that had so often nudged me with affection and admiration. At length he turned, disbelieving, and slowly swam to Mark, who helped him aboard the launch. I heard my friend gun his engine as he turned the prow north. I paddled south with every fiber of my body. Had I ever looked back, I would never have left the northland. (p. 198)

In spite of their informal nature and necessary reliance on only one or a few animals, these unselected stories have a compelling commonality. All authors describe the playfulness and curiosity of the cubs that extended well beyond the early months of maternal dependence (note that three extensively discuss relations with a specific dog). All describe the affectionate, even trusting, behavior that developed between human and bear, especially with the main care giver. And all came to the same realization that black bears grow large and, unlike dogs, are ultimately incompatible as human companions. Yet the parting scenes I have quoted, literary license aside, show a truly emotional attachment of the human to the animal that is as intense as any I have come across. All four of these hand-rearing experiences, by the way, seem to have taken place without knowledge of the others, adding to their validity.

One can, of course, adduce reasons for these people's attachments to bears, such as a childhood fascination with teddy bears, the bipedal abilities of bears, and general anthropomorphic considerations (cf. Lawrence, 1990). But these are not enough. Although their backgrounds differ, all four writers were adult men with considerable insight into the dangers of anthropomorphism and sentimentality concerning animals, but were nonetheless deeply affected. Certainly all voluntarily made a commitment in time, energy, and devotion that became a major focus of their lives for some years. Also, in all cases the bears' entrance, as orphans, into these persons' lives, was unplanned, indeed serendipitous.

I had no inkling of any of this when I too became involved in raising orphaned black bear cubs. But my very ignorance eventually convinced me that, with black bears, the bond is both inevitable and enduring.

Black bears in the southern Appalachian Mountains

Shortly after I moved to the University of Tennessee in 1968, a public controversy developed in the nearby Great Smoky Mountains National Park. Black bears were becoming emboldened panhandlers in campgrounds, and especially along roads, where "bear jams" could cause considerable delays. As the number of tourists increased to millions a year at this most heavily visited national park, traffic delays, property damage, and personal injuries were increasing and a bewildered park administration was desperately looking for solutions. A several-point plan, developed without the benefit of much research, was publicized. One idea was the use of bearproof garbage cans (eventually developed and now in widespread and effective use). Another, which concerned me, was to instill a "natural fear of humans" in bears by having park rangers smack offending bears with baseball bats. This appeared problematic both scientifically and ethically. After the park superintendent graciously visited the university to give a joint zoology–psychology departmental seminar and subsequent extensive consultations with the National Parks Administration, a research plan developed.

A colleague in wildlife research, Mike Pelton, had just arrived at the University of Tennessee, and we decided that he and his students would concentrate on ecological aspects (food habits, demography, movements) and my emphasis would be on behavior. The relationship of knowledge about black bears, and attitudes toward them, to visitor demographic characteristics was documented, along with preferred responses to intrusive bears (Burghardt, Hietala, & Pelton, 1972).

Our behavior work began with observations of bear–human interactions at campgrounds and picnic areas, but in April 1970 our plans shifted when a park ranger found two orphaned bear cubs about 10 weeks old. The bear project was contacted, and we decided to keep them for a while to observe and study their behavior. The two female cubs (named Kit and Kate) were kept temporarily by a wildlife graduate student working on the bear project while I had a playpen-like enclosure built. In May the cubs were moved to my house, where we kept them in our dining room (when not at large). That summer they were moved to a 18.3 m × 18.3 m chain link enclosure and observation tower built in the Tremont area of the national park (see Jordan & Burghardt, 1986). They were studied in this enclosure for the next 4 years.

The research

During the course of this project, a diversion from my devotion to reptile behavior, several students participated in the maintenance and study of the animals. My graduate student Ellis Bacon moved to a trailer in the park, 40 miles from campus, commuting for most of his graduate education. Two dissertations (Bacon, 1973; Pruitt, 1974) were based almost exclusively on data gathered from Kit and Kate, and another dissertation (Jordan, 1979) and master's thesis (Ludlow, 1974) utilized these two bears extensively. Our studies necessitated a close working relationship with the bears to accomplish careful physical and behavioral measurements and experimental manipulation. An overview of our work has been published

Figure 23.1. Kate sucking on Kit's ear after bottle feeding had been discontinued.

(Burghardt, 1975), as has a general review of communication processes (Pruitt & Burghardt, 1977).

The primary care givers for the bears were initially Lori Burghardt and myself and later Ellis Bacon and Cheryl Pruitt. Robert Jordan and Jeanne Ludlow also cared for the animals and worked with them extensively; joining the project later, they did not develop the same confidence and rapport with the bears that the others did. It is probable that intensive interaction with young bears is necessary for maximal bonding by both bear and human. In any event, Cheryl and especially Ellis worked with the bears inside the enclosure, while others observed from outside and went inside primarily to feed, water, and clean.

The work we carried out can be divided into three areas: maintenance, descriptive, and experimental. The first was a constant concern, especially when the bears were young. Bottles had to be prepared and given at least every 4 hours day and night (Figure 23.1). Diets had to be researched, cleanliness maintained, records kept. The first months in captivity, including the period in my house, were most exciting, and the developmental changes in food habits, size, and behavior were dramatic (Burghardt & Burghardt, 1972). The onset of bipedalism much earlier than recorded previously and the ontogeny of play, aggression, tree climbing, and other basic responses were all noted (see Figure 23.2).

This early phase of our project was devoted largely to establishing a

Figure 23.2. The author and proverbial "dancing bear" (*Ursus americanus*).

routine from which further work could be accomplished. From the very beginning the bears were totally dependent on us for food and shelter; however, they were also very rapidly developing individual personalities and a host of aggravating behaviors to test their care givers.

Ellis took on primary responsibility for the bears when they were about 6 months of age and about 10 kg in weight. He quickly realized that although little bears are extremely cute they were also (1) highly self-centered, particularly concerning food; (2) very aggressive, even in play, covering his legs and arms, as mine had been, with bites, scratches, and bruises; (3) incredibly strong; and (4) extremely clever and curious. The last caused an amazing and amusing array of problems. For a few months there was a true test of wills pertaining to diet (what the cubs preferred vs. what they should eat) and their containment, perhaps related to the shift from one "parent" to another. Ellis developed a true admiration and respect for the animals (Figure 23.3). After about 2 months, he noted a curious and welcome phenomenon. They acted as if they discovered he was different from a bear: There was a rapid decline of inadvertent scratching and biting, and they distinguished clothing (which was still fair game) from flesh. From then on, with the exception of occasional dominance testing, Ellis considered himself physically safe in their presence. In contrast, the cubs were still very rough with each other in play or fights. In retrospect, this was the beginning of Ellis's true bonding with Kit and Kate.

Figure 23.3. Bears interacting with Ellis Bacon.

Pruitt (1974, 1976) observed the bears extensively, focusing on social behavior, especially fighting and play. Rough and tumble play with people and each other differed in both the kinds of play invitations used and the intensity of play "bites," which, as already noted, were eventually inhibited with, at least, familiar people. Cheryl's husband, Bob, suffered a deep puncture wound in his hand while playing with Kit when she was about 6 months old. This motivated us all to be more attentive to behaviors anticipating agonistic play and to appreciate the cubs' strength. As the bears matured, the level of vigorous play declined and involved extensive "jaw play." The danger to humans, who were not as strong, with thin skin, less endurance, and no loose folds of fur or long ears to tug or gnaw on, also increased as rough play could turn serious rapidly. Ellis reports that his attachment to the bears increased as they became more solitary and less vigorously playful.

A basic ethogram of movements and postures was developed that was later elaborated by Jordan (1976, 1979) and Jordan and Burghardt (1986). Again, the familiarity with the individual bears was an asset. However, the location of the animals in a remote area of the park in which their primary human encounters were with the researchers did lead to some difficulties. Careful observation showed that the bears' behavior changed over an observation period (observer effect) in a way not seen in same-aged bears observed at a tourist attraction in Pigeon Forge, Tennessee, outside the park (Jordan & Burghardt, 1986). It was impossible for the bears to be observed unnoticed.

Experimental studies were carried out primarily by Ellis Bacon. These included the development of a modified WGTA (Wisconsin General Test Apparatus), originally designed for use with nonhuman primates, to study preferences for native and nonnative food items and their seasonal change (Bacon & Burghardt, 1983). Curiosity, manipulation, and exploratory behavior were studied by presenting a standardized set of novel objects based on Glickman and Sroges's (1966) seminal comparative study, which did not include bears. Our black bears were well within the "higher" primate range at all three testings (at 16, 20, and 26 months of age) and above all other carnivores (Bacon, 1980). It was remarkable how they used single claws and their highly mobile lips to lift, manipulate, and carry small objects. This "delicate" aspect of their behavior, including "greeting kisses," was another contribution to the bonding between the primary care givers and the bears.

Sensory abilities were analyzed using a test paradigm involving having the bear move from one section of the cage to another where objects of different colors or marked with different forms were placed. It was remarkable to see a large bear lumber up to an array of stimuli and gently turn over only the "correct" one for a reward of one or two raisins. In this way it was first demonstrated that black bears have color vision and good short-range visual acuity (Burghardt, 1975; Bacon & Burghardt, 1976a,b).

These experimental studies could not have been carried out without a mutually close, trusting, and playful relationship between experimenter and bear. We had to try out many techniques and apparatus to find ones that both worked and were relatively indestructible. The task had to be one that the bear voluntarily undertook. The ability of the bears to learn to make the discriminations rapidly was probably enhanced by their concentrating on the task at hand and not being distracted by unfamiliar people and situations. In this sense the relationship was essential to the work.

As the bears grew, the special relationship between Cheryl, Ellis, and the bears became more apparent (Figures 23.3 and 23.4). Keep in mind that the bears were never tame in the sense that a domestic animal is. Typical agonistic displays to or flight from persons visiting the research area were frequent. Kate, in particular, developed certain dislikes of later researchers. She would display to them an impressive array of charges, snorts, slaps, and chomping of the teeth. As she approached 140 kg, these displays were effective indeed.

However, with Cheryl and Ellis in the everyday routines of testing and observation, there was very much a sense of calm. It was unfamiliar human intruders who caused agitation and unease in the bears. One of Ellis's quiet

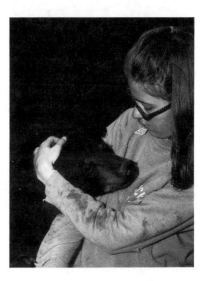

Figure 23.4. Cheryl Pruitt and research subject.

delights was to sit on the roof of the bears' den watching the forest and the rushing mountain stream with one or both of the bears as company.

But it was not all calm. Both bears tended to test Cheryl and Ellis in what appeared to be a dominance assessment. The latter developed a policy of never backing down. Once, Kate, then at least 100 kg, lunged at Cheryl's chest and almost tore off a loose-fitting jacket. After her initial scream, Cheryl promptly slapped Kate and scoldingly chased her away. On another occasion, during testing, one of the bears lunged and slapped Ellis on the leg, knocking him down. With the only thing he had in his hands, a pencil, he immediately chased her about the enclosure, swatting her on the nose as she bellowed and ran. That testing session ended and the subject pouted for the rest of the day.

The ending

Kit, the most adept, learned how to climb up a tree and escape the enclosure over the sheet-metal walls, but always hung around and was easily coaxed back. However, in 1973 she climbed out when under the supervision of a student less comfortable with the bears, who left and went for help. Kit eventually strolled (or was frightened) away from the area and, in spite of much searching, was never seen again. Kate was given to the

Knoxville Zoo in 1974. It was traumatic, especially for Ellis, to see her under the care of keepers who could not relate to her as the individual she was, nor play the games we shared. Kate was eventually send to the Hershey Park ZooAmerica in Pennsylvania. Cheryl visited her in 1987 after an interval of 13 years. Although Cheryl was not permitted to be closer than 5 meters from the cage, Kate came when she called and watched her intently for 15 minutes. She never responded to any ZooAmerica staff in this manner. Healthy and long-lived (19½ years), she was by all reports a loner and never mated, rebuffing all males. This probable imprinting on people shows, in an unfortunate way, the reciprocal nature of the bond and underscores ethical dilemmas in hand rearing wild animals.

The bears have been gone from the enclosure in the national park for more than 15 years now. No nonhuman animals have been such intimate and intense partners in my life, even if all too briefly, as Kit and Kate. Ellis was devoted to the bears he studied and lived with for several years. He commissioned a large painting of Kate by a local wildlife artist that was released as a limited edition print (one is displayed in my home). He ended the acknowledgment section of his dissertation with this sentence (Bacon, 1973): "Ironically, the greatest thanks is given to the two girls with whom I have integrated my life, and who cannot accept the appreciation, Kit and Kate" (p. iii). Cheryl ended her acknowledgment section this way (Pruitt, 1974): "Finally to Kit – wherever she is – and to Kate, this all would never have been possible without them" (p. iv). She also had a dedication: "To Bob [her husband] and all the experiences the girls have given us." Only now have I realized that neither Jordan or Ludlow wrote anything comparable in their theses, confirming their weaker, less salient bond. Cheryl, Ellis, and I keep in touch and plan an occasional reunion. We rarely talk to one without bringing up the other. Through Kit and Kate we have formed mutual bonds that endure.

Although outwardly bears of all ages generate anthropomorphic reactions in people (Lawrence, 1990), intimately known bears engender quite different responses that nevertheless counter those who view bears as unpredictable or even vicious. The affective world of the black bear does not mirror ours. Nor is it a parallel alien universe, but one running obliquely to our own. If we have the privilege of being individually bonded to bears, their world touches us at their convenience, and we are given partial access to it through a darkened glass, but never face to face. Through the years, I have come to consider the efforts of von Uexküll (1934) to grasp the inner life of other species as quite profound, ultimately incomplete, yet well worth pursuing (cf. Griffin, 1976).

There is a danger here, of course, recognized by Baynes. Was the bond nothing but projection? I think not. The bears did treat people differently. No one but Ellis could accomplish the learning trials. There was no doubt that the bears were both attached to and accepting of us; occasional threats, bad moods, forgetfulness, even anger is normal for relationships between children and their parents and should not be unexpected with animals either.

Once, when Kate got sick (at about 50 kg) and severely dehydrated, she was carried to Ellis's trailer, where she laid on the floor, quiet, accepting of our care, and like a young child when sick, regressing to patterns of an earlier age. By this time the bears were not interested in sitting quietly in our laps. But Kate, with the look of a sick child who quietly appreciates being waited on, enjoyed an ice cream cone, her baby bottle, and lapping up the gruel and baby food of her youth. Although lethargic and in discomfort, she also appeared contented and childlike, drawing out care-giving behavior from all who saw her, and especially those who had reared and loved her. As Kate recovered, she spent three days on the ground floor of our observation tower next to the enclosure. Ellis and Cheryl alternated sleeping with her. Kate maintained close physical contact all night.

Of course, few of these events or feelings made it into the scientific papers. It is undoubtedly proper to keep certain aspects of behavior research objective and clear, free from the taint of mentalism and emotion. Yet we may thereby unwittingly do science, and especially young researchers, a disservice. Standard ethological tradition calls for a genuine liking and respect for one's animals in order to do effective and creative research. But can or should the line really be drawn there? Is not a *critical* anthropomorphism the next important step we are too wary to embrace openly, but covertly use (Burghardt, 1991)? Recall the emotional rush that the bear hunters undoubtedly experienced when they discovered that they or their partners were now the quarry. So too, but more gradually and stealthily, does the phenomenology of more complex and enduring bonds grow and shape us. These we must also respect and try to comprehend.

Acknowledgments

Support for preparing this chapter was provided by a research grant from the National Science Foundation (BNS-8709629). The research itself was supported by the University of Tennessee and research grants from NIMH and NSF from 1970 to 1978. The cooperation of the National Park Service, Hershey ZooAmerica, and the Tremont Environmental Center is also gratefully acknowledged, as are Mike Pelton and all the people who worked with Kit and Kate or commented on the manuscript. Ellis Bacon and Cheryl Pruitt added important information from their notes.

References

Bacon, E. S. (1973). Investigations on perception and behavior of the American black bear. Unpublished Ph.D. dissertation, University of Tennessee.

(1980). Curiosity in the American black bear. *International Conference on Bear Research and Management, 4*, 153–7.

Bacon, E. S., & Burghardt, G. M. (1976a). Ingestive behaviors of the American black bear. *International Conference on Bear Research and Management, 3*, 13–25.

(1976b). Learning and color discrimination in the American black bear. *International Conference on Bear Research and Management, 3*, 27–36.

(1983). Food preferences in the American black bear: An experimental approach. *International Conference on Bear Research and Management, 5*, 102–5.

Baynes, E. H. (1929). *Jimmie: The story of a black bear cub.* New York: Macmillan.

Burghardt, G. M. (1975). Behavioral research on common animals in small zoos. In *Research in zoos and aquariums* (pp. 103–33). Washington, DC: National Academy of Sciences.

(1982). Comparison matters: Curiosity, bears, surplus energy, and why reptiles don't play. *Behavioral and Brain Sciences 5*, 159–60.

(1991). Cognitive ethology and critical anthropomorphism: A snake with two heads and hognose snakes that play dead. In C. A. Ristau (Ed.), *Cognitive ethology: The minds of other animals* (pp. 53–90). Hillsdale, NJ: Erlbaum.

Burghardt, G. M., & Burghardt, L. S. (1972). Notes on the behavioral development of two female black bear cubs: The first eight months. *International Conference on Bear Research and Management, 2*, 255–73.

Burghardt, G. M., & Herzog, H. A., Jr. (1989). Animals, evolution, and ethics. In R. J. Hoage (Ed.), *Perceptions of animals in American culture* (pp. 129–51). Washington, DC: Smithsonian Institution Press.

Burghardt, G. M., Hietala, R. C., & Pelton, M. R. (1972). Knowledge and attitudes concerning black bears by users of the Great Smoky Mountains National Park. *International Conference on Bear Research and Management, 2*, 207–20.

Craighead, F. C., Jr. (1977). Foreword. In W. H. Wright (1909/1977), *The grizzly bear* (pp. v–vii). Lincoln: University of Nebraska Press.

Glickman, S. E., & Sroges, R. W. (1966). Curiosity in zoo animals. *Behaviour, 27*, 151–88.

Griffin, D. R. (1976). *The question of animal awareness*, New York: Rockefeller University Press.

Haynes, B. D., & Haynes, E. (1966). *The grizzly bear: Portraits from life.* Norman: University of Oklahoma Press.

Hediger, H. (1950). *Wild animals in captivity.* London: Butterworth.

(1955). *Studies of the psychology and behaviour of animals in zoos and circuses.* London: Butterworth.

Jordan, R. H. (1976). Threat behavior of the black bear (*Ursus americanus*). *International Conference on Bear Research and Management, 3*, 57–63.

(1979). *An observational study of the American black bear (Ursus americanus).* Unpublished Ph.D. dissertation, University of Tennessee.

Jordan, R. H., & Burghardt, G. M. (1986). Employing an ethogram to detect reactivity of black bears (*Ursus americanus*) to the presence of humans. *Ethology, 73*, 89–115.

Kuckuk, E. (1937). Tierpsychologie beobachtungen an zwei jungen braunbaren. *Zeitschrift für Vergleichenden Physiologie, 24*, 14–41.

Lawrence, E. A. (1990). The tamed wild: Symbolic bears in American culture. In R. B. Browne, M. W. Fishwick, & K. D. Browne (Eds.), *Dominant symbols in popular culture* (pp. 140–53). BowlingGreen, Ohio: BowlingGreen State University Popular Press.

Leslie, R. F. (1971). *The bears and I.* New York: Ballantine.

Leyhausen, P. (1948). Beobachtungen an einen jungen Schwarzbaren. *Zeitschrift für Tierpsychologie, 6*, 433–44.

Ludlow, J. C. (1974). A preliminary analysis of the activities of captive black bears (*Ursus americanus*): Locomotion and breeding. Unpublished M. A. thesis, University of Tennessee.

Pelton, M. R., & Burghardt, G. M. (1976). Black bears of the Smokies. *Natural History, 85*(1), 54–63.

Pelton, M. R., Scott, C. D., & Burghardt, G. M. (1976). Attitudes and opinions of persons experiencing property damage and/or injury by black bears in the Great Smoky Mountains National Park. *International Conference on Bear Research and Management, 3*, 157–67.

Pruitt, C. H. (1974). Social behavior of young captive black bears. Unpublished Ph.D. dissertation, University of Tennessee.

(1976). Play and agonistic behavior in young captive black bears. *International Conference on Bear Research and Management, 3*, 79–86.

Pruitt, C. H., & Burghardt, G. M. (1977). Communication in terrestrial carnivores: Ursidae, Mustelidae, Procyonidae. In T. A. Sebeok (Ed.), *How animals communicate* (pp. 767–93). Bloomington: Indiana University Press.

Rogers, L. L., & Wilker, G. W. (1990). How to obtain behavioral and ecological data from free-ranging, researcher habituated black bears. *International Conference on Bear Research and Management, 8*, 321–7.

Roosevelt, T. (1893/1983). A man-killing bear. In P. Schullery (Ed.), *American bears: Selections from the writings of Theodore Roosevelt* (pp. 117–20). Boulder: Colorado Associated University Press.

Seton, E. S. (1900). *Biography of a grizzly.* New York: Century.

(1904). *Monarch: The big bear of Tallac.* New York: Scribner's.

Shepard, P., & Sanders, B. (1985). *The sacred paw.* New York: Viking.

Tate, J. (1983). A profile of panhandling black bears in the Great Smoky Mountains National Park. Unpublished Ph.D. dissertation, University of Tennessee.

Uexküll, J. von. (1934). Streifzüge durch die Umwelten von Tieren und Menschen. Berlin: Springer. Translated in *Instinctive behaviour* (1957), Ed. C. H. Schiller. London: Methuen.

Whiten, A., & Byrne, R. W. (1988). Tactical deception in primates. *Behavioral and Brain Sciences, 11*, 233–73.

Wright, W. H. (1909/1977). *The grizzly bear: The narrative of a hunter naturalist.* Lincoln: University of Nebraska Press.

(1910). *The black bear.* New York: Scribner's.

Yerkes, R. (1925). *Almost human.* London: Jonathan Cape.

Scientist–animal bonding:
some philosophical reflections

Hugh Lehman

Editors' introduction

The book's final chapter examines the relevance of our central theme to ethical concerns and the philosophy of science. Lehman, a philosopher, surveys definitional issues, examines problems related to anthropomorphism, and discusses the range of effects bonding might have on research conclusions. Finally, Lehman examines the ethical implications of the scientist–animal bond. Does the scientist have a special obligation to the animal by virtue of that bond and, assuming it is possible, should scientists avoid research in which bonding with animals is likely to occur?

The nature of bonding

Do animals that serve as research subjects in scientific experiments bond to the people who handle them – to scientific researchers or their assistants? The title of this book, *The Inevitable Bond*, suggests an affirmative answer. It may be doubted that such bonding occurs in every case, for every animal in every research project; nonetheless, as many of the chapters in this volume indicate, many scientific claims imply that bonding is a regular occurrence in research settings. There are some reasons for being skeptical about such claims. We shall discuss these reasons briefly in the second section and again, though on different grounds, the third section. In the fourth section we shall assume that research animals (at least sometimes) become bonded to scientists and discuss a number of ethical issues that arise as a consequence of bonding.

First, it is necessary for us to consider the nature of bonding. One way to proceed at this point would be to develop a definition of *bonding* on the

383

basis of a discussion of definitions or explanations of this term in the literature. No doubt such a review would be useful and would suggest many
interesting lines of research; however, we will not follow that course here.
Rather, we shall formulate one definition based on the work of Gubernick
(1981) and subsequently take note of some questions that may be raised
with respect to this definition. In light of these questions it is not unlikely
that some ethologists, psychologists, and others would opt for alternative
definitions. Nonetheless, we need to consider this (or a similar) definition
to give sense to the epistemological and ethical questions that we pursue.
The definition suggested here should turn out to be compatible with a
range of theories about bonding. Further, the term, as here understood,
is applicable to behavior that serves a range of functions, for example,
mating and protection of young. Of course, theoretical developments in
psychology or other sciences may motivate the formulation of alternative
definitions. As the etiology of the behavior that serves these functions is
more precisely understood, alternative definitions of *bond* may well be
required.

The definition is as follows: A bond is a stable relationship among two or
more individuals, *A* and *B*, or between an individual and some object,
provided that three conditions are satisfied. These are (1) *A* feels affection
or some similar emotion toward *B* (or to the object or place); (2) the affection is directed specifically toward *B* (or to a specific object or place) rather
than toward all closely similar individuals (or objects or places); and (3) the
bonded individual displays preferences in its behavior for the individual (or
object) to which it is bonded.

We will not here explore the nature of relationships or attempt to define
relationship. It does not appear that such exploration would help to clarify
the points we are considering in this chapter.

The reference to feelings of affection is, of course, somewhat vague. Our
language for describing feelings is not precise. Presumably, since a bond is
a stable relationship, the feelings referred to here are not merely brief
occurrences, but rather enduring psychological states. Presumably also
there are a range of feelings that may loosely be called feelings of affection.
These may include such mild feelings as a sense of security and liking,
as well as much more forceful sentiments. In light of the range of feelings
that fall within the concept of affection, we may expect that there is a
comparable range of behaviors that are taken as indicative of the presence
of a bond. While this vagueness increases the difficulty of the scientists
concerned with describing or explaining behavior, it does not appear that
the notion of bonding or attachment is so loose as to be useless to psychologists, ethologists, and others. Feelings of affection can frequently be reliably
distinguished from other feelings such as anger, boredom, and fear.

This definition of *bond* is clearly not an operational one. The inadequacy of operational definitions has long been known (Hempel, 1965a). Nor do we think it promising to try to define *bond*, as some well-demarcated collection of observable behaviors. In claiming that the term *bond* cannot be operationally defined, we are not saying that it is not possible to spell out observational criteria for application of the term. Rather we are saying that such criteria cannot exhaustively determine the content of the term. Rather, in distinct contexts, distinct observational criteria for application of the term are required. A researcher who proposed to investigate bonding would have to give some indication of which forms of behavior were to serve as indicators of the presence of bonding in the context of his or her investigation. Presumably this context would vary with respect to many factors, including species, sex, age, local environment, behavioral capacities of the individuals, and so on. However, identifying a bond with the behaviors that indicate the presence or strength of the bond is, we believe, an error, as great an error as identifying a mother's affection for her infant with her giving the infant a large number of caresses or performing other related behaviors. Our position on this question is in agreement with that of Hinde (1979), who said, "An account of interpersonal relationships solely in terms of overt behavior will not suffice" (p. 22).

That terms such as *bond* cannot be operationally defined renders the goal of attaining reliable generalizations more complex. While an operational definition would simplify the task of distinguishing bonds from other forms of stable relationships (as well as the task of distinguishing the consequences of bond formation from the consequences of the presence of other relationships that may occur simultaneously with bonding), the absence of an operational definition in this case does not mean that reasonably clear distinctions cannot be maintained.

The notion that a bond is a stable relationship does not imply that it is an unchanging relationship. Saying that a bond is stable is compatible with allowing that it develops, becomes stronger or weaker, passes through several phases, and so on. Saying that the relationship is stable implies, we believe, that the relationship persists over a relatively extended period of time, that it does not consist merely of a few interactions. Indeed, we take it that this definition implies that the bond exists even during intervals within which the bonded individuals are not presently interacting. For example, a lioness may be bonded to her cub even when both are sleeping.

Further, we take it that the definition does not imply that an individual can be bonded to only one individual at a time. In saying that "the affection is directed toward specific individuals," say from a mother to her cub, it is not implied that the mother cannot be bonded to other cubs or to the father of the cubs or to other, less closely related felines. We take it that what is implied by this phrase is that the relationship reflects a discrimination of particular individuals within a collection of individuals. If one goose is bonded to another goose, it is not thereby bonded to all geese or even to all geese within the flock of which it is a member. Very likely, characteristics of the particular individuals, not present in the same combinations in other members of the flock, species, and so on, give rise to and elicit behaviors that are indicative of the presence of the relation.

Two more points concerning the definition may be worthy of comment. First, in discussions leading up to the writing of this chapter, the question was raised as to whether the bonding relation is symmetric. That is, does the fact that individual *A* is bonded to individual *B* necessarily imply that *B* is bonded to *A*. Some confusion over this matter may arise in consequence of thinking of bonding as analogous to cases in which objects are tied or glued to each other. Clearly, if *A* is glued to *B*, then *B* is glued to *A*. That relation is symmetric.

The definition we have given implies that bonding is not symmetric. (Of course, this does not mean that the relation is asymmetric; it is possible for *A* to be bonded to *B* and for *B* to be bonded to *A*.) For *A* to be bonded to *B*, *A* must feel some affection (or similar emotion) for *B*. However, it is possible that *A* feels affection for *B* but that the affection is not reciprocated. Thus, the relation is not symmetric. Of course, if *A* is bonded to *B* (and *B* is a living being rather than a place or thing), *B* normally will modify its behavior in relation to *A*. But *B*'s feelings or emotions with regard to *A* need not be affection; *B* may feel aversion for *A*. Where *A*'s bonding to *B* causes *B* to modify its behavior toward *A* we may speak of a degree of mutuality in their relationship (Hinde, 1979).

Second, the definition leaves open the possibility that an individual can be bonded to an object or a place or some other object as well as to other living creatures. This is consistent with common usage in which, for example, it is sometimes suggested that cats bond to places and human infants to objects such as blankets.

Reductionism, physicalism, anthropomorphism

According to the definition of *bond* that we have adopted, to attribute a bond to a pair of individuals is to say that at least one of the individuals feels affection (or some similar emotion) for the other. Some scientists may be disinclined to accept such attributions because of a reluctance to accept references to theoretical or inferred entities that are based on empiricist or physicalist preconceptions. We will not pursue this matter at length here. Clearly, many of the authors of chapters in this volume feel comfortable with such theoretical assumptions (among others, Estep & Hett, Chapter 2;

Dewsbury, Chapter 3; Oden & Thompson, Chapter 13; Caine, Chapter 22). Under the pressure of sustained criticism, many of the preconceptions of rigid empiricism have been abandoned by most philosophers concerned with the formulation of principles of scientific methodology (Scheffler, 1964; Hempel, 1965b). Even the assumption that there is a clear distinction between observational data and scientific theory has been challenged (Quine, 1953, 1970). It should perhaps be noted that Hinde defends the use of such mentalistic concepts as "affection" on the grounds that such terms refer to "emergent" properties, as well as by pointing out that the use of such concepts is "essential in practice" (Hinde, 1979, p. 34.) Baker (1987) has also criticized philsophical physicalists (also known as "eliminative materialists") such as Dennett on the grounds that they surreptitiously appeal to mentalistic concepts in defending their views. Perhaps the best reason for abandoning such preconceptions is that they do not appear to accord with the best scientific practice, as indicated by many excellent analyses of historical developments in science (Kuhn, 1962).

Even though some reference to inferred entities is inevitable, many scientists are uncomfortable with references to states of emotion, feeling, cognition, and so on in animals. One often finds scientists using forms of words that do not fit comfortably together, that is, words that have incompatible connotations. The nonharmonious forms of words indicate both ontological and epistemological concerns. For example, Gubernick (1981) appears to accept the idea that the term *bond* refers to "certain aspects of the relationship" (p. 244), but he also suggests that the term is a "shorthand descriptive label that refers to preferential responding" (p. 246). "Aspects of the relationship" suggests that the term bond refers to something that is not identical to the specific behaviors from which one may infer its presence. However, the phrase "shorthand descriptive label that refers to preferential responding" suggests that Gubernick does not mean the term *bond* to refer to anything distinct from observable behaviors.

If the term *bond* can be eliminated, as Gubernick suggests, by spelling out the longer, more cumbersome descriptions, his definition of *bond* should indicate how this is to be done. It does not do this. In his discussion he goes on to list "criteria of attachment." These criteria do not provide a method for replacing the term *bond* by a set of purely observational terms that exhaust the content of the term. In some of his criteria, Gubernick refers to emotional aspects of the relationship – for example, by referring to the attachment figure as a "secure base." He also refers to the bonded individual as "seeking . . . proximity to" the attachment figure. These criteria do not appppear to serve as a basis for eliminating reference to emotional states in favor solely of items of observable behavior. Of course, Gubernick's discussion is not as clear as it could be. He has not said precisely what items are observable. Furthermore, some of the criteria he mentions are rather vague.

Not all scientists are reductionists. While Gubernick seems to favor what might be called a reductionist perspective, that is, one which maintains that reference to feelings, emotions, and so on is reducible to reference to observed behavior, Hinde may be described as a realist; in other words, he understands himself to be providing a framework for discussing structures. Structures are entities that are not to be identified with particular behavioral manifestations (Hinde, 1976, p. 8). However, while Hinde is self-consciously a realist with regard to structures, he expressed qualms about references to affective states. For example, he said, "Where the relationships of others are concerned, the only access to such feelings is through behaviour" (Hinde, 1976, p. 6). Presumably this statement implies that affective states of others are not observable. This implication is an assumption that is apparently widely made without any indication that it is subject to doubt (Quine, 1953, 1970). In spite of differences among psychologists on these matters concerning realism and reductionism, the questions raised are too big to be pursued in a brief chapter, and so we will not comment further on them. There are, however, some more manageable epistemological issues to consider.

Where references to affective states are made, it is not unusual to hear them criticized as involving some form of "anthropomorphism." Anthropomorphism, we are led to believe, is an error of some sort. Since the attribution of a bond to animals involves, we believe, the attribution of affection, if all such attributions are erroneous or unwarranted, the validity of much scientific work on bonding is called into question. If all such attributions are unwarranted, then the definition of *bonding* that we have adopted would be useless, since attributions of affection would never be warranted. Let us consider this criticism. (We should note perhaps that sometimes the term *anthropomorphism* is used without any implication that an error has been committed. In such uses the term may refer to a tendency of some people to treat animals much as they treat other people.)

It is not entirely clear exactly what error is committed when one attributes a human affective, emotional, or cognitive capacity to animals. Two possible errors come to mind. On the one hand, one commits a factual error if one attributes a characteristic to an object which that object does not possess. On the other hand, one commits a logical or evidential error if one attributes a characteristic to an object in the absence of sufficient evidence to warrant the attribution. However, if we assert that animals feel affection for one another or for humans, that assertion is not necessarily an error of either type. An animal may indeed feel affection for a specific individual; for example, a pair of geese may feel affection for each other, and we sometimes indeed have adequate evidence to warrant such an attribution, that is, to make it reasonable to believe the claim that the animals feel affection.

A full defense of this claim that scientists sometimes have adequate evidence to warrant the attribution of affection or other emotional states to animals is beyond the scope of this chapter. To defend this assertion we would first argue that scientific methodology is in accord with the hypotheticodeductive method, which has been elaborated by Hempel and others (Hempel, 1966). Then we would point out that hypotheses which use concepts of feeling or emotion have often been tested and confirmed by observation and experiment. Hinde has noted that research using such mentalistic concepts is no less scientific than research that uses only purely physicalistic or mechanistic concepts (Hinde, 1979).

Sometimes one gets the impression from the words that some scientists use that they think that in order to avoid anthropomorphism scientists must free themselves from all of their human or mammalian preconceptions and simply describe a behavior as it is in itself or as it would appear to an observer who had no preconceptions. Such Baconian conceptions of science have, as noted earlier, been largely rejected (Holton & Roller, 1958; Hempel, 1966). Preconceptions are unavoidable and any attempts to avoid them all would stifle all intellectual activity, including science. Nonetheless, assumptions made in the course of an inquiry may be subjected to experimental testing and revised or abandoned if observational evidence indicates their inadequacy.

Under what circumstances is it reasonable to postulate that bonding between a pair of individuals exists? Presumably, the postulate is warranted, to some degree, when one or both of the individuals display behavior that is most reasonably explained by reference to some pleasant feeling such as affection. Such behaviors may include those that tend to preserve proximity between the pair or to reestablish proximity when the pair is separated, behaviors that indicate distress upon separation, behaviors that indicate a special protectiveness among members of the pair, and so on. The postulate that a bond exists is warranted provided that the behaviors are better understood by reference to affection than to other feelings and when the behaviors are displayed specifically in reference to the bonded individuals and not generally – for example, if a mammal displays solicitous behavior toward her own offspring but not toward any other comparably young species members.

We do not mean to say, of course, that a scientist who attributes a bond (or some other psychological state) to an animal is always correct or always justified on the basis of the evidence. A scientist may indeed be guilty of anthropomorphism in either sense. The attribution of the bond to an animal may be mistaken, or the evidence that would make such an attribution reasonable (evidentially warranted) may be lacking. Clearly, a scientist who attributed a bond merely on the basis of facial expression would be guilty of anthropomorphism. Indeed, we should be cautious in our attributions of psychological states to animals. Except for people who have considerable experience with animal behaviors, such as may be the case with animal trainers, pet owners, ethologists, and other scientists who study animal behavior, most of us have very little evidence for attributing such states to animals. Further, some people, like pet owners, who have more experience with animals, do not normally methodically test their assumptions in a scientific way. Consequently, many such assumptions are not strongly supported by observational evidence.

Scientists who criticize others as having committed anthropomorphism seem to think that they are being particularly good scientists because they are being (appropriately) rigorous with regard to the evidence. However, such scientists are not necessarily being exceptionally good scientists – even in this one respect. It is possible to be too demanding with regard to the evidence required to show that a belief is warranted. (Consider a scientist who refused to accept a hypothesis regarding causal relations unless his data disconfirming the null hypothesis were significant to a far higher degree than usual. Such a scientist might pass up many truths because of his or her exceptionally rigorous standards.) Further, a scientist's disinclination to be guilty of anthropomorphism could lead to inadvertent errors if she or he were unaware that the animal research subjects had formed bonds to the researcher or research assistants. The scientist might fail to recognize that the bond was affecting the behavior under investigation.

This last point suggests that we need a term to refer to errors that are, in a sense, opposite to anthropomorphism. There are two such errors, namely, the denial that organisms have feelings or emotions when, in fact, such states are present and the failure to draw the conclusion that a bond or some other state involving feelings or emotions is present when such a conclusion is supported by adequate evidence. We might call these errors "disanthropomorphism."

The chapters in *The Inevitable Bond* are concerned with cases of alleged bonding of animal research subjects to human researchers. It appears that such bonding is more likely when the animals are mammals or even birds, than it is when the animals are amphibians, reptiles, or invertebrates. Investigations of this question would be complicated by the concern that in attributing affective states to amphibians, reptiles, or invertebrates investigators would be more likely to attribute such states on the basis of inadequate evidence. In some reptile species, adults display parental care toward their young. But does bonding among reptiles occur in other contexts? Is parental care among reptiles bonding in the sense of the term we have specified? One would have to display great caution in making such a judgment. One would have to consider whether there are reptilian behaviors that indicate distress on separation, whether the care-giving behavior is sufficiently specific, and so on. In general, we wonder whether there are patterns of behavior manifest among reptiles that cannot be adequately characterized or explained without reference to emotional states. Physiological investigations might shed some light here. One wonders whether the neurological or glandular structures on which affective states depend are present in reptiles.

Perhaps a more important reason for being skeptical about the attribution of bonds in phyla or orders far removed from mammals or birds has to do with the limits to which concepts involved in the description or explanation of affective states can be stretched. We indicated earlier that the concept of "affection" is vague but that its scientific application is dependent on taking certain observed behaviors as indicators of its applicability. There must be limits to the range of vagueness if the term is to retain any usefulness for making distinctions. If we are to be justified in speaking of bonds in reptiles there must be some behaviors that we recognize as being indicative of the requisite affective state. It may well be that in the case of reptiles or invertebrates, the limits of vagueness for terms signifying emotions have been exceeded.

Further doubts

It seems clear that not all research that requires observations of animals will lead to bonding of the animals to the researcher. Some of the contributors to this volume note that sometimes it is possible to observe an animal without the animal being aware of the observer's presence. Animal behavior can be videotaped, for example. If a videotape is used, not only is the human not seen, he or she is not heard, smelled, and so on. In such cases, it seems unlikely that the animal could become bonded to the researcher.

For various reasons, it may be impossible for a researcher to observe the animals that are serving as research subjects without the animals also observing the researcher. In such a case, is bonding inevitable? Estep and Hetts (Chapter 2, this volume) suggest a negative answer to this question. In some cases the animal might perceive the human observer, but if the animal is not predisposed to find humans biologically significant, for example, as prey or as predators or as care providers, then it may fail to recognize the same human even after repeated exposures. (See their discussion of the scientist as a socially insignificant part of the animal's environment.) In such cases bonding might not occur. It seems likely that such cases exist. Some animals are, I believe, harmed because they do not seem to care about the presence of humans, for example, certain island species such as are found on the Galapagos Island. Furthermore, even if the animal regards the human as significant and so is capable of bonding to the human, it may be that the bonding will occur only under certain circumstances, circumstances that are not necessarily present in the research situation. In such cases also the animal will not become bonded to the researcher.

Let us ask another question. Suppose we are investigating an animal that perceives us and does not ignore us when it does so. Suppose further that we are unable to conduct our research in any way that will prevent the animal from bonding to the observer. Such a case could occur in situations such as those described by Crowell-Davis (Chapter 20) or Caine (Chapter 22). What bearing will this have on our conclusions?

Clearly, as others in this volume have noted, it is possible that some of the conclusions we draw from our studies will be too general. This could occur if the animal's bonding to the researcher influenced the animal's subsequent behavior. Such a generalization might lead researchers to have expectations that would not be met in situations in which the animal had not bonded to a human (Caine, Chapter 22).

However, in a situation such as we are here considering, is skepticism inevitable? Do we have premises in this situation which imply that a general understanding or knowledge of an animal's behavior in this situation is impossible? It appears not. (It should be noted that bonding, as we have conceived it, is not the only animal–scientist relationship in which this sort of question may be raised. A scientist's research results could be invalidated as a result of social relationships with the scientist other than bonding.) Suppose that we are investigating some type of behavior in which bonding is likely to occur. It is possible that even though the animal subjects become bonded to the researcher (e.g., recognize the researcher and modify some of their behavior in light of that recognition), the animal behavior that is modified as a result of this bonding is not the behavior under investigation.

However, even if the bonding affects the behavior under investigation, it does not follow that the researcher cannot investigate and come to understand this effect. For example, suppose that the bonding affects the behavior of the animal subject in the presence of its nest. The researcher may develop hypotheses concerning the effect of the bonding on this behavior and then develop observational protocols for testing these hypotheses. Suspicion that the influence of bonding cannot be investigated may arise as a result of the assumption that in order to investigate this phenomenon scientifically one would have to contrive experimental circumstances in which bonding was absent and compare the animal behavior in those circumstances with its behavior in circumstances which were identical except that bonding was present. Very likely it would be impossible to contrive circumstances that differed only in this one factor, that is, the presence of bonding. However, observational or experimental studies of the effect of bonding do not require that one contrive circumstances that differ in only this one respect. Consequently, it does not appear, even in cases in which bonding of the animal subject to the researcher is inevitable, that this implies some inherent limitation in human understanding of animal behavior.

We could ask at this point whether the bonding of an animal to a scientific researcher ever gives rise to a relationship that interferes with the capacity of the scientist to fulfill her or his scientific role competently. One can certainly imagine scenarios in which as a result of bonding the scientist's data are distorted. Estep and Hetts (Chapter 2) and Caine (Chapter 22) call attention to that possibility. As they suggest, error could arise because the scientist ignores the effect of bonding (or some other relationship) on the animal's behavior. Error could also arise because the bonding gives rise to a relationship that causes the scientist to fail to attend to some factors that are pertinent to the study or to assign more weight to some factors in the data than a strict disinterested statistical analysis would warrant.

Ethical questions

There has been considerable discussion in recent years concerning the moral acceptability of the use of animals in scientific research, especially if the animals suffer pain or are killed (Singer, 1976; Rollin, 1981; Regan, 1983; Sapontzis, 1987). We will not review or discuss that general issue here. To sort out and discuss all the pertinent questions regarding values and moral obligations that arise in consideration of the use of animals in scientific research requires much more space than is appropriate here. Rather, it might be interesting to consider whether bonding between an animal subject and a researcher raises any special ethical issues and, if it does, whether any moral injunctions may be derived. We shall consider two ethical questions. The first is whether we need to take the fact that animal subjects become bonded to scientists into account in determining our obligations to those animals. The second is whether scientists ought to eschew research in which the animals develop bonds to the researcher.

For the sake of this discussion we shall start by assuming that in general we ought not to add to the burden of suffering or frustration of others. We shall also assume that we ought, where it does not cause excessive suffering or harm to ourselves or others, take steps to add to the pleasure and satisfaction of others. These obligations are only prima facie ones. That is, we are obligated not to add to others' suffering or frustration unless there are strong, morally relevant reasons for doing so – and there may well be such reasons. Similarly, there may be strong countervailing obligations that outweigh our obligation to enhance the quality of animal lives.

Bonding of the animal to the researcher often facilitates the research. As a result of the bonding, the researcher can approach the animal more easily and when this occurs the animal will not become overly excited. Excitement could influence the variables under study, and the reduction or elimination of the excitement can help make investigations more easily controlled. If the scientist can approach the animal more easily, this may save time and reduce or eliminate injuries to the scientist or the research subject. Thus, bonding is often of significant benefit both to the researcher and to the research subject.

The other side of the matter is that bonding may in some circumstances create additional burdens for the animal. In addition to whatever other suffering the animal may endure, it may suffer some frustration if expectations that formed as a result of bonding to the scientist are not met. Given that bonding can have consequences that make an animal's life either better or worse and given the moral assumptions that we have made, it is clear that in considering our obligations to research animals, we must take bonding into account. The answer to our first question is affirmative. As in other matters, we can say that if our research questions can be answered while causing less suffering or harm by modifying the research strategy, then it ought to be modified. Thus, when we have good reason to believe that satisfactory research results can be achieved by methods that prevent that animal from bonding to humans and where the severance of bonds, if they were formed, would cause suffering to the animal, then those methods should be followed. Similarly, if modifications to the research method can enable animal subjects to have more interesting or pleasant experiences, without reducing the efficacy of the method, then these modifications should be made.

Estep and Hetts (Chapter 2) note that there is a widespread belief that relationships between animal subjects and scientists should be minimized because such relationships are an additional factor that may not always be present. This variability complicates the interpretation of scientific results in various ways. Nonetheless, this additional complexity is only one value (or rather disvalue) that we must consider in determining our obligations.

It is conceivable that in many circumstances standardization of experimental arrangements is not sufficiently important to sacrifice the quality of animals' lives by eliminating the opportunity for bonding to the researcher. Since bonding may be of significant benefit to animals (as well as to researchers), the answer to our second question is that sometimes the bonding of animals to researchers should be encouraged.

In general, it appears that bonding is a relatively minor factor to be taken into account in determining our moral obligations to animals. However, in certain cases, this factor could assume much larger proportions. In some circumstances the severance of a bond could be the only cause of significant suffering inflicted on an animal either in the course of or at the end of a scientific research program. Furthermore, there is reason to believe that the suffering associated with breaking a bond can be of major proportions. I am thinking, in particular, of research on nonhuman primates. There is evidence that such creatures experience severe stress from the severance of bonds to human beings (Oden & Thompson, Chapter 13, this volume). This could also be a factor that should be taken into account in evaluating proposals to do research on dolphins or other cetaceans. Before undertaking research in which it is likely that the animals will form bonds with their human associates, one should give serious consideration to the steps that might be taken to ensure the quality of the animal's life if and when it ceases to be of use as a research subject.

References

Baker, L. R. (1987). *Saving Belief: A Critique of Physicalism*. Princeton, NJ, Princeton University Press.

Gubernick, D. A. (1981). "Parent and Infant Attachment in Mammals," in *Parental Care in Mammals*, eds. D. J. Gubernick & P. H. Klopfer. New York, Plenum, pp. 243–305.

Hempel, C. (1965a). "A Logical Appraisal of Operationism," in *Aspects of Scientific Explanation and Other Essays in Philosophy of Science*. New York, Free Press, pp. 331–496.

(1965b). "The Theoretician's Dilemma: A Study in the Logic of Theory Construction," in *Aspects of Scientific Explanation and Other Essays in Philosophy of Science*. New York, Free Press, pp. 173–224.

(1966). *Philosophy of Natural Science*. Englewood Cliffs, NJ, Prentice-Hall.

Hinde, R. A. (1976). "Interactions, Relationships and Social Structure," *Man, 11*, 1–17.

(1979). *Towards Understanding Relationships*. New York, Academic Press.

Holton, G., & Roller, D. H. D. (1958). *Foundations of Modern Physical Science*. Reading, MA: Addison-Wesley.

Kuhn, T. S. (1962). *The Structure of Scientific Revolutions*. Chicago, University of Chicago Press.

Quine, W. V. (1953). "Two Dogmas of Empiricism," in *From a Logical Point of View*. Cambridge, MA: Harvard University Press, pp. 20–46.

(1970). "Grades of Theoreticity," in *Experience and Theory*, eds. L. Foster & J. W. Swanson. London, Duckworth, pp. 1–17.

Regan, T. (1983). *The Case for Animal Rights*. Berkeley and Los Angeles, University of California Press.

Rollin, B. E. (1981). *Animal Rights and Human Morality*. Buffalo, NY, Prometheus Books.

Sapontzis, S. F. (1987). *Morals, Reason, and Animals*. Philadelphia, Temple University Press.

Scheffler, I. (1964). *The Anatomy of Inquiry*. London, Routledge & Kegan Paul.

Singer, P. (1975). *Animal Liberation: A New Ethics for Our Treatment of Animals*. London, Jonathan Cape.

Index

Affection, 250, 252–253, 260, 370, 372, 384–391
Affiliative behavior, 21, 39, 43, 100–103, 250, 252
Age, 18, 105, 162–166
Aggression, 13, 50–51, 100, 171, 232, 244–246, 321–323, 328, 344, 350–354, 358, 367–371, 374–376, 378
Agonistic behavior, see Aggression
Alarm, 138, 142, 359, 362
Anecdotes, 4, 17–18, 35, 44, 58, 61, 150, 228, 229, 251, 253–254, 287, 367–370
Animal welfare, 253, 264–265, 272, 277, 288
Anthropomorphism, 15–16, 19, 23, 31, 261, 365, 372, 379, 380, 386, 389–390
Antipredator behavior, 11, 20, 252, 357–362
Applied animal behavior, 291, 317–318, 323–326
Approach; 266–268, 288, 290, 319–320; see also Avoidance
approach-withdrawal, 246–247, 266, 280, 285, 289, 299–300
Arousal, 125, 132, 135–150, 227
Assimilation tendency, 15, 188
Attachment, 8, 14–17, 22–23, 72–73, 76, 79, 83, 275, 334–335, 349, 354, 372, 376, 387; see also Bond
maladaptive, 84
operational definition of, 385
site, 76, 79, 83, 89
Attention, 223–225
Avoidance, 50, 232–237, 252, 266, 279, 280; see also Approach
lever-press, 234, 237

Baconian conceptions of science, 389
Biting, 233, 242, 352–353
Bond, 1–5, 8, 13, 20–23, 29, 38, 45–59, 66, 72–73, 93, 98, 105, 218, 223, 230, 251, 334–335, 349, 354, 367, 370–375, 379, 383–387, 390–395
definition of, 384
Breed, 95–97

Cardiac responses, 114–126
Caretaker effect, 34–35
Clever Hans effect, 2, 219–220, 225, 325–326
Coaction effect, 35
Cognition, 2–3, 195–196, 218–220, 228, 229, 353
Communication, 16–17, 19–23, 55–56, 62–66, 81, 101–102, 104, 178, 180–181, 185–186, 192, 195, 252
Comparative psychology, 28–29, 31, 38, 182, 228
Comprehension, 193, 195
Conceptual matching, 225
Constraints in learning, 52, 59
Context, 49–50, 52, 60, 67, 181, 183, 185, 189, 190–192, 219–220, 227, 229
Cooperation, 101, 172
Corticosteroids, 154, 161, 163–164, 168, 175–176, 266–272, 280, 304–305
Cortisol, see Corticosteroids
Critical period, 30, 74–75, 77, 80–82, 89; see also Sensitive period

Defensive behavior, 171, 252, 254–260, 358
Development, 30–31, 47–49, 62–64, 74, 184–186, 322–323, 325, 335; see also Ontogeny

Index

Disanthropomorphism, 390
Discrimination, 27, 34–35, 103, 254–255, 258, 321
Displacement behavior, 192, 276, 285
Displays, *see* Communication
Domestication, 18, 96–98, 102–103, 106, 264, 286, 296
Dominance, 20–21, 44, 80, 86, 100, 115, 157, 161, 166, 251, 322, 350, 354–355, 378
Duetting, 190, 192

Early experience, 30–32, 36–37, 258–260, 296, 311–312, 319, 324
Electroencephalogram (EEG), 136–137, 141
Emotion, 56–57, 227, 278, 380, 384, 386–391
Empiricism, 386–387
Endocrine responses, 153
Endogenous rules, 74
Ethics, 383, 393
Ethology, 178–182, 184
Experimenter
  attitude, 276–277
  behavior, 86, 88
  biases, 3, 220, 252
  effect, 36, 324, 362–363
  expectancies, 3, 219
  gender, 86, 87, 94, 103, 326
  movements, 51–52, 55–56, 64, 87, 102, 259
  skills, 225–226, 259

Familiarity, 17, 220–221, 266
Fear, 18–20, 22, 50, 84, 264–281, 285, 288
  response, 10, 74, 75, 76, 84, 266, 285, 288
Feral dogs, 97
Feral horses, 317–319
Flooding, 235–236
Following response, 76, 79, 287
Footshock, 232
Functionality, 188, 190

Genetic predispositions, 19–20
Genetic variation, 74, 77, 82, 86, 97–98
Gentling, 10, 33–34, 36–37
Guide dogs, 94, 97

Habituation, 9, 11, 156, 244, 266, 278, 281, 289, 291, 299, 318–321, 362–363, 369
Handling, 10, 31, 33–34, 36–37, 87, 102, 104, 153–154, 156, 158, 161–168, 171–176, 233, 256–260, 268, 270–277, 289–290, 324–325
Handrearing, 47–62
Hedonistic learning, 113
Hierarchies, *see* Dominance
Hunters, 366–367, 371, 380

Immune function, 154–155, 164–167
Imprinting, 16, 73, 83, 253, 275, 286–289, 334–355, 379
Individual differences, 54–55, 65, 82, 87, 245–246, 256–257, 299, 301, 304, 310
Individual recognition, 34, 250–251, 253, 254–261, 347
Infantile stimulation, 273
Instrumental response, 110–113, 117
Integration, of behavior, 64–66
Intelligence, 52, 194, 240
Interaction, *see* Attachment; Bond; Relationships
Invertebrates, 240–248

Labelling, 192–193
Lactation, 137, 307
Latency, 112
Longevity, 79, 85

Match-to-sample, 222, 224, 225
Mechanism, 389
Mentalism, 380, 387, 389
Mentor, 218, 221–229
Milk ejection, 307–309
Mimicry, 185–186, 352
Mitogens, 155, 165
Modeling theory, 182–187, 189, 191, 196
  model/rival, 185, 188–190, 192
Moral obligations, 393–395
Motivation, 110, 219–220, 222, 228–229, 266, 340

Neophobia, 279–280
Nesting behavior, 290
Neuronal activity, 133–150
Novelty, 223–225

Objectivity, 365, 370
Object matching, 225–226
Observational learning, 225
Observer effect, 45, 226–227, 285, 291, 318, 362–363, 376
Ontogeny, 362–364, 374, 380; *see also* Development
Ontology, 387
Open-field test, 31–32, 35, 79, 80
Opiates, 76, 84, 116, 126
Organizational processes, 75–76
Orientation response, 114–115
Overtraining, 111

Panhandling, 369, 372
Parasympathetic activation, 121–123
Parental care, 20, 251, 391
Parental investment, 358

Index

Partnership, social, 12–13
Pavlovian conditioning, 2–3, 95, 114–116
Perception, 9–10, 14, 19–23, 369
Petting, 106, 110–113, 123
Physicalism, 386–387, 389
Physiological psychology, 28
Pituitary adrenal responses, 269, 304–307, 310–311
Play, 51–53, 218, 220–229, 322–323, 353, 370–372, 374–377
Playmate, 218, 221–229
Popularity, 222–223
Predation, 10–11, 17, 20, 97, 241–242, 286, 357–363; *see also* Antipredator behavior
Productivity, 265, 272, 274, 318
Punishment, 17–18, 22, 83–86, 97, 279; *see also* Reinforcement
Pygmalian effect, 3

Quality of life, 394

Reactivity, 297, 310
Realism, 388
Reductionism, 386, 388
Referentiality, 182–183, 186–192
Reinforcement, 17, 22, 32–34, 110–111, 125, 172, 228; *see also* Punishment; Rewards
Relationships; *see also* Attachment; Bond
  conspecific, 12, 13
  definition of, 8
  inevitability of, 20
  insignificant, 11
  interspecific, 29
  predator-prey, 10–11
  scientist-animal, 1–5, 7–8, 19, 23, 220, 223, 232, 241, 377, 380
  symbiotic, 11–12
REM sleep, 137–138
Reproduction, 264, 270–272
Restraint, 175, 288
Rewards, 17–18, 80, 83, 87, 106, 109, 279; *see also* Reinforcement
  appetitive, 109
  extrinsic, 191
  food, 109–113, 172, 222, 369
  intrinsic, 191–192
  social, 32–34, 110–111, 222

Same/Different, 193–194
Scaffolding, 188, 191

Sensitive period, 18–19, 22, 74, 77, 81–82, 350; *see also* Critical period
Sensitivity to environmental events, 50, 53, 55–56, 61, 96, 105
Sensory abilities, 21, 252, 370, 377; *see also* Perception
Sensory contact, 17, 20
Separation
  distress, 14, 76–77, 80
  studies, 157–158
  syndrome (kennel-dog syndrome), 77, 84, 99–100
Shuttlebox, 233, 236–237
Signals, *see* Communication
Skepticism, 383, 392
Sleep, 137, 250, 252
Social behavior, 28–29
Social facilitation, 35
Socialization, 13, 62, 98–99, 102, 186
Social milieu, 219–220, 222, 227
Song, 184–185
Spatial abilities, 105
Species specific defensive reactions (SSDR), 234, 237
Stimulation, 31, 33, 37
Stimulus generalization, 279
Stress, 142, 166–167, 264, 269–270, 272, 365
Suckling, 141–142
Surrogate mother, 81, 334–350, 352, 354
Symbiosis, 11
Sympathetic activity, 121, 156, 305

Tactile stimulation, 109–110, 120
Taxonomies in behavior, 45–47, 54, 57
Teaching, 206, 213, 218
Temperament, 98, 99, 275, 295–310, 321
Temporal discharge patterns, 133–134, 139, 144, 146, 149
Time (as variable), 75
Timidity, 246, 302–305
Tongue flicking in snakes, 252–253, 258, 260
Tool use, 58
Two-process theory of avoidance, 233

Venipuncture, 158, 171–176
Vocalization, 49, 55, 65, 101, 183–187, 189–191, 347–348

Work, 218, 221–229

Zoomorphism, 15